& fork

fork

Phaidon Press Limited
Regent's Wharf
All Saints Street
London N1 9PA

Phaidon Press Inc.
180 Varick Street
New York, NY 10014

www.phaidon.com

First published 2007
© 2007 Phaidon Press Limited

ISBN 9780714847689

A CIP catalogue record for this book is
available from the British Library.

Designed by Jannuzzi Smith

Printed in China

Preface

With the publication of *&Fork* we have added another piece of cutlery to the Phaidon design table. After the success of its predecessor, *Spoon*, *&Fork* follows the same formula and presents the work of 100 of the world's most interesting product designers to have emerged during the last five years.

We asked 10 internationally respected figures in product design to select 10 contemporary designers each. The 100 designers they chose come from all over the world: from the Americas to the Far East and Australia, passing through the Philippines, Europe and the Middle East. The result is a comprehensive and extensive survey of contemporary product design.

All 10 curators are highly established figures in the design world; they are experienced museum or gallery curators, renowned critics, and sharp talent scouts. They have an extensive knowledge and are expertly-informed about the international design scene but they are also increasingly aware of more local design endeavours. They are constantly searching for the rising stars of tomorrow and the designers they have chosen here represent the very best in today's cutting-edge product design.

What became most apparent during the book's creation was that our current MySpace/iPod generation is becoming more and more concerned with social responsibility and the ethics of design. There is a widespread use of recycled materials, and many contemporary designs recycle existing objects that might otherwise have been thrown away. Sometimes this has been done by transforming used or old items into other new objects (a used bottle becomes a watering can, for example) and sometimes, broken objects are restored using new flashy materials. Young designers are also concerned with the lack of space in our increasingly overcrowded world, and they have reacted to this global problem by inventing objects that can be folded, dismantled, and easily stored.

Inevitably, new product design caters for our increasing dependence on portable and personal technology, such as mobile phones, MP3 players and laptop computers. Designers are looking at how these objects affect our interpersonal relationships, and they are designing with a mind to future generations and their inevitable interaction with technology, creating products that are sometimes specifically aimed at familiarizing children with the technology on which they will be dependent when they grow up. It might be said that technology is responsible for opening up the world, allowing easier access to knowledge and communication, and throughout this book it becomes apparent that contemporary product design reflects a world that is coming together as a whole, a world that is more conscious of its many parts than ever before.

The current generation may have grown up with computer technology present in all aspects of life, but we are also fascinated and inspired by the beauty of nature. For example, contemporary designers can be seen using the most sophisticated technologies to precisely reproduce floral patterns and natural structures in their work.

It is also significant that many of this generation's young designers are connected to each other: they know one another's work, they exchange experiences, they don't believe in working in an isolated environment, and very often they team up in groups where there is no leader, with a number of designers collaborating equally between themselves.

Providing an insight into the production processes of designers, from ideas to manufactured objects, and illustrated with early design sketches and photographs of both prototypes and the finished products, *&Fork* gives a worldwide view of what's happening right now and the best of what can be expected in years to come. With each curator delivering a professional commentary on the designers they have chosen, we are offered a fascinating glimpse into what top design experts most admire in the work of the next generation. Cataloguing new work from 100 of the most exciting and important young product designers, detailing advancements in production methods, and showing new ways of using existing materials, *&Fork* is an essential and comprehensive volume, featuring innovative work from the cutting-edge of international product design.

In addition to their choice of ten designers, each of the curators has also been asked to choose and write about an object that they think fully represents the concept of 'good design'. This section, which concludes the book, gives the curators another opportunity to describe what constitutes good design to them and to explain to us why it is so important.

Good design has never before been so readily available, whether in products that make our lives simpler, or simply in those that make us smile. *&Fork* is an invaluable reference book on an industry that continues to shape the way we live.
Emilia Terragni

Curators

Tom Dixon

Self-taught maverick Tom Dixon is one of the UK's most dynamic and influential designers. Famous for products such as the 'S' Chair and Jack Light, his appointment as creative director of Habitat in 1997 further increased his influence over the direction of British design. His company, TOM DIXON, was launched in 2002, developing its own collection of contemporary lighting and furniture, which has entered the international design major league through shows at venues such as the Milan International Furniture Fair and London Design Museum. Dixon was awarded the OBE for services to British Design in 2000, and in 2005 became creative director of Artek.

Maria Helena Estrada

Maria Helena Estrada is a design critic with a background in philosophy and journalism. In the 1980s she worked as design editor of *Folha de São Paulo* and was influential in bringing the Campana brothers to international attention, curating their first exhibition in 1988–9 at the Nucleon-8 gallery in São Paulo. She is now the publisher of the magazines *Arc Design*, specializing in architecture and design, and *SustentAção*, focusing on sustainability. Estrada is also currently the director of design affairs at the Museu da Casa Brasileira (Brazilian House Museum) Friends Society, and finds time to curate design exhibitions and participate in design seminars in Brazil and in Italy.

Pierre Keller

Since graduating with first-class honours in graphic design from the Ecole Cantonale d'Art de Lausanne, Pierre Keller's multifaceted career has included work as an artist, graphic designer, publisher, art consultant, curator and lecturer. Following his appointment in 1995 as director of ECAL in Lausanne, the school has undergone a process of radical change, evolving from a regional school of applied arts into an institute of European significance. From the outset, Keller has been working to create a new environment characterized by the breaking of conventions and the stretching of limits. In 2005 he was appointed professor at the Swiss Federal Polytechnic University of Lausanne (EPFL).

Sang-kyu Kim

Originally trained as an industrial designer at Seoul National University, Sang-kyu Kim now works as a curator at the Hangaram Design Museum in Seoul, an organization that explores design culture while seeking a 'Korean design identity'. The exhibitions that he has curated include 'No design No style, Droog Design' (2003), 'Design Culture in Korea 1910-1960' (2004) and 'The New Vision from László Moholy-Nagy' (2005). Kim is constantly researching diverse design subjects for new exhibitions, focusing on issues generated by contemporary visual culture, and striving to communicate them and their historical context to the visiting public.

Didier Krzentowski

Didier Krzentowski is managing director and founder of Galerie Kreo, one of the most influential contemporary design galleries in Europe, representing internationally renowned designers such as Ronan & Erwan Bouroullec, Hella Jongerius, Jasper Morrison and Marc Newson, and also showcasing the work of emerging young designers. Galerie Kreo is dedicated to exploration and analysis – a 'research laboratory' where designers' most experimental work gives birth to remarkable limited edition pieces. An expert in design and contemporary art, Krzentowski has worked on many significant exhibitions, including 'Ronan & Erwan Bouroullec' at the Design Museum, London (2002) and the Museum of Contemporary Art, Los Angeles (2004), and 'Marc Newson' at the Groningen Museum, The Netherlands and the Design Museum, London (2004).

Julie Lasky

Julie Lasky is editor-in-chief of *I.D.*, the award-winning magazine of international design. A widely published writer and critic, she won a journalism fellowship from the Medill School of Journalism in 1995 and her subsequent writings on the cultural life of post-war Sarajevo gained her the prestigious Richard J. Margolis award. She has contributed to the *New York Times, Metropolis, Dwell, Architecture, Slate, Surface, The National Scholar* and *NPR*, and is also the author of two books: *Borrowed Design: Use and Abuse of Historical Form* (written with Steven Heller) *and Some People Can't Surf: The Graphic Design of Art Chantry*.

Guta Moura Guedes

Guta Moura Guedes is co-founder and chair of Experimenta (Lisbon's cultural association devoted to promoting design). As director of the international biennale ExperimentaDesign, her aim is to nurture design talent, both from Portugal and abroad, and take design and culture to a wider audience. Acting as a designer herself, in 1992 she founded the Elementos Combinados design studio, and from 1997 onwards she has worked on interdisciplinary design projects, acting variously as designer, writer, lecturer, consultant and curator. In recognition of her important contribution to the design world, the French government awarded her in 2005 the honorary title of Chevalier de l'Ordre des Arts et des Lettres.

Brian Parkes

Brian Parkes is a key figure in the Australian design world with an interest in making design accessible to all. He has been Associate Director at Object: Australian Centre for Craft and Design since January 2000, and manages Object's artistic programs including exhibitions and publications. His major curatorial projects include: 'Akira Isogawa' (2002); 'Dinosaur Designs' (2003); and 'Design Island' (2004), a survey of contemporary Tasmanian design exhibited at Sydney Opera House. His most recent exhibition, 'Freestyle: New Australian Design for Living' (2006), is a ground-breaking collaboration between Object and the Melbourne Museum, and the first major survey of contemporary Australian design. Parkes managed the development of Object magazine from 2002–2004 and is a regular contributor to international design publications.

Francesca Picchi

Francesca Picchi is an architect, design writer and curator living and working in Milan. Currently the editor of the influential Italian design magazine, *Domus*, she is at the cutting edge of developments in international product design. Picchi has also worked as a researcher at the Barragán Foundation in Basel and as visiting professor of design history at Politecnico di Milano's design school. Her curatorial work includes 'Enzo Mari: il Lavoro al Centro' (Barcelona and Milan, 1999), with Deyan Sudjic 'The View from Domus. Fotografie/Photographs 1928–2002' (Milan, 2002) and 'Kuramata's Tokyo' (Milan); her published books include *Enzo Mari* (1999) and *Alberto Meda* (2003). She also contributed to *KGID*, a monograph on Konstantin Grcic published by Phaidon Press in 2005.

Chieko Yoshiie

Chieko Yoshiie is the editor-in-chief of *Casa Brutus*, Japan's regularly published magazine focusing on architecture and design founded by Tokyo's Magazine House in 1998, and also a design critic. The magazine introduces its ever-increasing readership to the most innovative new designers from Japan and around the world, whilst also celebrating legendary architectural and design giants such as Tadao Ando, Le Corbusier and Sori Yanagi. Yoshiie has written for some other design publications and was one of the distinguished design figures who contributed to a monograph on the Japanese retail designer Masamichi Katayama, which was published in 2003. She was also on the jury for the 'Open Living Container' prize for Tokyo Designers' Week in 2004.

Contents

5 Preface

10 Curators

6   Tom Dixon
    Maria Helena Estrada
    Pierre Keller
    Sang-kyu Kim
    Didier Krzentowski
    Julie Lasky
    Guta Moura Guedes
    Brian Parkes
    Francesca Picchi
    Chieko Yoshiie

100 Designers

10 5.5 designers
14 ANTEEKSI
18 Antenna Design
22 Atelier Oï
26 Maarten Baas
30 Suyel Bae,
     MILLIMETER MILLIGRAM.INC
34 Manuel Bandeira
38 Yves Béhar/fuseproject
42 Mathias Bengtsson
46 bernabeifreeman
50 Mana Bernardes
54 Big-game
58 Steven Blaess
62 BLESS
66 Jörg Boner
70 Fernando Brízio
74 Stephen Burks,
     Readymade Projects
78 Sam Buxton
82 Elio Caccavale
86 Louise Campbell
90 Leo Capote
94 Jennifer Carpenter,
     TRUCK Product Architecture
98 PIERRE CHARPIN
102 Lucas Chirnside
106 Clemens Weisshaar & Reed Kram
110 Kenneth Cobonpue
114 Paul Cocksedge
118 D-BROS
122 DEMAKERSVAN
126 Stefan Diez
130 Florence Doléac
134 Doshi Levien
138 david dubois
142 Piet Hein Eek
146 Martino d'Esposito
150 FRONT
154 FUCHS + FUNKE
158 Martino Gamper
162 Alexis Georgacopoulos
166 Adam Goodrum
170 graf:decorative mode no.3
174 Tal Gur
178 Ineke Hans
182 Jaime Hayon
186 Simon Heijdens
190 Jackson Hong
194 Junya Ishigami
198 Ichiro Iwasaki
202 ixi

206 Trent Jansen
210 Patrick Jouin
214 Chris Kabel
218 Lambert Kamps
222 Yun-je Kang,
     Samsung Electronics Co., LTD.
226 Meriç Kara
230 Bosung Kim
234 André Klauser
238 Tetê Knecht
242 korban/flaubert
246 Joris Laarman
250 Nicolas Le Moigne
254 Simone LeAmon
258 Sang-jin Lee
262 Mathieu Lehanneur
266 Arik Levy
270 Julia Lohmann
274 Alexander Lotersztain
278 Xavier Lust
282 Cecilie Manz
286 Tatsuya Matsui
290 Matthias Aron Megyeri
294 Marcelo Moletta
298 Nada Se Leva
302 Khashayar Naimanan
306 nendo
310 Nido Campolongo
314 Nódesign
318 Johannes Norlander
322 Ken Okuyama
326 OVO
330 Flávia Pagotti Silva
334 Zinoo Park
338 Russell Pinch
342 POLKA
346 ransmeier & floyd
350 Adrien Rovero
354 Alejandro Sarmiento
358 Inga Sempé
362 Wieki Somers
366 Alexander Taylor
370 Carla Tennenbaum
374 TONERICO:INC.
378 Peter Traag
382 Maxim Velcovsky
386 voonwong&bensonsaw
390 Dominic Wilcox
394 WOKmedia
398 Shunji Yamanaka
402 Noriko Yasuda
406 zuii

10 Curators' Choices

412 Armchair 41 'Paimio' chosen by Tom Dixon
413 Poltrona Mole (Sheriff) armchair chosen by Maria Helena Estrada
414 Ulmer Hocker chosen by Pierre Keller
415 Sella stool chosen by Sang-kyu Kim
416 Lampadaire '1063' chosen by Didier Krzentowski
417 Schwinn Sting-Ray chosen by Julie Lasky
418 Paperclip chosen by Guta Moura Guedes
419 Dilly Bag chosen by Brian Parkes
420 TP1 (T4 & P1) chosen by Francesca Picchi
421 Steel Shelving Unit chosen by Chieko Yoshiie

422 Artists' Biographies
439 Index
443 Credits

100

Designers

1

2

3

1–4

Furniture to Garden
Garden furniture
2006
b-ton design
Concrete

This garden furniture is designed so that the plants within the furniture can absorb water from the ground. The playful series essentially functions like a series of plant pots that you can sit on.

5·5 designers, Paris

Picture a hospital with trolleys and operating tables. A team of white-coated specialists attends to patients suffering from broken backs and missing limbs. Soon the injured are made whole and ready for use again – as armchairs and coffee tables. This scene, which has been staged half a dozen times on two continents, is part of Réanim, a project by the Paris-based design group 5.5 to extend the life of battered furniture. Using prefabricated pieces of glowing green plastic, the team replaces maimed table legs, covers distressed chair seats, and sutures the fronts of bland cabinets. The point isn't to restore these items to their original condition, the designers say, but to find a systematic way of reinventing them – one that anyone can accomplish.

Though Réanim represents just a slice of 5.5's work, it bares several themes that preoccupy the firm's four young principals – Anthony Lebossé, Jean-Sebastién Blanc, Claire Renard and Vincent Baranger (a fifth departed shortly after the studio was founded in 2003). They are intent on prolonging the use of products in a culture of waste, honouring an object's defects and battle scars, and turning design into a participatory activity, if not a spectacle. Recently invited to porcelain manufacturer Bernardaud's factory in Limoges, 5.5 collaborated with the workers to create mutant dishware that fits together in oddly conjoined arrangements and sometimes even erupts into nests of china worms. For the 2006 Rayons Frais arts festival in Tours, the group staked out a corner of the old town and invited passers-by to band together in building an enormous edifice from cardboard bricks. Every visit to the beach appears to be another opportunity for 5.5 to build easy chairs out of sand using a variety of prefabricated plastic moulds. Says member Anthony Lebossé, the studio even has a slogan to convey its dynamic, unorthodox processes: 'We have to design a way to design.' **Julie Lasky**

4

5

9

6

5–8

Graft Kit
2005
5.5 designers
Painted plastic

Part of the Réanim project, this is one of four ingenious furniture kits designed by the group. This Graft Kit can be used to connect different pieces of specially designed furniture together.

7

9–10

Seating Prosthesis
2005
5.5 designers
Plastic PMMA

The Seating Prosthesis can be used to 'cure' chairs with a broken seat by attaching an ingeniously designed seat substitute.

10

11–12

Crutch
2005
5·5 Designers
Painted steel

A metal crutch which can be used to fix a chair with a broken leg.

8

11

12

13–15

Cup Elements
Coaster
2005
Bernadaud
Porcelain

5.5 designers used the handles of cups to create different table accessories – in this case a table coaster.

16

Salad Bowl
2005
Bernadaud
Porcelain

Part of the Workers-Designers project, the bowls are created using a technique called giggering, which allows for the creation of large pieces from semi-soft clay.

17–23 & 26

Creamer Casting
2005
Bernardaud
Porcelain

Part of the Workers-Designers project, an 'upside down' creamer becomes a 'hoofed' mug. Created using a slip-casting moulding technique the excess hoof-shaped slip becomes the base of this unique mug.

24–25

Sauce-Boat
2005
Bernadaud
Porcelain

The Sauce-Boat inverts traditional porcelain design by handpainting the delicate gold pattern onto the object's interior, thus making it an irreplaceable hidden treasure.

17

22

18

23

13

24

14

19

25

15

20

16

21

26

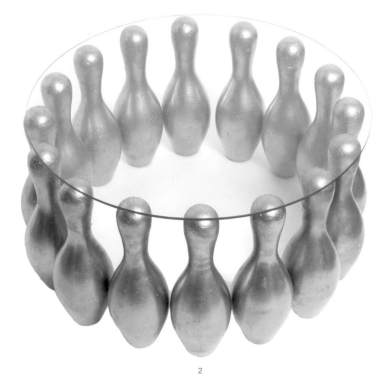

ANTEEKSI, Helsinki

While other designers might work together for comfort and greater clout in the design world, ANTEEKSI just seems to work together to be creative. The Helsinki-based, 18-member group (the desginers are currently: Selina Anttinen, Malin Blomqvist, Johanna Hyrkäs, Erika Kallasmaa, Jussi Kalliopuska, Maippi Ketola, Johannes Nieminen, Vesa Oiva, Johan Olin, Pauli Ojala, Anna Oksanen, Anne Peltola, Tuomas Siitonen, Aamu Song, Janne Suhonen, Mari Talka, Tuomas Toivonen and Nene Tsuboi) is perhaps one of the most exciting and innovative of the growing number of design collaborations and collectives. What's more, they seem to be having a lot of fun.

The ANTEEKSI office houses 14 separate workspaces and is also used as a place where members can work on individual, more commercial projects. The collective's emphasis, however, is decidedly non-commercial. Experimental and crazy, it casts aside the confines of mainstream design and mines the ignored in-between spaces. The ANTEEKSI designers run events, produce books, give lectures and hold two shows a year: one a product design show, the other a fashion show.

The shows are in some ways a 'mock' version of mainstream events, with furniture fairs set in street kiosks and amateur fashion shows displaying items such as a coat festooned with stuffed animals. The group is made up of architects, graphic designers, product designers and others, so that architects might work on fashion, or graphic designers on product design, etc. Using this mix of professions and disciplines in such a free way is what gives its work its experimental edge and appealingly playful aesthetic. The fact that members often come and go also ensures that the group's design identity is constantly in flux.

It seems apt that anteeksi in Finnish means 'excuse me' as the 18 members are concerned with cheerfully pushing past established design boundaries as easily as we might pass someone in the street. Their design spans the whole cultural landscape, overlapping with conceptual art. They bash out numerous ideas, of which maybe two or three will be useful. That is what is exciting about their work: they are, first and foremost, a prolific ideas bank. **Tom Dixon**

1

2

3

1

LightButton
Light
2005
ANTEEKSI
Plastic buttons, film, metal,
light bulb

*Designed by Johanna Hyrkäs,
the buttons give the lamp
a reclaimed feel.*

2

MutsisoliFaijas
Table
2004
ANTEEKSI
Glass, plastic, wood, paint

*Designed by Vesa Oiva,
this kitsch table is made
out of recycled bowling
pins painted gold.*

3

Christmas Tree
Card holder
2004
ANTEEKSI
Metal postcard stand, postcards

*Designed by Erika Kovanen,
this stand houses the
designer's own postcard
collection for all to see.*

4

Iltama
Chandelier
2006
ANTEEKSI
Steel wire, paper lottery
coupons

*Designed by Mari Talka, this
chandelier is made of lottery
coupons that create a delicate
light piece.*

5

FormFoam
Chair
2004
ANTEEKSI
Foam plastic, pearl-head pins,
birch

*Designed by Aamu Song,
this chair's shape and visual
upholstery is created by
sticking pins into foam plastic.*

6

Tellut
Pendant lamp
2006
ANTEEKSI
Plastic globes, light bulbs

*Designer Erika Kovanen used
globes to make a fun pendant
light that shows you the world!*

4

5

6

7

Jeesus Furniture
Chair
2006
ANTEEKSI
Old chair, duct tape

Designed by Johan Olin, this chair has been covered with duct tape.

8

FlightCaseChair
2005
ANTEEKSI
Plywood, foam, aluminium

Designed by Tuomas Toivonen, this chair is specially designed for touring musicians.

9

Milky
Pendant lamp
2004
ANTEEKSI
Milk cartons, light bulbs, paint

Malin Blomqvist creatively utilizes milk cartons to create these quirky pendant lamps.

9

8

7

16

10

10

Rubber Chair
2004
ANTEEKSI
Wood, rubber paint

Designed by Janne Suhonen, this chair is completely weatherproof.

11

Lidl by Lidl
Chair
2004
ANTEEKSI
Wood, beer cans

Designer Jussi Kalliopuska has transformed an old wooden chair by covering it in beer cans sold by the Lidl supermarket.

12

Pöpö Stool
2004
ANTEEKSI
Plastic, paint

Nene Tsuboi's stool has its nickname written on it.

13

White October
Light
2006
ANTEEKSI
Plastic, paint

Designed by Pauli Ojala, this lantern with a flexible arm was designed to be placed in a park or small gazebo.

14

Tuoli Mummilleni
Chair
2005
ANTEEKSI
Old sofa,
plastic artificial flowers

Designed by Tuomas Siitonen, this is an exuberant 3-D floral decoration.

12

13

11

14

1

Antenna Design, New York

New York City is famously populated by non-natives contributing to every level of culture. So it should come as no surprise that Antenna Design, a Manhattan studio whose work affects the daily lives of millions of New Yorkers, is run by a Japanese-born man and an Austrian-born woman, who are themselves a couple. What is astonishing is that Masamichi Udagawa and Sigi Moeslinger – much-lauded designers of subway cars and ticket-vending machines, of biometric security cards and desktop data terminals, of public artworks and mobile phone technology – represent two-thirds of Antenna's staff. Tiny, yet monumental in scope and originality, the studio, which was founded in 1997, excels at making connections.

Udagawa, an industrial designer previously employed by Apple Computers, and Moeslinger, trained in interactive design, bring particular comfort to the meeting of human and machine. They are especially good at manipulating the invisible medium through which data flies, allowing users to extract and share information from personal digital assistants or JetBlue Airways' automated check-in machines or the media company Bloomberg's information kiosks as easily as breathing. The partners are equally at home in the analogue world, where interaction may simply mean a comforting word to a stranger. Recently, at a New York art gallery, they presented such socially minded concepts as traffic signals that flash exercise routines so that waiting pedestrians can participate in a burst of group calisthenics; sculptures fashioned communally from discarded chewing gum; and street furniture combining the forms of a chaise-longue and an upright chair, designed to encourage strangers to indulge in spontaneous analytical sessions. 'Some people have the urge to talk, others have the urge to eavesdrop,' Moeslinger noted. 'This would bring the two halves together.' **Julie Lasky**

1

Exercise Stop –
Sidewalk Series
2006
Concept project

A traffic signal becomes an exercise 'instructor' making valuable use of waiting time for pedestrians.

2

Escape Loft –
Sidewalk Series
2006
Concept project

A way of escaping the sidewalk for a few moments to chat, smoke or just reflect.

3

Gum Sculpture –
Sidewalk Series
2006
Concept project

This sculpture allows for an artwork to be created from used chewing gum.

4

Hugging Tree –
Sidewalk Series
2006
Concept project

In a bustling city, sometimes everyone needs to be comforted. The Hugging Tree proposes a way to provide this unusual service.

5

Shrink Bench –
Sidewalk Series
2006
Concept project

The shrink bench encourages complete strangers to play out the roles of therapist and patient.

2

3

4

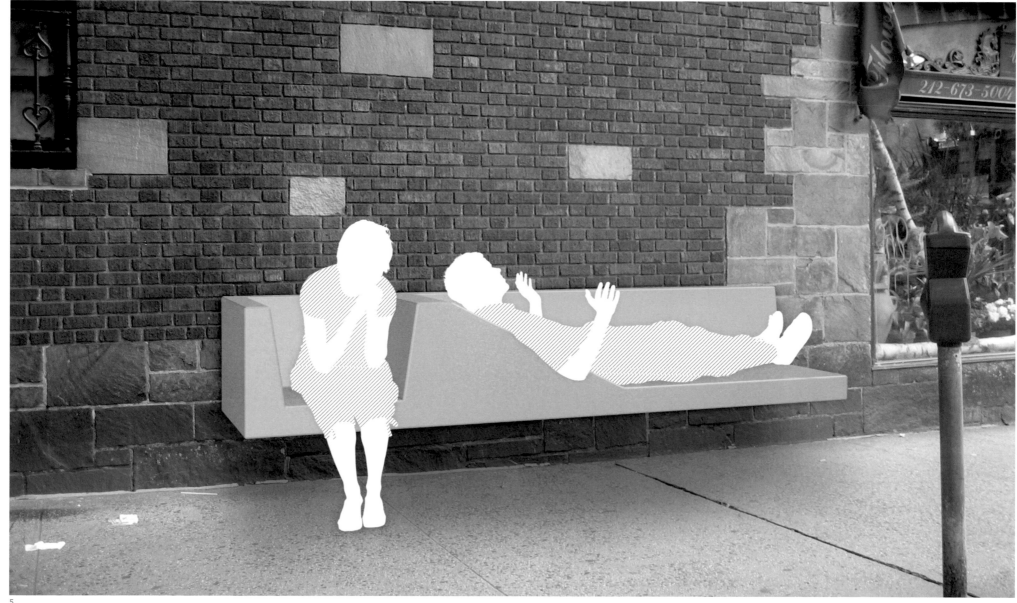

5

6

8

7

9

6–9

Fujitsu 23/6 Mobile PC
2005
Fujitsu
Die-cast aluminium alloy,
OLED screen

*A multifunctional
portable computer that
can be used in both
game and computing/
communications mode.*

10–11

Google Cab
2006
Antenna Design
New York Inc.
Ultra-light MDF, paint

*An object based on
a satellite image from
Google Maps of a New
York City taxi cab. Each
pixel has been translated
into a 2-inch cube.*

12–13

Civic Exchange
2004
Prototype
Steel, glass mosaic, LED
displays, LCD displays,
solar panels

*Useful portal to local
information, and a public
space for social interaction.*

12

10

13

11

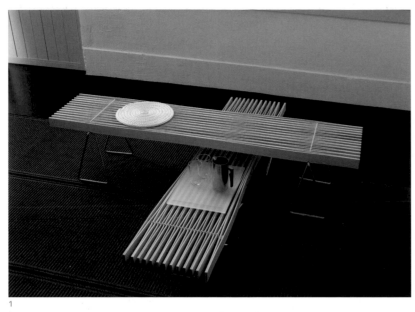

Atelier Oï, La Neuville

Atelier Oï, founded in 1991 in La Neuveville (Switzerland) by Aurel Aebi, Armand Louis and Patrick Reymond, takes its name Atelier Oï from *troïka* (Russian, a trio) and conjures up the dynamism of this group. Its practice located at the intersection of Switzerland's two main linguistic regions, Oï is resolutely interdisciplinary and collective. The designers extend their research from individual objects to environment matters, and their commissions range from product design to architecture, by way of set design and town planning. As well as developing products for brands such as Wogg, Ikea, Swatch, B&B Italia and Desalto, they have also designed private homes, the arteplage for the national Expo.02 exhibition at Neuchâtel, exhibition pavilions and street furniture.

While maintaining a direct and intuitive relationship with materials, Atelier Oï likes to upset scale and subvert the way in which an element is traditionally used, adapting it to contemporary use. Thus the Wogg 24 bed (1999) combines a mosquito net, a canopy and a tent into a single, portable object; the Mesofär lamp (1999), for Ikea, deconstructs the shade into a system of rings diffracting light; and the Pause series (2003), developed for the town of Bienne (Biel), transforms the austere public bench into a comfortable chaise longue. With a great deal of precision and intelligence regarding the forms and opportunities afforded by the materials they use, Atelier Oï develops objects by analysing their uses and the behaviour engendered by the settings in which they are deployed. This approach is epitomized by the Composition of Cords collection, which represents Atelier Oï's current experimentation and was presented at the 2006 Milan International Furniture Fair.

Atelier Oï's pragmatic and hands-on experimentation with materials and their technological processes have become their trademark. **Pierre Keller**

1

Bank Plus
Table/bench
2005
Roethlisberger
Oak, rope, stainless-steel
foot

Trendy multifunctional light-weight table/bench combo.

2–4

A Composition for Cords
Sculptural lighting
2006
Foscarini
Aluminium tubes

A lamp installation at the Swiss Cultural Centre during the Milan International Furniture Fair in April 2006.

2

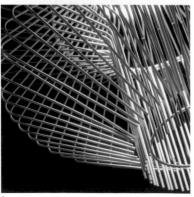

3

5

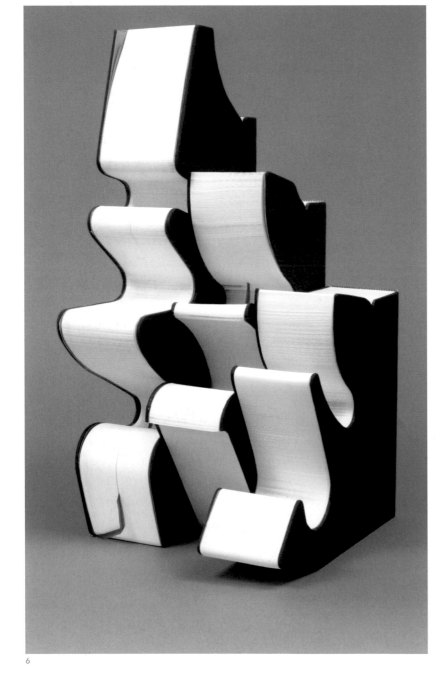

6

7

8

5–6

Lamp Tome
2005
Mouvements Modernes
Paris
Paper

*Compact booklets that fold
out into beautiful table
lamps like paper lanterns.*

7–8

Rigicordes
2005
Atelier Oï,
Rope, stainless-steel rod

*Unique hanging device that
really makes a statement.
Perfect for clothes, towels
or umbrellas.*

9

Poche-Cul
Stool prototype
2002
ECAL

*Handy stool that fits inside
your pocket to allow you to
take a seat whenever you
feel the need.*

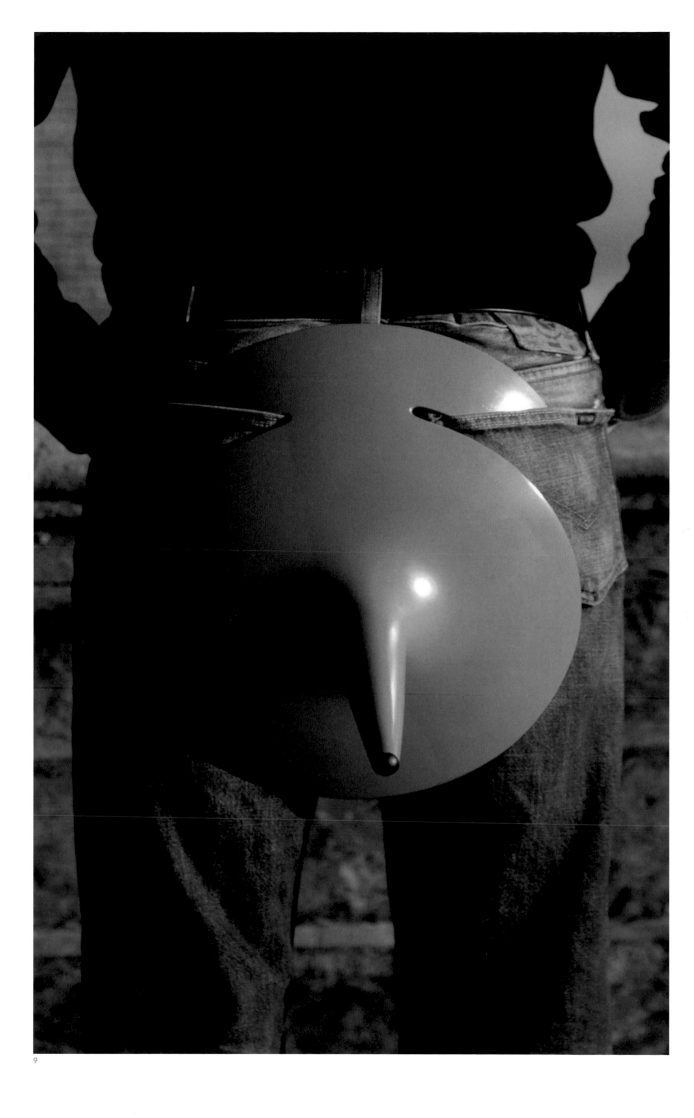

9

Clay Chair
2006
Baas & den Herder
Industrial clay, metal

These handmade, hand-moulded chairs are all unique: the clay moulding process produces a slightly different chair form each time.

Maarten Baas, Eindhoven

Maarten Baas began his career with an acetylene torch and a taste for bravado. While still a student at the Design Academy, Eindhoven, from which he graduated in 2003, he scorched ordinary wood furniture until the pieces blackened and began crumbling. Before long, the collection, called Smoke, caught fire. Marcel Wanders began selling it through his design company, Moooi, and retailer Murray Moss commissioned one-offs – charred versions of Mackintosh, Mollino, Eames and Sottsass, each possessing a haunting delicacy.

For Baas the project reinforced a paradoxical lesson: surfaces contain depths. Seeing no point in creating a new shape for a chair, he sought instead to baptize existing forms by fire, causing them to be reborn as artefacts that are both elemental and extraordinary. 'I wanted to make something by changing the idea of it,' he says.

At other times Baas has used found objects to conduct playful formal experiments. His Hey, Chair, Be a Bookshelf! (2005) is a jumble of seats, tables, lamps and planters stacked chaotically and painted a uniform colour. Objects are deposited in any cranny that will accept them. His Flatpack Furniture (2005) provides the materials and instructions for building a table, gathered together from Ikea's Ringo stool and Stefan chair. 'Why wouldn't you see a box of flatpack furniture as a kind of Lego toy?' he wonders.

Nothing if not Promethean, Baas recently turned his attention from fire to earth. Clay, a set of furniture launched at the 2006 Milan International Furniture Fair, features pieces hand-moulded from clay in colourful, elongated shapes, like life-size versions of a child's crafts project. Although the irregular surfaces recall Smoke, Clay is something of an inversion: Baas isn't giving up control to an unpredictable force of destruction, but having his way with a malleable tool of creation. **Julie Lasky**

1

3

2

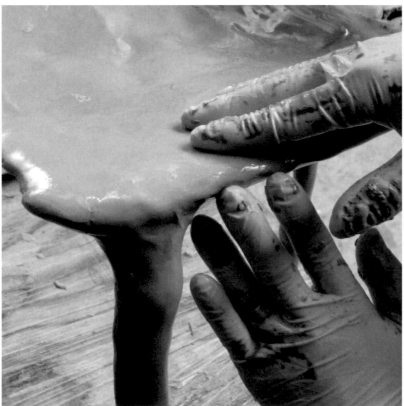

4

5

6

6

Smoke ZigZag chair
Charred version of Gerrit
Rietveld's ZigZag Chair
2004
Cassina/Maarten Baas

*A limited edition of
25 pieces for Moss NY,
this chair is part of a series
of works, 'Where There's
Smoke', in which Baas takes
famous design works
and burns them to create
a new, challenging piece.*

7

Treasure Dining Chair
2005
Baas & den Herder
MDF

*Part of the Treasure
Furniture series, this
dining chair is made from
discarded MDF sourced
from furniture factories.*

7

Suyel Bae, MILLIMETER MILLIGRAM.INC, Seoul

Suyel Bae is a designer and an executive who wants to create useful things. He has always been interested in expressing his design vision and in 1999, he established MILLIMETER MILLIGRAM.INC as a means of cultivating communication.

Bae believes small things, such as a little notebook, can serve as a link between designer and user. His company name – a combination of 'millimeter' (length) and 'milligram' (weight) – proclaims its intention: attention to detail is the starting point for larger design concerns and projects.

Bae's designs, with their stable forms and simple graphics, use familiar or affordable imagery rather than provocative images and shapes. His earlier study of metal craft taught him to build longevity into his designs for everyday products.

Products such as Mr Green (2001), Capsule Letter (2000) and Messenger Band (2006) propose new ways of communicating. He has extended his communication platform from products to exhibitions. At his MILLIMETER MILLIGRAM.INC exhibition, which took place during 'London Designer's Block' in 2003, he displayed notebooks that people had used, exhibiting the content, rather than the product itself.

Bae also likes using design to make people think, and his graphically striking Everyday Calendar (2004) lends a philosophical and timeless appeal to the normally practical desk calendar as it shows only the number of the day in the month – and nothing else.

Bae is always interested to know how his designs 'get along' with people in their everyday lives, and prefers not to present himself as a designer or an artist. For him, the design process is what creates opportunities for communication.
Sang-kyu Kim

1
Everyday Calendar
2004
MILLIMETER MILLIGRAM.INC
Paper, wire, PVC

A flip calendar showing only the number of the day in the month.

2
Everyday
Annual calendar
2005
MILLIMETER MILLIGRAM.INC
Paper, wire, PVC

A colourful desktop calendar that clearly shows both the day and the date.

3–11
Baby Holder
Keyring
2006
MILLIMETER MILLIGRAM.INC
Rubber, metal

Variety of smiley colourful keyrings to brighten up any key chain.

3

4

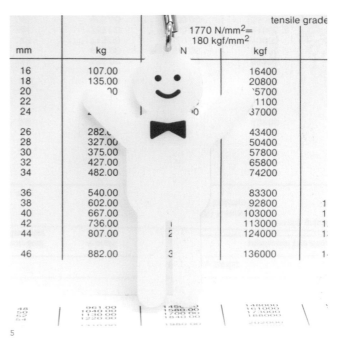

5

6

7

8

9

10

11

Suyel Bae,
MILLIMETER MILLIGRAM.INC

12

13

14

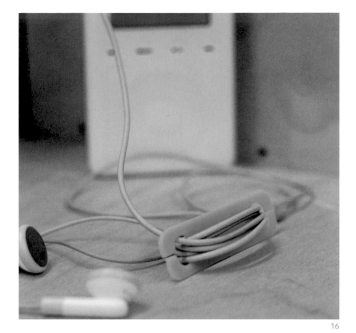

12–14

Pen Smile Man; Pen Cloud;
Pen Kindergarten; Pen Peace
Pen clips
2005
MILLIMETER MILLIGRAM.INC
PVC

*Colourful and fun ballpoint
pen clip collection designed
to put a smile on your face
while you work.*

15

Messenger Band
2006
MILLIMETER MILLIGRAM.INC
Rubber

*Simple bracelet which takes
its inspiration from the dove
as a messenger of peace.*

16

Roll Up
2006
MILLIMETER MILLIGRAM.INC
PVC

*An attractive cable tidy
ideal for loose wires or
headphones.*

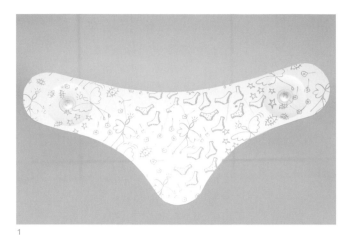

Manuel Bandeira, Salvador

Manuel Bandeira has become used to receiving awards. As an architecture student he won prizes in both Italy and Brazil for the prototype of the Contorno chair, inspired by an avenue in the city of Salvador da Bahia. After graduating in 1998, one of his projects was selected and he received an honorary mention at the XII Museu da Casa Brasileira Awards, São Paulo.

Bandeira obtained his master's degree in industrial design from the Domus Academy, Milan, where five of his projects received awards, including his packaging for Baci Perugina chocolate and a furniture piece for the Promosedia exhibition in Udine, Italy. He became assistant professor at the Domus Academy and took up an offer to work at designer Marc Sadler's Novus studio, where he designed home appliances for Moulinex. In 2001, having returned to Brazil, Bandeira created the Água chair (2001) in close collaboration with the furniture industry. His most intriguing projects are Sequinha® (an ingenious underwear dryer) for its creative concept and simplicity, and the Levita Chair, both designed in 2003 and given Museu da Casa Brasileira Awards.

Bandeira does not have a defined style and is not concerned with having one. His products always reflect the search for a coherent design, be it from the formal, functional or constructive point of view. 'I do not seek an aesthetic result. It appears from an equation involving the constructive process, its use and my fantasy. As a formula for my creations I combine simplicity and practicality resulting in forms expressing my life experience,' he explains. Bandeira's confessed influences are his passion for sailing boats, and his admiration for Marcel Duchamp and Achille Castiglioni.

With seductive, provocative and refined designs, Bandeira attributes new values to traditional products, or simply teases us by questioning certain habits or manners.
Maria Helena Estrada

1–3

Sequinha®
Underwear dryer
2003
Ultradesign
Polypropylene

This ingenious underwear dryer attaches to smooth surfaces enabling the user to easily and stylishly dry their knickers whilst reducing unsightly clutter.

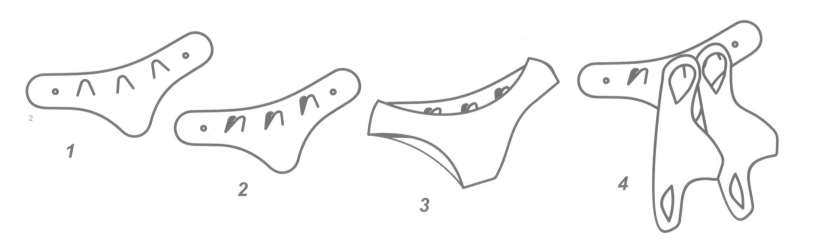

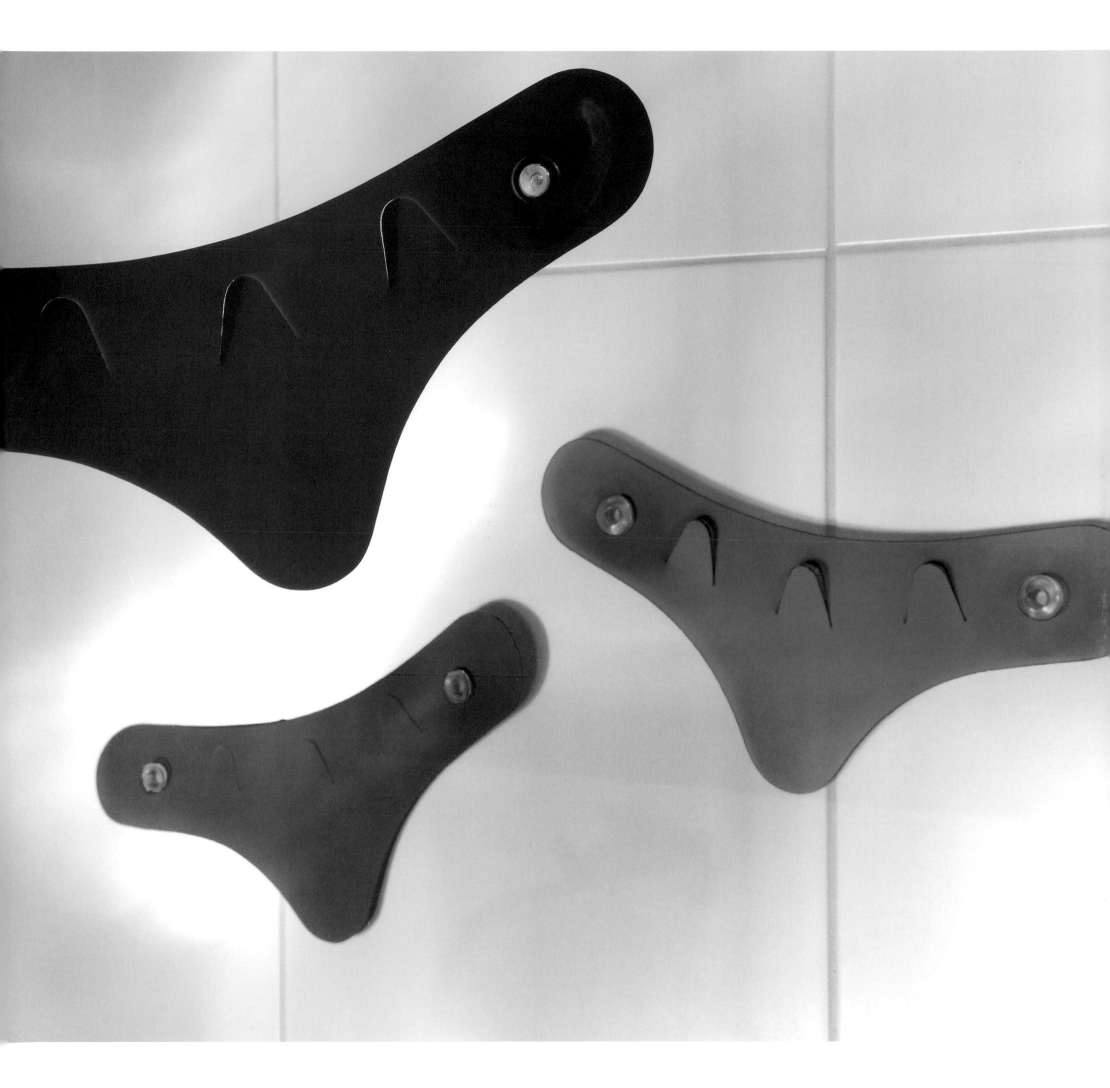

Manuel Bandeira

4

Euripedes Hanger
2006
Kenison Furniture
Eucalyptus wood

The Euripedes Hanger is a practical household clothes hanger that also playfully references the erotic.

5–6

Stair Chair
2003
Ultradesign
MDF, steel

This utility design allows a stylish, simple chair to double up as a useful household stepladder.

7–8

Levita Chair
2003
Ultradesign
Steel, Cordura© canvas

Sitting in the soft material seat of the Levita is, as the name suggests, like floating on air.

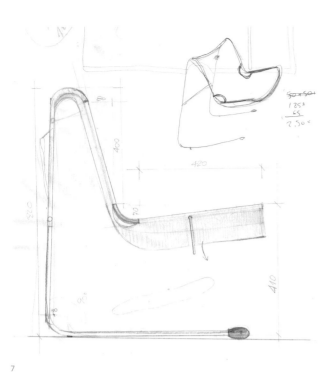

7

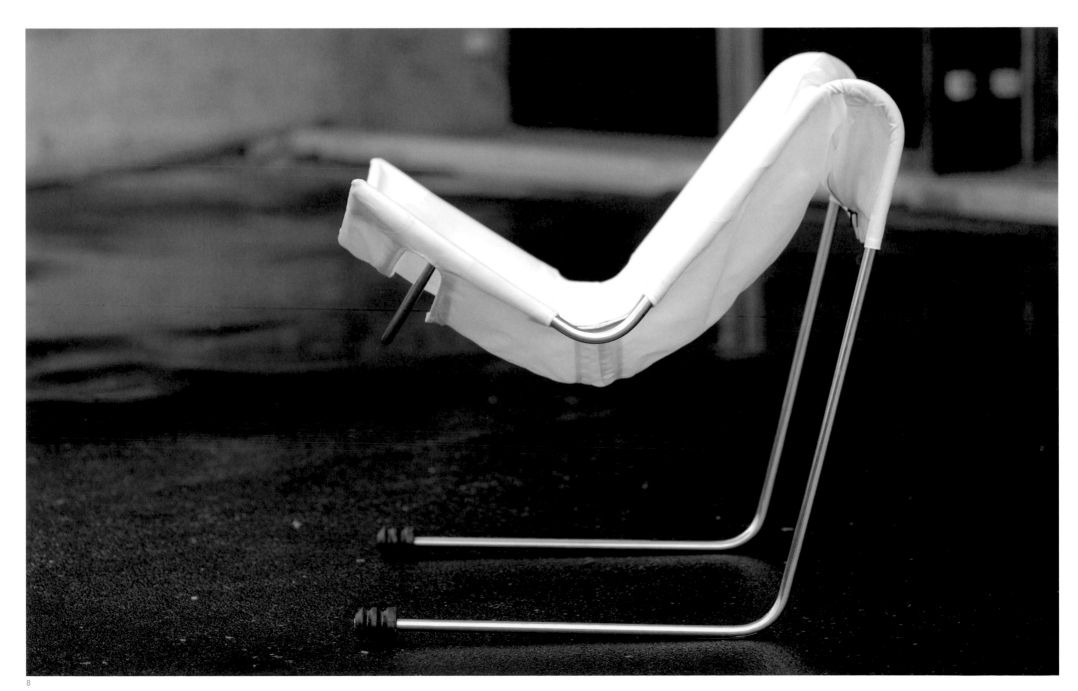

8

Yves Béhar/fuseproject, San Francisco

Yves Béhar, the founder of the fuseproject design studio in San Francisco, is a 'classical' designer. Observing the new needs created by the technological changes that are occurring in our daily lives, he solves the problems that are thrown up by our adaptive behaviour. For example, working with Danese, Béhar has designed a multifunctional side table – Kada (2006) – and a chair with pockets – Farallon (2006) – to allow us to store the myriad technological 'prostheses' we carry around with us: mobile phones, mini-computers, MP3s, etc. Noting that good manners have been eroded by gadgets, Béhar counteracts with elegant objects that are designed to minimize the impact on our behaviour of new technologies.

With Leaf (2006), the lamp distributed by Herman Miller and now in the permanent collection of New York's Museum of Modern Art, Béhar successfully combines elegant design with LED technology and low-energy consumption. The colour of the light is modulated using cutting-edge technology and Béhar stuck to his principle of the 'right form' allied with sustainability, which is what this lamp, likely to become a benchmark, demonstrates with panache. Similar environmental and technological concerns can be seen in his reinterpretation of the 'Birkenstock', known as the Birkis (2004). Made from the ultra-light plastic material EVA (ethylene vinyl acetate), each shoe is manufactured from a single piece without joints or glue. They are washable, comfortable, durable and they convey the spirit that has epitomized these shoes from the start: comfort, freedom and outdoor activity.

Having already had a solo show at San Francisco's Museum of Modern Art in 2004, Béhar's young studio, not yet a decade old, is destined to become even more influential in years to come. His highly successful designs encompass a diverse range of products and display his desire to connect society with the possibilities of the future. **Pierre Keller**

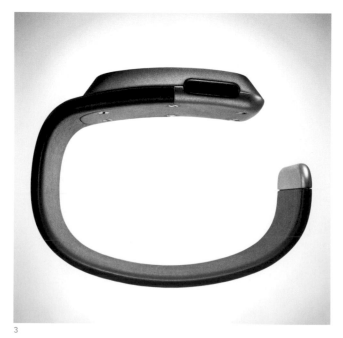

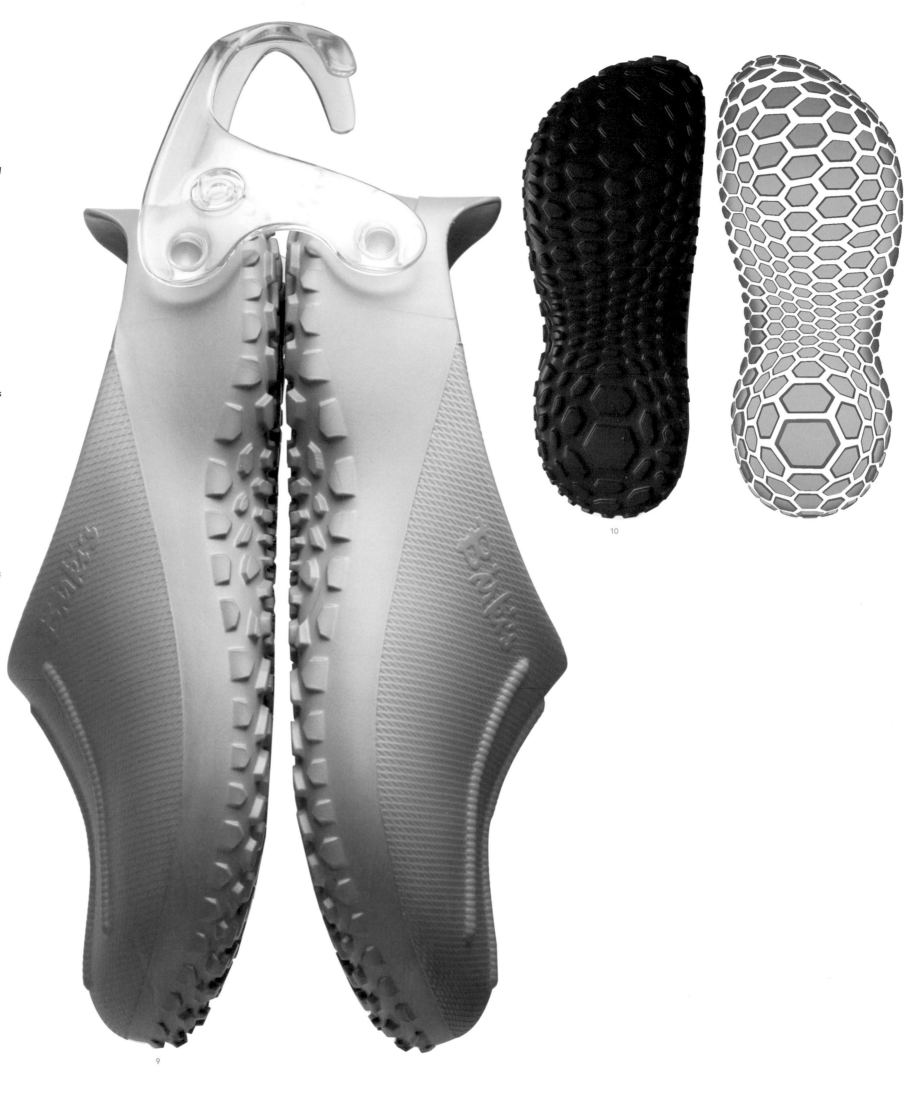

1–3

MINI_motion watch
2004
MINI/BMW Group
TPU, stainless steel

The watch has a striking, but very soft, sculpted strap that is is strengthened and shaped by a spring steel core.

4–8

KADA
Table/stool/power-hub
2006
Danese
Industrial neoprene, laminated wood panels
In collaboration with Youjin Nam and Naoya Edahiro

KADA is shipped flat and acquires strength when folded. Its different surfaces allow for varied usage as a stool, low table or tray.

9–10

Birkis
Slip-on clogs
2004
Birkenstock
Biodegradable TPU, EVA

Each Birkis is made of a single piece of material. There are no joints, stitches or glues used and they can be worn either indoors or outdoors.

9

10

11

12

13

14

15

16

17

11–13

Jawbone Air
2006
Aliph
ABS Plastic,
elastomer, metal
In collaboration with
Youjin Nam

*The only Bluetooth headset
with a voice activity sensor
touching the skin of the
wearer, which distinguishes
the human voice
from background noise.*

14

Toshiba Laptop Transformer
2003
Toshiba
Injected-moulded ABS

*The laptop is designed
with a sliding 'hinge' to
allow an easy transformation
from laptop computer
to flat-screen home
entertainment centre.*

15–17

One Laptop Per Child
(OLPC)
2006
One Laptop Per Child,
Quanta
PC/ABS, elastomer

*The cheap and efficient
OLPC brings learning
to children in the
developing world.*

18–20

Leaf
LED table light
2006
Herman Miller
Stamped aluminium, TPU,
injection-moulded ABS

*Leaf burns 40% less energy
than traditional LED lights,
and more than half of the
energy required to operate
a comparable halogen or
fluorescent lamp.*

18

19

20

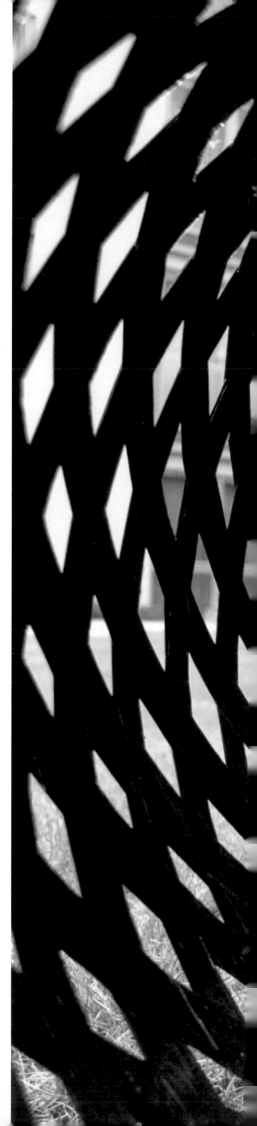

Mathias Bengtsson, London

Many designers are comfortable with both industrial technology and handicrafts, but Mathias Bengtsson presents an extreme case of divided loyalties. A Dane who makes his home in London, Bengtsson laboriously assembles furniture by hand from slices of wood, aluminium, acrylic and foam cut by a computer-programmed laser. Depending on the material, the rippling forms might resemble petrified wood, eroding sandstone or the exoskeleton of an insect. Other pieces are produced by a robot that spins a single strand of carbon fibre into unimaginable shapes, a method developed for the construction of lightweight rockets but which reminds Bengtsson of a spider's craft.

Bengtsson believes the urge to exploit industrial techniques and materials while honouring the lyrical imperfections of the hand is only natural for a designer who came of age at the threshold of the digital revolution. 'When I started my career, people were still wearing sandals and smoking pipes and spending three months making a nice paper drawing and then a prototype in wood,' he recalls. As a homage to fellow countryman Vernon Panton's sinuous plastic chair, Bengtsson reproduced the piece in cardboard, rebuilding it in a form that was both recognizable and his own. This act of bravado represented the anatomy of an icon but also a statement of Bengtsson's place in the daunting tradition of Danish design. Adding even greater resonance to the project, completed at London's Royal College of Art, was the fact that the original model had been donated by his instructor Ron Arad, an exemplary member of a generation of designers with whom Bengtsson and his peers would soon be contending.

Today, Bengtsson's struggle is not with the past but with a future that may have to be scoured of technology's false promises. He points out that, owing to our inventions, 'we have new problems': environmental strain, endless hardware and software glitches, and the flattening of original ideas into a parade of lookalike objects and deadening clichés. So, while he is besotted with advanced tools and materials, he is also programming flaws into his processes to ensure that no creation looks robotically perfect. **Julie Lasky**

1–2

Spun Carbon Fibre Bench
2005
Bengtsson Design Ltd
Carbon fibre

This 17-metre (55-foot) bench is spun from one continuous carbon-fibre tow.

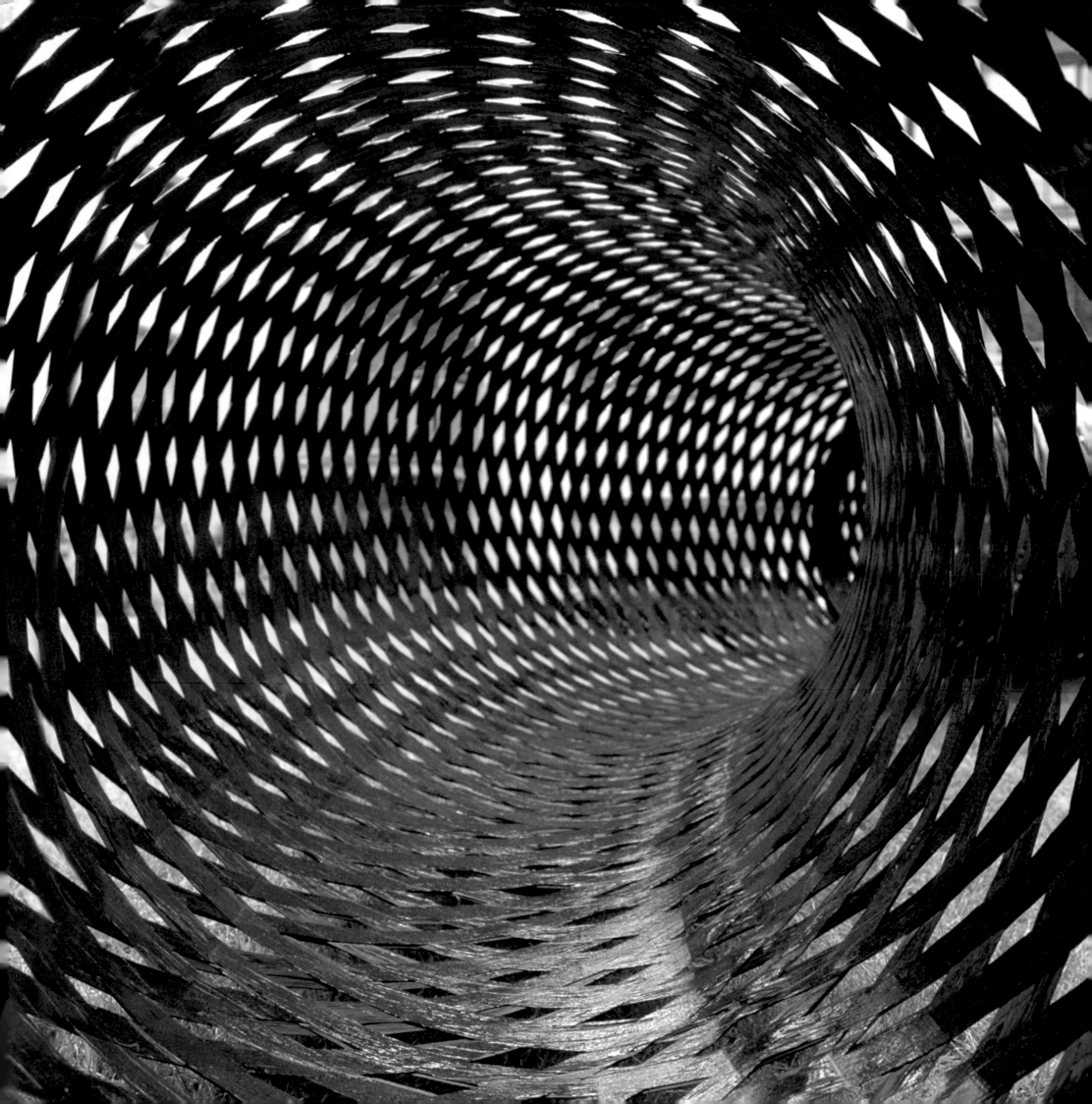

3

Spun Carbon Fibre Chair
2003
Bengtsson Design Ltd
Carbon fibre

*The geometric, structurally
transparent design of
the chair makes it a striking
furniture piece.*

4–5

Magnetic Jewellery
2004
Bengtsson Design Ltd
Magnetized silver

*This modular jewellery
system means that
the wearer can create
necklaces, rings and
bracelets from the same
piece of jewellery.*

bernabeifreeman, Sydney

bernabeifreeman was established in 2002 by industrial designers Rina Bernabei and Kelly Freeman. The partnership specializes in decorative lighting for commercial and domestic interiors, and has recently expanded into furniture. Bernabei worked in design studios in Sydney and Milan before becoming a lecturer in industrial design at the University of New South Wales in 1996. A former student of Bernabei's, Freeman graduated in 1998 and now also lectures at the same university.

bernabeifreeman describe its design sensibility as overtly feminine and decorative within a predominantly masculine industry. While Bernabei and Freeman draw heavily on the rich history of women's domestic handicrafts (particularly textiles), it is crucially important to them to use purely industrial materials and processes in the production of their work. Their Peony chandelier (2003) is made from a series of punched and laser-cut anodized aluminium panels that are suspended in three tiers. The perforations allow light through, creating an overall image based on traditional Japanese floral textile design. The Leaf pendant light (2003) applies the same principle: the dappled shadows cast by the William Morris-inspired leaf pattern create a wonderful soft ambience. In more recent work bernabeifreeman has taken the delicate patterning of lace through computer design and CNC milling processes to create products such as the Ema Pendant (2006) and the Cloth Table (2006), in which the traditional doily tablecloth becomes the tabletop itself.

As academics, Bernabei and Freeman regularly present papers and undertake research projects. The development of their innovative Shaali Light Scarf (2002) – a scarf sewn with conductive thread and jewel-like LEDs, which becomes a light when placed on its coat stand – has led them to investigate new technologies. **Brian Parkes**

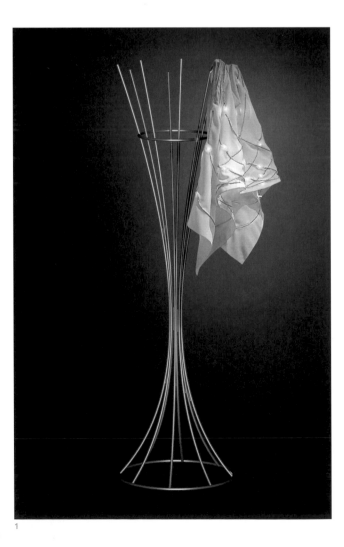

1

3

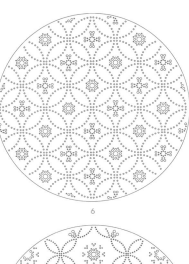

4

Cloth Table
2006
bernabeifreeman
Powder-coated stainless
steel

*The Cloth Table's perforated
top has a check design and
mimics a traditional tablecloth. The table top can
also be reversed to form
a serving tray.*

5–7

Ema Pendant
Light
2006
bernabeifreeman
Powder-coated aluminium,
hand-printed fabric

*The broderie anglaise
design of the diffuser is
appealingly continued in
the shade by the use of
hand-printed fabric.*

Mana Bernardes, Rio de Janeiro

Mana Bernardes is a jewellery designer and visual artist, who is also a member of a well-known creative family – her grandfather was the famous Brazilian architect Sergio Bernardes. Her professional story began when, at the age of seven, she created her first necklace. By the time she was 11 she had already produced pieces for a Rio de Janeiro store and headbands for a TV soap opera.

Bernardes invents objects in order to express a single concept: 'the power of transformation is the human being's jewel'. She uses small glass bottles, PET bottles, toothpicks, hair clips, pearls, silver and gold to create a modern and challenging look. For example, glass balls are transformed into the Glass Ball Necklace (2003), a coffee spoon becomes the Teasing Necklace (2004) and collars with mirrors display You're Something on Me (2004). Her pieces are sold in select fashion stores in Brazil and at the Collette store in Paris.

In 2004 Magnomomento (2004), a weightless magnetic clasp made with transparent plastic, was her first product manufactured on an industrial scale. Since then she has designed products such as Nokia Sphere (2005), a mobile phone pendant for Nokia. In 2005, she was invited by the Campana brothers to participate in the exhibition 'J'en Rêve', at the Cartier Foundation, which gathered together young artists from around the world. As part of this same project she contributed to a commemorative exhibition with the British magazine *i-D*.

Since the age of 14 Bernardes has held workshops, and has already trained over 300 teenagers in art and jewellery-making through a social project in Rio de Janeiro. Currently she is developing a methodology called 'Life Story through the Object, Story of the Object through Life', offering training to artisans in northern Brazil and to students of the European Design Institute in São Paulo. Bernardes' designs have been highlighted in the book *Momentum: Design Carioca* (2004), about Rio de Janeiro designers, and have been exhibited in the Biennale de St Etienne, France. **Maria Helena Estrada**

1–2

Nokia Sphere
Mobile phone pendant
2005
Tatil Design,
Fernanda Saboia
Plastic ball, nylon, leather

A mirrored ball with leather fringes.

3

Glass Ball Necklace
2003
Suelen Figueira
Glass balls, plastic net, plastic magnetic latch

Glass balls are transformed into an elegant floating necklace.

4

Urban Indian Bracelet
2006
Tatiana Barbosa
Acetate, elastic

Pieces of acetate are hand threaded with acetate threads.

3

4

5

Brazil Necklace
2006
Fernando Vitória
Leather, plastic

*Plastic circles are bent
and hand-placed, forming
geometric shapes.*

6

Fan Necklace
2005
Juliana Lima de Silva
Bamboo toothpicks,
pearls, nylon,
plastic magnetic latch

*Bamboo toothpicks
separated by pearls come
together to form this
fan-inspired necklace.*

7

**You're Something On Me
Necklace**
2004
Erivelto Rodrigues
Acrylic, mirror, nylon,
plastic magnetic latch

*A necklace which
reflects the person you
are talking to.*

8

Serpent Necklace
2004
Juliana Lima da Silva
Flower-cut PVC sheets,
waxed nylon, pearls

*Individual flower-shaped
PVC sheets are
hand threaded with
waxed thread.*

9

Stirrer Necklace
2004
Juliana Lima de Silva
Plastic coffee stirrers,
leather, crystals

*An unusual combination
of materials come together
to form a graceful
and translucent necklace.*

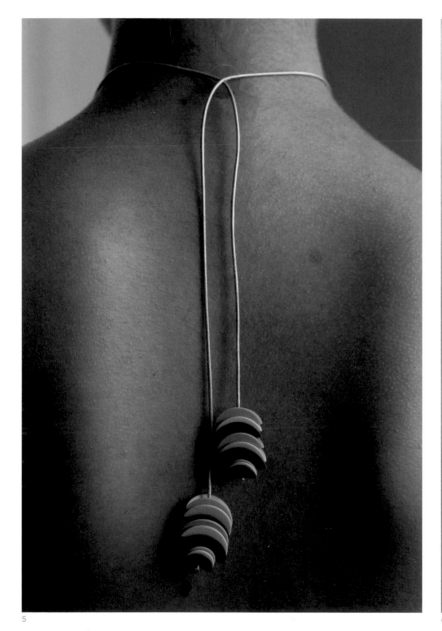

5

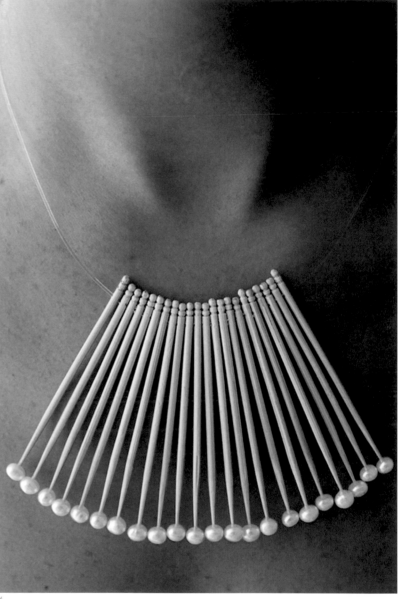

6

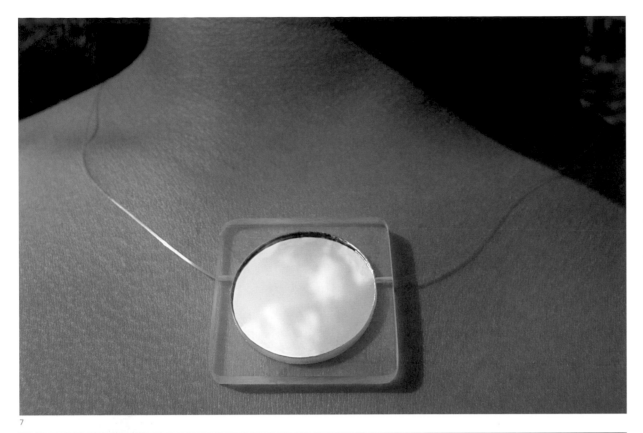

7

8

9

Big-game, Brussels and Lausanne

Big-game, founded only two years ago, resulted from the meeting of Elric Petit, Grégoire Jeanmonod, Adrien Rovero and Augustin Scott de Martinville at ECAL – École cantonale d'art de Lausanne, where they were studying industrial design (although Rovero has now left the group). Drawing on their diverse backgrounds and the dynamics of creativity and communication emanating from the school, the three members demonstrate tremendous ambition and different approaches to product design.

Big-game's first collection, HERITAGE IN PROGRESS (2005), plays with the concept of pitting traditional ideas against our contemporary way of life. It humorously reinterprets hunting trophies, transforms industrial lamps into domestic objects and converts old tables into trestle systems. Conceived with industrial production in mind, these objects demonstrate Big-game's technological expertise and realization of economic potential. In fact, most of its products are now in manufacture.

Big-game's second collection, New Rich (2006), explores an ironic take on the accessory as viewed by the fashion and luxury worlds. Taking the opposite path from a traditional design approach, the team here ventures into a more exclusive and refined field. Thus the BIC ballpoint pen, an icon of democratic design, is given a 'new look' with a solid gold cap. Similarly, the push-button of the BIC lighter is adorned with gold, Swatch watches become luxury objects, and even a simple drawing pin, transformed by the Midas touch, becomes a precious item of jewellery.

PACK, SWEET PACK (2006), Big-game's latest collection, reinvents the domestic world using the vocabulary of industrial packaging: a tufted wool carpet assumes the form of an unfolded cardboard box; a lamp made of expanded polystyrene is fixed by adhesive tape; a stool made of thermo-lacquered aluminium borrows the techniques used in folding cardboard; and a polystyrene and cardboard vase cleverly merges the object and its packaging. These are just some of the new items produced by this group, who seem destined to have a bright future. **Pierre Keller**

1–2

BOX stool
2006
Big-game
Powder-coated aluminium

The BOX is based on cardboard folding techniques, which have been adapted for use with aluminium sheets.

3

PACK, SWEET PACK
Home Item Collection
2006
Big-game
Various

This collection takes advantage of the formal and constructive qualities of packaging, to create products for domestic use.

3

4

Big-game

4

HERITAGE IN PROGRESS
Home Item Collection
2005
Big-game
Various

*A collection of objects
born from the
clash between a certain
old-fashioned bourgeois
culture and the realities
of today's way of life.*

5

Treteau
Trestle
2005
Ligne Roset
Lacquered MDF

*A simple and functional
trestle becomes the
shadow of a turned-wood
Regency-style table leg.*

6

Moose
Wall-mounted ornament
2005
Vlaemsch
Beechtriplex

*The traditional idea of the
hunting trophy has been
playfully transformed into
a very modern wall piece.*

5

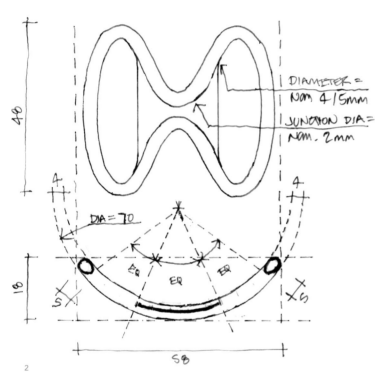

Steven Blaess,
Dubai and Sydney

Steven Blaess's childhood was spent moving around with his family through the vast and extreme landscapes of South Australia including the Nullarbor Plain and the coast of the Great Australian Bight. These early experiences informed and stimulated his creativity, and he eventually studied interior architecture and furniture design at the University of South Australia, Adelaide.

After graduating in 1994, he worked with several of Australia's leading architecture, interior and graphic design companies on high-end residential and very large-scale commercial projects. Blaess has developed a significant multidisciplinary design practice, encompassing furniture, product, lighting, interior, architectural and transportation design. With a growing interest in designing interiors for private jets and yachts, Blaess relocated to Dubai in 2006. The move has provided him with first-hand experience of the intriguing blend of East and West.

Blaess's first international success came in 1999 when pinkusblaess, the partnership he had established with Kendra Pinkus, was engaged by global company Johnson & Johnson to develop plastic carry cases for sanitary products. Well over ten million copies of its 3pac Digital Tampon Container (1999) have since been produced, and several other items have been added to the range. That same year Blaess began developing product ideas for Alessi, although it was not until 2005 that Alessi put one of his designs into production – the highly successful Marli Series bottle opener (2005), which doubles as a tactile and compact key ring.

Blaess takes his cues from colour, texture, light and form in nature and attempts to inject personality and character into each of his products and interiors. He has enjoyed significant success at the Milan International Furniture Fair since first exhibiting there in 2001 with his Meditation Pod Seat (2001) for Edra. And he continues to develop lighting and furniture concepts for several other European manufacturers, including Artemide. **Brian Parkes**

1–3

Marli Apribottiglie
Bottle opener
2005
Alessi
Investment-cast high tensile
18/10 stainless steel

The design is based on the rotation of the human wrist to twist open a bottle cap, rather than the lever action of traditional bottle openers.

4–9

Marli Ice Tongs
Prototype
2005
Alessi
Investment-cast high tensile 18/10 stainless steel, injection co-moulded silicon grip

The 1-piece moulded form is narrowed at the return radius enabling a flexible cast hinge within the form. The integration of injection co-moulded silicon gives grip when serving ice.

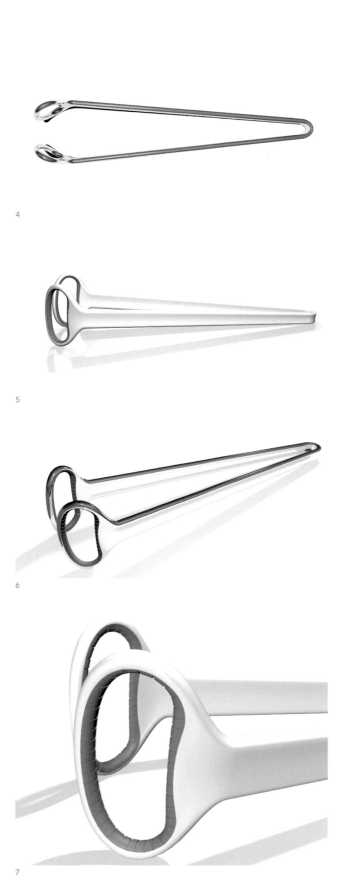

4

5

6

7

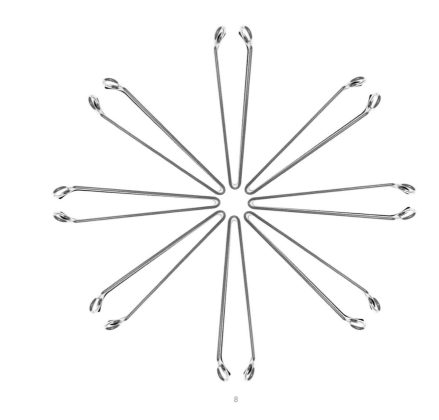

8

9

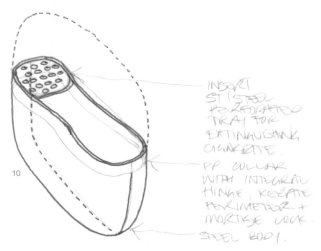

10

10–16

EXT Square
Cigarette butt case
Prototype
2005
Alessi
Machine-pressed 18/10
stainless steel, injection-
moulded polypropylene

*Cigarette butts are often
carelessly discarded on
the street. The EXT allows
smokers to get rid of
their butts in a sealed
case, to be disposed of
properly later.*

17–19

Shaker
Cocktail shaker prototype
2004
Fratelli Guzzini
18/10 stainless steel,
enamel

*This prototype shaker
has a measure in the lid
and a strainer in the main
body to help you make
the perfect cocktail.*

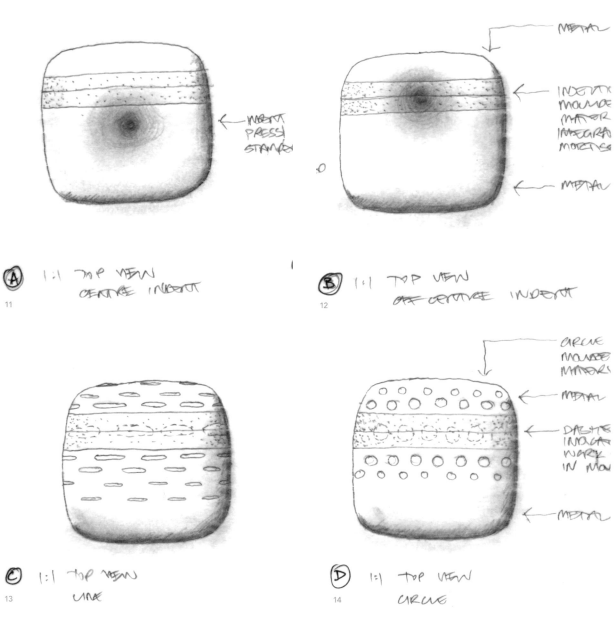

Ⓐ 1:1 TOP VIEW
CENTRE INDENT
11

Ⓑ 1:1 TOP VIEW
OFF CENTRE INDENT
12

Ⓒ 1:1 TOP VIEW
LINE
13

Ⓓ 1:1 TOP VIEW
CIRCLE
14

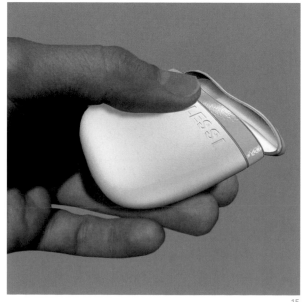

15

16

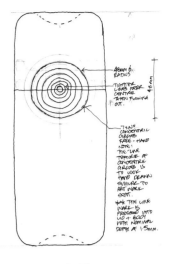

section

18

ELEVATION FRONT
BATES ©2004

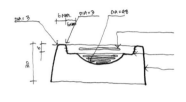

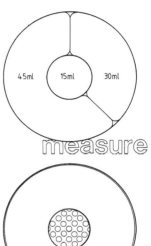

45ml   15ml   30ml

measure

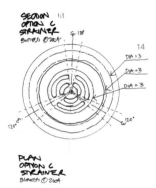

SECTION
OPTION C
STRAINER
BATES ©2004

14

strainer

19

PLAN
OPTION C
STRAINER
BATES © 2004

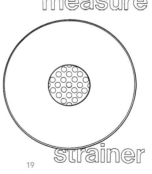

1

BLESS, Berlin and Paris

Based in Paris and Berlin, Desiree Heiss and Ines Kaag, known as BLESS, have rapidly made a reputation for themselves after their first product, a wig made from recycled fur, was selected by Martin Margiela for one of his most legendary collections (Winter 1997–8). Since then, they have created their own collections, allocating them sequential numbers, each product being produced in a limited edition of between 20 and 2,000 pieces.

However, despite having presented regular fashion collections since 2002, BLESS cannot be easily categorized under any particular discipline. This has given it the freedom to follow its changing interests, producing unexpected results. It can subvert the rules, adopting comfort or a certain functionality as the basis for an alternative. The objects, dresses and accessories are in fact always the expression of a recognizable functionality, albeit unexpected and unpredictable.

The BLESS duo has developed a particular sensitivity to the 'subversive' potential of its design. The subversion is conveyed with subtle irony through concrete solutions for everyday problem areas, no matter how insignificant or mundane, such as dressing, shopping and travelling. In Bagcollector (2001), part of their No.14 Shopping Supports project, the bag adopts the shape of a leather envelope, with simple hooks to hold together plastic or paper bags accumulated during shopping. Shopping bags can be added or removed modifying the shape and colour of each Bagcollector assuring the comfortable transportation of new belongings on the way home whilst staying very close to, and almost protecting, the body.

Despite the diversity of the fields addressed in their projects such as furniture from No.22 Perpetual Home Motion Machines (2004), hammocks from No.28 Climate Confusion Assistance (2006) or cable accessories from No.26 Cable Jewellery (2005) – certain themes are recurrent in BLESS's work: a fascination with the recycling of materials, and with traditional handicrafts such as embroidery, knitting and crochet; an aptitude for deconstruction; and a penchant for adopting assemblage as an aesthetic practice. **Francesca Picchi**

2

8

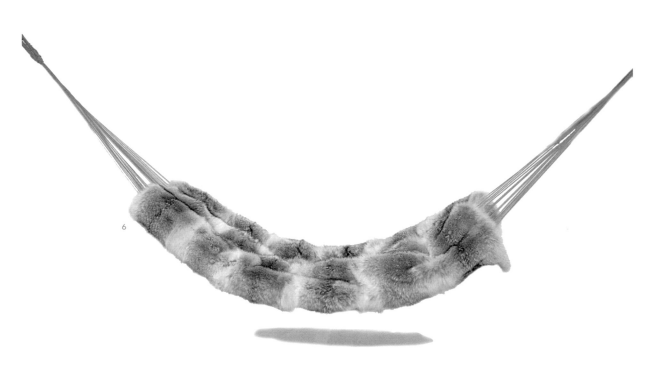

6

9

7

10

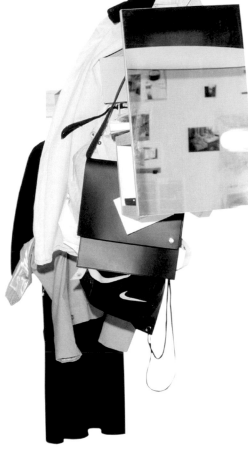

11

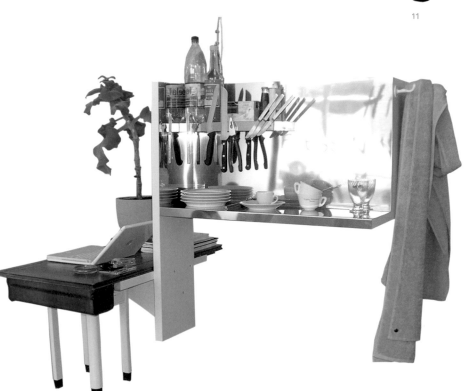

12

6

Fur Hammock 2
2006
BLESS
Fox or coyote fur, or
sheepskin, cotton

7

Stylefree
Signature Pin
2006
Fireproof nylon,
polyester stuffing

*Part of the No.28 Climate
Confusion Assistance
collection, these items
playfully provide ways of
dealing with the cold.*

8–9

4 Zipper Purse
2006
BLESS
Calf leather

*Part of the No.27 Eased Up
collection, this purse has
four individual zips so the
user can separate different
currencies, cards and other
personal items with ease.*

10–12

Mobile
2004
BLESS
Wood

*Part of the No. 22
Perpetual Home Motion
Machines collection, these
Mobiles arose from the
question of how to keep
garments tidy and on
display at the same time,
and plays with ideas of
the public and private.*

Jörg Boner, Zurich

Jörg Boner lives and works in Zurich, where he has been operating on a freelance basis since 1999. After studying product and interior design at the Schule für Gestaltung, Basel, he won a number of prizes for his furniture projects, which were put into production by companies such as Team by Wellis, Wogg and Nils Holger Moormann. As designer, author and lecturer at ECAL – École cantonale d'art de Lausanne, Boner's approach embodies mobility, creativity and openness, which undoubtedly makes him one of Switzerland's most important designers.

In his work Boner successfully brings together two opposite design trends: product design and communication design. The clarity of his concept and meticulous implementation through appropriate use of traditional materials bring him close to a family of designers for whom modernism is still alive. However, his use of objects as vehicles for anecdotes, memories and emotions places him among that generation of designers for whom the object must, above all, communicate with its user.

This intimate link between the object and its user can be seen in his Ajax writing desk (2000), manufactured by ClassiCon: it is an up-to-the-minute adaptation of a classic piece of furniture, yet loses none of the private and practical qualities of a working table. Made of plywood with birch-faced veneer and chromium-plated steel tubing, the desk incorporates an energy-saving lamp that is concealed beneath the raised storage, providing an even, glare-free light. When not in use, Ajax can be folded and stored flat. Similarly, the demountable Dresscode (2005), a wood and fabric cupboard distributed by Moormann, while employing cutting-edge modular design, conjures up memories of camping and student life, combined with elegance, ration-ality and lightness.

Another example is the BoCu armchair (2004) for Team by Wellis. Produced from classical materials, the design features a basic sculptural idea. Seat and back are each formed by geometrical planes transformed into three-dimensional objects through two-dimensional distortion. The lower frame provides a visual reference to classical modernism. In this instance, material and technical aspects fade into the background. BoCu is first and foremost a sculptural achievement.

Boner's work is distinguished by clear references and interactions between classic materials and new approaches.
Pierre Keller

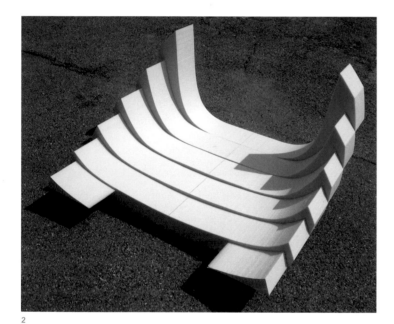

1–7

BoCu
Armchair
2004
Team by Wellis
Cellular foam core, leather,
shiny chrome

This stylish chair with
swivelling base and four
cruciform feet, also
comes in a fabric cover
with a choice of steel,
matt varnish or shiny
chrome for the stand.

Jörg Boner

8

9

8–9

Who's gonna drive you
home tonight?
Bench
2002
Gallery Spazio Opos
Galvanized steel, coloured wood

*This limited edition bench
is shaped like a car to draw
attention to both the car's
domination of public spaces,
and also to remind us that
a public bench instantly creates
a universal public space, even
in a car park.*

10

Cookuk2
Cooking studio
2006
Steel, wood, fabric

*The Cookuk2 cooking studio
can be used as a classroom,
restaurant, or location for
private or business events. The
space aims to enhance the
good taste of the meals, with
the good taste of the décor.*

10

1

2

Fernando Brízio, Lisbon

Fernando Brízio is undoubtedly one of the most thought-provoking practitioners in design today. Born in 1968, in Portugal – a country that for several years was distanced from the international industrial design epicentre but that has been carving out an increasingly prominent place for itself in the last decade – Brízio stands out as an indisputable and consistent reference. The density and richness of his work and his outlook on the design world have progressively led to the creation of a myriad of projects whose standing goes far beyond the production of simple functional and formal solutions. The real purpose of Brízio's research and his objects is to probe the more conceptual dimensions of design tied to material production and the interaction between the user and the product.

Brízio's explorations not only reveal his unique vision, underpinned by a sharp sense of humour and a need for self-questioning, but also examine production processes, material behaviour and the 3-D semiotics of objects. Brízio is interested in developing creative practices in which he is deliberately not in complete control and where the outcome of the final piece is unpredictable: for example, his Painting a Fresco with Giotto vase and jar (2005), where we are unable to anticipate the final display of colours, the Journey ceramic pieces (2005), where the form is decided by their voyage inside a car as the ceramic pieces are drying, and the Sound System lamp and jar (2003), where the designs are made from the result of the shapes sound waves take when the words 'lamp' and 'jar' are spelt.

In addition, his desire to challenge limits and dynamics and confront the user with new questions has led him to such pieces as the unstable Bowls with Pin, the HB Drawing Shelf, Table Layered Carpet and Drawing Table/Telephone Table. The irony in works such as Eaten Plate, Tablecloth Tableshirt, the Casa lamp and the Pata Negra stool reveal his engaging, enticing spirit, which also surfaces in his exhibition design projects and other more specific areas, such as set design.

Brízio is a researcher who uses design as a way of making sense of the world and as a medium for questioning it, leaving us, the users, to experience his exciting challenges.
Guta Moura Guedes

1–2

Painting a Fresco
with Giotto #1
Vase
2005
Fernando Brízio Studio
Faïence, felt-tip pens

These faïence pots absorb the ink from the felt-tip pens to create one-off colourful designs.

3–8

Journey
2005
Fernando Brízio Studio
Porcelain

Pre-fired ceramic pieces are taken for a bumpy ride to create weird and wonderful forms; they are then fired to become an unusual pottery range.

3

4

5

6

7

8

Sound System
2003
Fernando Brízio Studio

*The Sound System records
the sound waves created
by a certain word and then
translates these oscillations
into a 3-D form to create
an actual object.*

1

2

Stephen Burks, Readymade Projects, New York

Since completing his studies in architecture (at Illinois Institute of Technology and later Columbia University's Graduate School of Architecture) and product design (at IIT's Institute of Design) Stephen Burks has unleashed his hyperactive energy and unconventional ideas onto the design world. Burks brings a distinctive character to his work, believing that 'under the best circumstances design communicates with immediacy' and his products demonstrate this aim through their boldness, functionality and intelligence.

Burks' Cup Chairs (2006) for Modus engages with the idea of surfaces and planes, as the two-part system visually separates the seat from the support of the chair into two distinct planes that can be upholstered in different fabrics such as wool or leather. Burks' laminated plywood Work Station for E&Y (2002) is also adaptable to the moment: it can be formed into desks, varying sizes of shelving unit and even comes with optional additional parts such as a wine rack. It is sleek utility furniture that encourages change and can be constantly rebuilt as required by the owner and their living/working space. Other Burks designs such as the Ambrogio magazine rack (2004) and Gio clothes dryer (2004) also show the designer's desire to combine style with functional simplicity.

This is just a taste of the incredible amount of work Burks has designed. Perhaps one of the reasons that this talented, instinctive designer is not a household name is the sheer diversity of projects he has undertaken with his New York-based studio, Readymade Projects. His work embraces graphics, product design, retail interiors, packaging and concept design, with clients including Estée Lauder, Missoni, Cappellini, Vitra and Idee.

In his contribution to the Conductor's Baton design project organized by the École cantonale d'art de Lausanne, the stem of the delicate gold stick he designed suddenly bursts into uncontrollable swirls and curves as if echoing the conductor's movements from the past and anticipating those that are to come. Its key quality is its dynamism and the same could be said of Burks. **Tom Dixon**

1–2

Light Frame
Shaded prototypes
2005
Readymade Projects
Laser-cut acrylic, polycarbonate, nylon mesh fabric, zipper

A nylon mesh cover is zipped onto the existing Light Frame (2004) allowing the clear acrylic structure to be produced in various colours.

3–5

Parallel Shelving System
2006
Modus
Lacquered medium-density fibreboard, rubber-coated steel

These practical shelves fold along concealed hinges and flat-pack for easy shipping and storage.

3

4

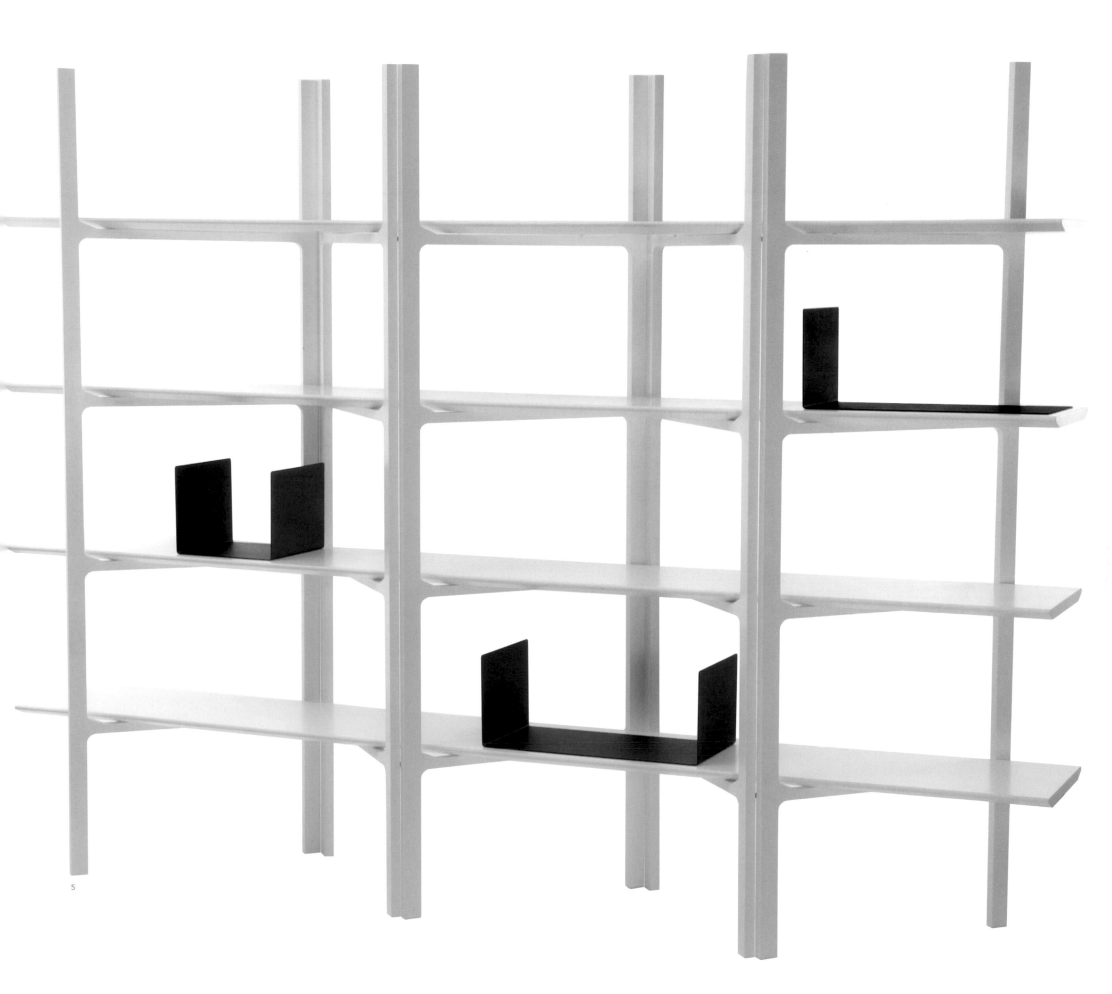

6

7

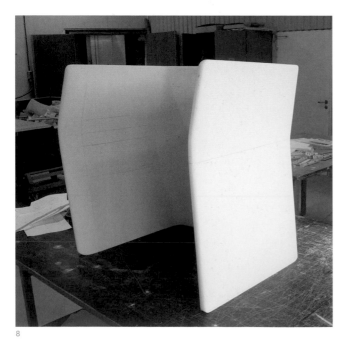

8

Stephen Burks,
Readymade Projects

10

9

6–10

Cup Chairs
2006
Modus
Cold-cured polyurethane
foam, internal steel frame,
wool or leather upholstery

*The chairs utilize a 2-part
system that visually
separates the seat from
the rest of the chair so that
they can be upholstered in
different fabrics.*

11–13

Ambrogio
Magazine rack
2004
Zanotta
Moulded plywood, stainless
steel

*A horizontal access
system for face-out or
spine storage of books,
CDs and magazines.*

14

Gio
Clothes dryer
2004
Zanotta
Stainless steel

*A simple frame with
fixed hinge points
whose essential form
communicates its function.*

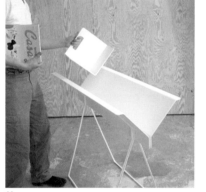

11

12

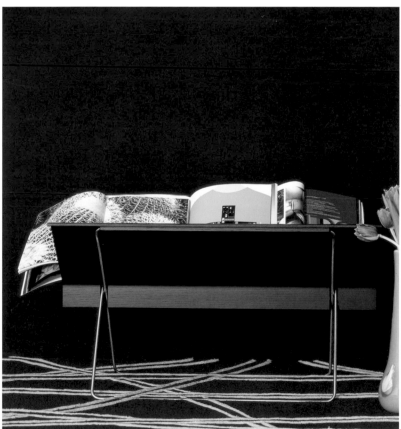

13

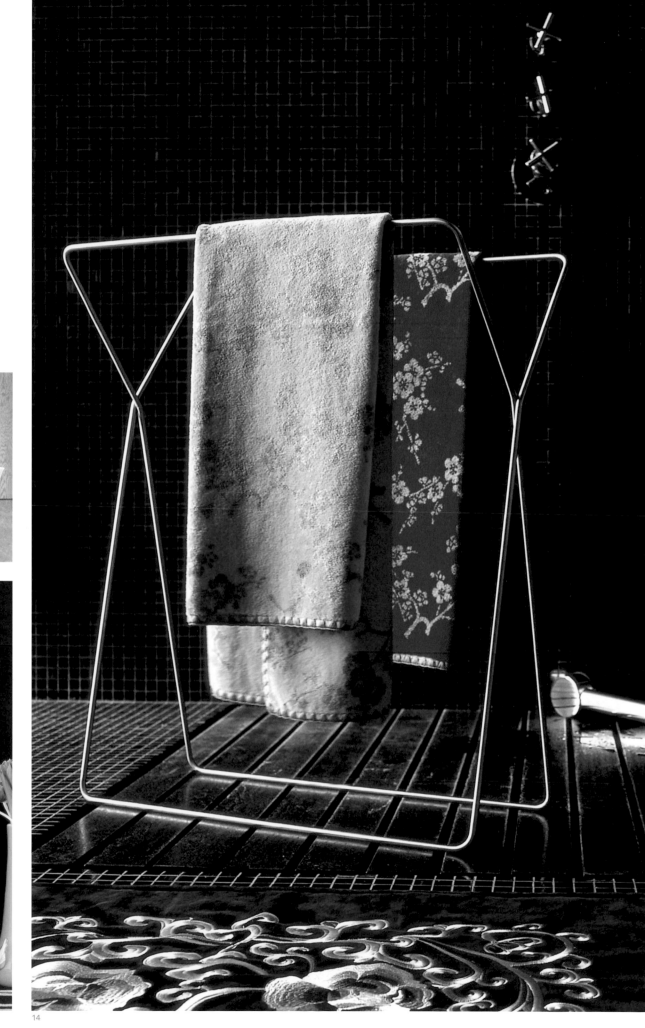

14

Sam Buxton, London

Sam Buxton is an entrepreneurial British designer who aims to work on as diverse a range of design projects as possible. His latest endeavours are incredibly varied: a clothing and footwear project for Reebok; an ongoing collaboration with Vauxhall Motors UK; and a clock that tells the time with illuminated words. Buxton graduated from the Royal College of Art in London in 1999, and since then has continually experimented with industrial technologies and materials, often using them to explore relationships between objects and the human body. He is most interested in creating work that crosses the perceived boundaries between art, science and design.

Buxton came to prominence in 2001 when he exhibited MIKRO-Man (2001), his imaginative chemically etched metal business card, at the 'Design Now London' exhibition at London's Design Museum. The card folded out to create a 3-D representation of himself working at his desk with a laptop computer. A UK-based company approached Buxton to manufacture the card as a commercial product and it became the first of what is now a growing family of highly successful MIKRO products, all designed by Buxton. In 2004 he developed the idea even further to create the 15m² MIKRO-City, an extraordinary installation for the Design Museum.

Buxton's eagerness to develop new modes of interactivity has fuelled an ongoing series of experimental products that he refers to as Surface Intelligent Objects (SIOS). He uses display technologies such as electroluminescence (conductive inks laminated onto transparent plastics that create a flat and flexible light source) to bring the surfaces of objects to life. For example, his SIOS Wristwatch (2003–6), where the entire strap communicates the time, or his Clone Chaise (2005), which graphically maps the systems of the human body – when approached, its heart begins to beat and its lungs appear to breathe. **Brian Parkes**

1

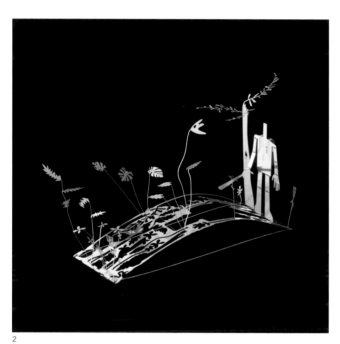

2

1
MIKRO-House
Fold-up sculpture
2003
Worldwide Co.
Acid-etched stainless steel

2
MIKRO-Man Jungle
Fold-up sculpture
2002
Worldwide Co.
Acid-etched stainless steel

Stemming from a fold-up business card, the burgeoning MIKRO collection of fold-up sculptures is one Buxton's most popular products.

3
Hub Commissioning Centre
Information point
2003
SB Studio
Laser-cut stainless steel,
laser-cut acrylic sheet

Situated at the Hub Commissioning Centre in Lincolnshire, the design is cut from the walls of a cube and houses two computer terminals where designers can access craft information.

3

Sam Buxton

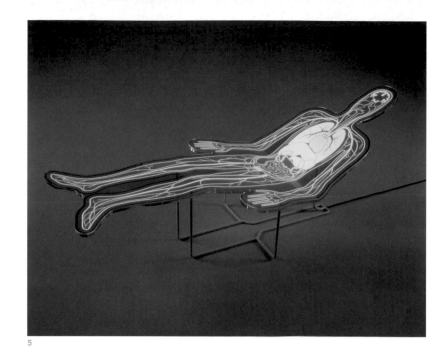

4

Timepiece
Clock
1999–2002
SIOS
Electroluminescent display,
sequencing electronics

*A clock that represents
the passing of time through
a series of animations
on the numberless panel.*

5–7

Clone Chaise
2005
SIOS
Electroluminescent display,
sequencing electronics,
acrylic, steel support

*As you approach the
Clone Chaise, the heart
illuminates and starts
to beat; the lungs begin
to breathe.*

8

Word Clock
2005–2006
SIOS
Curved electroluminescent
display, sequencing
electronics, acrylic

*Only the necessary words
are illuminated at any given
moment, to tell the correct
time in three time zones.*

9

SIOS Table
2003
SIOS
Electroluminescent display,
sequencing electronics,
acrylic, steel support

*66 individually controllable
areas create an intelligent
surface and communication
means that explores the
interface between human-
object and object-surface.*

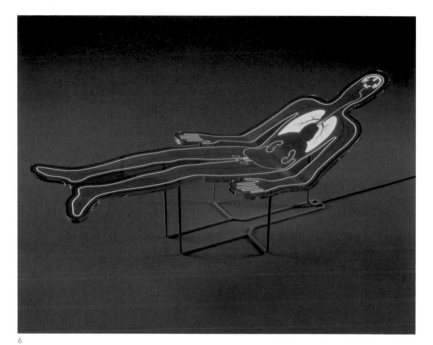

5

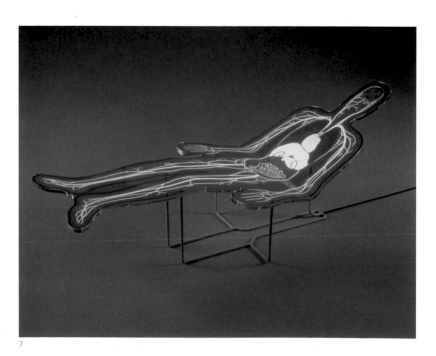

6

7

4

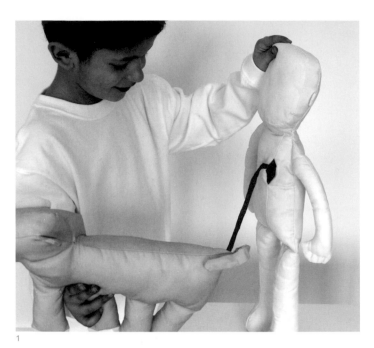

1

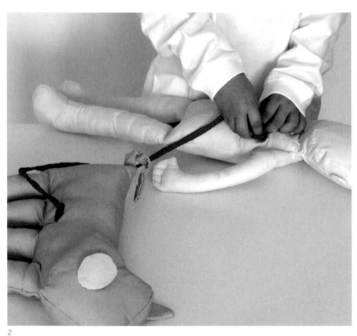

2

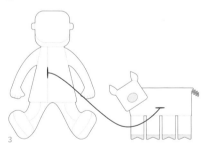

3

Elio Caccavale, London

Elio Caccavale graduated from London's Royal College of Art in 2003. He has pursued an intensely varied career combining consultancy, research, teaching and writing, and his design projects involve collaborations with scientists, social scientists and ethicists.

Straddling these different worlds, Caccavale considers separate disciplines to investigate the hybrid grounds where areas of expertise touch and overlap. His work participates in the debate on the 'critical' role of design in the face of the emerging needs of the contemporary world, involving, in particular, innovations introduced by advanced technology and biotechnology.

Biotechnology is for Caccavale an experimental ground where the most advanced technological innovation clashes with more human aspects and concerns such as ethics and social conduct. The transdisciplinary nature of his projects seeks an understanding of the role of design in creating a dialogue between science and everyday life.

Caccavale's Utility Pets (2004) project considers the possible donor-recipient/animal-human relationship. The organs of a pet piglet are genetically modified to make them perfectly compatible with those of its owner-recipient. A range of accompanying toys and accessories question the ethics and sentimental connotations of this complex relationship where the boundary between the animal and the human are blurred.

Using a similar 'narrative-based approach' Caccavale designed MyBio (2005), a range of transgenic dolls for experimenting with new teaching methodology. Using toys as a pedagogic tool he goes beyond verbal languages to assist children's approach to emerging biotechnologies and stimulate an easy understanding of complex concepts.

In both projects, Caccavale uses design to build familiar scenarios in order to present scientific information through the channel of the narrative. He is fascinated by our responses to the impact of biotechnologies, affected as they are by age-old fears that can make the concept of hybridity seem like frightening chimeras of our distant past. However, he sees contemporary advancement as an opportunity to re-examine, through design, the fundamental biological and physical properties that define human beings and their relationship with nature. **Francesca Picchi**

1–3
MyBio Xenotransplant
Concept product
2005
Batch Production,
Science Learning Centre
Fireproof nylon,
polyester stuffing

MyBio boy and MyBio pig demonstrate the physical transfer of the organ from the animal to the human.

4–5
MyBio Jellyfish
Concept product
2005
Batch Production,
Science Learning Centre
Fireproof nylon,
polyester stuffing

MyBio jellyfish glows bright green when illuminated with a UV light.

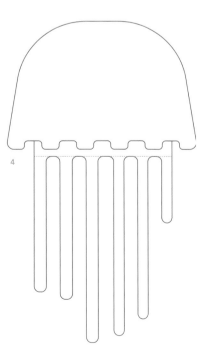

4

6

7

8

Toy Communicator
Concept product
2004
SLA-Epoxy

A pig toy with a microphone and a radio handset allows the owner to listen to the pig enjoying itself.

8

Comforting Device
Concept product
2004
SLA-Epoxy, pig snout

A psychological product made from the snout of the sacrificed pig, which serves as a memento after the xenotransplantation has been carried out.

9

Low-resolution Pig TV
Concept product
2004
SLA-Epoxy

A low-resolution TV exclusively for pigs that they can control by themselves.

9

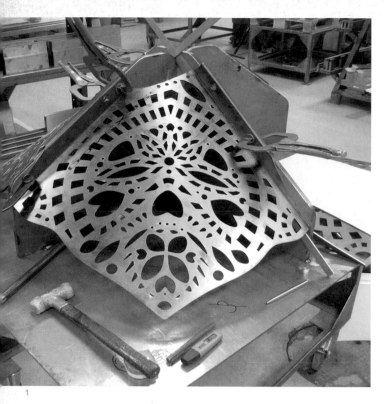

1–2

Prince Chair
2001 (launched 2005)
HAY
Laser-cut steel, water-cut neoprene rubber laminated with felt, powder-coated steel

The Prince Chair was created as part of a competition to design a chair for His Royal Highness, Crown Prince Frederik of Denmark. The chair represents the prince and his role in society by drawing inspiration from both historical precedents and cutting-edge modernity.

Louise Campbell, Copenhagen

Louise Campbell may be speaking the same design language as the current, popular aesthetic, but she has a unique sensibility that allows her to stand out in the overcrowded design scene. Her work is graphically strong, photogenic, sensual and highly desirable.

Campbell's design studio works to three golden rules: 1. always start from scratch; 2. everything is possible until the opposite has been proven; 3. there must be a good reason for every single decision made. These 'rules' highlight the importance of a systematic creative process, optimism, pragmatism and flexibility in her work, and, combined with the immediate visual appeal of her design, demonstrate her rigorous approach.

This combination of rationalism and imaginative delicacy is evident in some of her most recent work. The Veryround chair for Zanotta (2006) and Spiderwoman chair for Hay (2006) are both made from laser-cut steel and explore high-tech manufacturing processes. But they also play elegantly with ideas of light and shadow, and have a 'classic' design aesthetic.

After studying at the London College of Furniture in 1992 and then Denmark's Design School, in 1995 Campbell set up her own studio in Copenhagen in 1996 and since then has accrued an impressive number of clients such as Zanotta, Stelton and Louis Poulsen. For the latter she has created some stunning lighting pieces and for Stelton her Home Work range (2006) of desktop storage units is ergonomic and simple, showing her understanding of unfussy, strong design that really works.

There are a lot of designers who have a few good ideas, are 'one-hit-wonders' of design or follow the current trends a little too closely. However, there are telling signs that Campbell's distinctive style is going to keep pushing forward and progressing, and that we can expect a lot more from her in the future. **Tom Dixon**

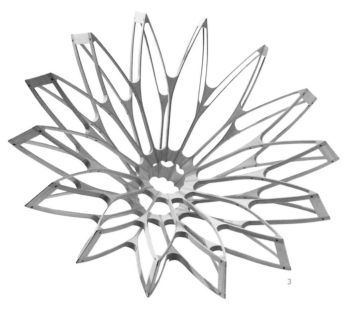

3

Billy goes Zen
Chair
2003
Ash

One-off chair designed for the 'Walk the Plank' design exhibition in 2003.

4–5

Home Work
Collection of storage units
2006
Stelton AS
3-D print plaster produced by rapid prototyping

This 14-container series intends to make life around the desktop easier – whether it be at home or at work.

6

Veryround
Chair
2006
Zanotta
Powder-coated sheet steel

The chair consists of 160 circles and focuses as much on shadow play as on formal construction.

7

Collage (detail)
Light
2004
Louis Poulsen
Laser-cut acrylic

Collage was inspired by the way light in a forest filters through thousands of layers before reaching the ground.

Louise Campbell

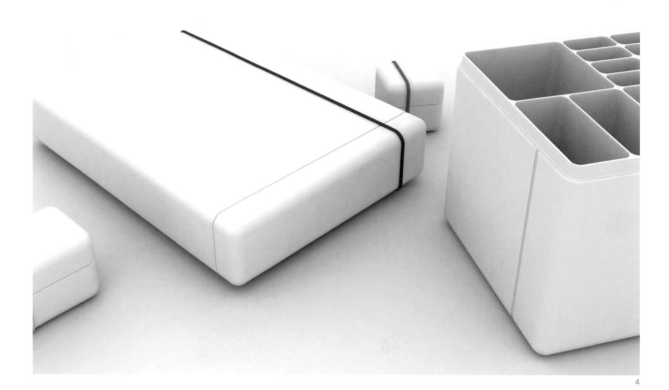

4

5

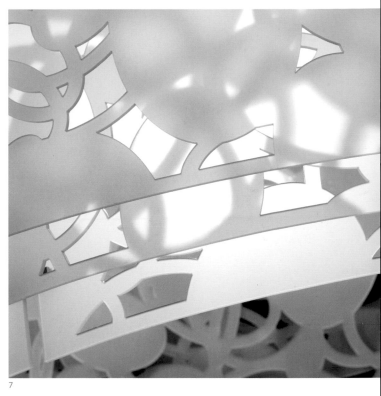

7

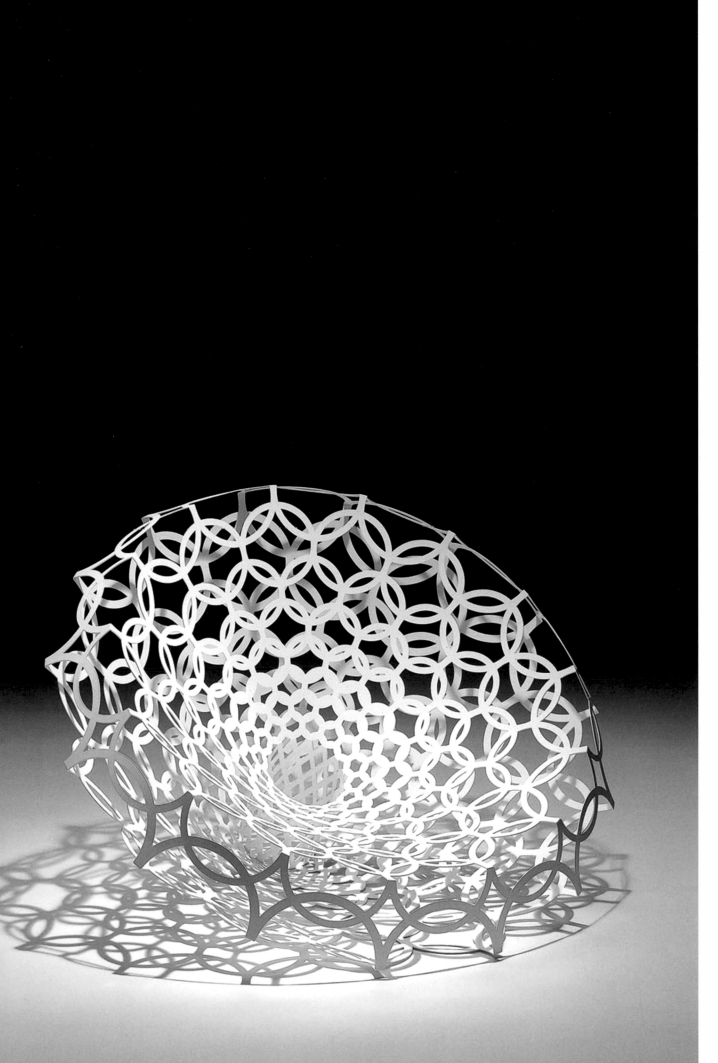

6

Leo Capote, São Paulo

São Paulo-born Leo Capote received his degree in industrial design from the city's Paulista University in 2002 and since then has worked at developing the concept of the functional reinterpretation of pre-existing objects for daily use.

In fact, this is an area to which many Brazilian designers have devoted their design energies. The reason for this focus is Brazil's small-scale retail market. Its limited industrial structure means that there are few companies that are willing and able to invest in a single product and the development of the tools, prototypes, production and marketing necessary for industrial product design. Perhaps for this same reason, Brazilian handmade design is gaining a greater market share worldwide.

Capote transforms cutlery, jugs, irons and vinyl records into chairs, lamps, photo-frames and soap dishes, stretching the capacity of his raw materials as far as they will allow. When he was a student, Capote completed his internship with designers Humberto and Fernando Campana, making the most of the opportunity to practise the technique of re-using materials. In his pieces, soup spoons became seats for armchairs – Spoon Chair (2001) – forks were spliced together to make photo-frames – Spider (2001) – while a glass jug formed the top of a lamp – Jar lamp (2003). With his Spade Chair (2005) a shovel is substituted for the seat and back of a chair and in Disk Chair (2003), remarkably comfortable and resistant considering the fragility of the material, two vinyl records provide the main parts.

Capote, like so many young Brazilian designers, closely guides the entire process of each of his pieces: he conducts the initial studies, develops the prototypes, produces limited editions and markets his own products. **Maria Helena Estrada**

2

1

Disk Chair
2003
Self-production
Stainless steel, vinyl disks

2

Disk Stool
2003
Self-production
Stainless steel, vinyl disks

The simple steel structure of the Disk chair and stool draw our attention to the playful use of LPs for the seat and back.

3–4

Spade Stool
2005
Self-production
Spade, steel, wood

5

Spade Chair
2005
Self-production
Spade, steel, wood

Capote creates these striking sculptural furniture pieces using a common garden spade.

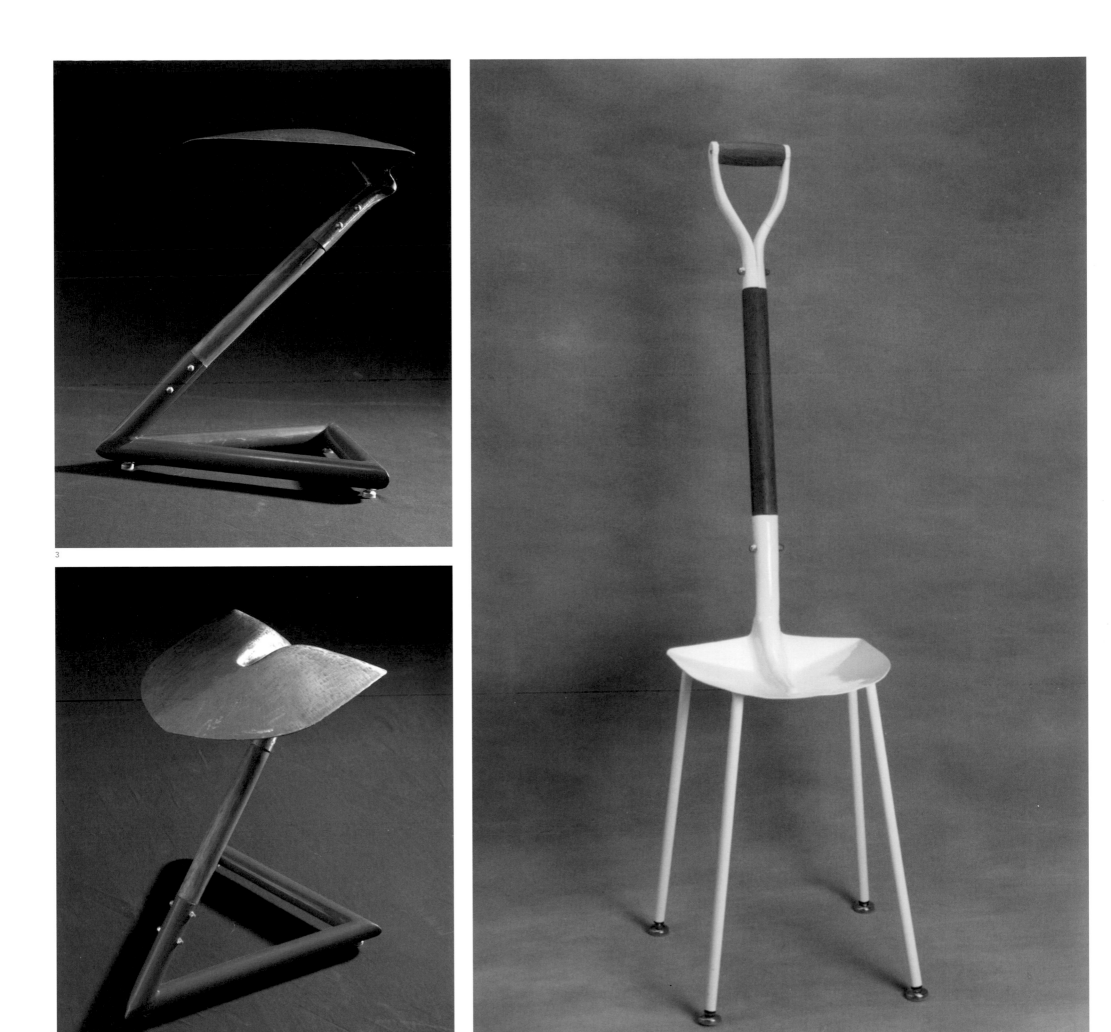

3

4

5

8

6

7

6

Spider
Picture holder
2001
Self-production
Metal fork

7

Spoon Simple
Picture holder
2001
Self-production
Metal spoon

Capote has cleverly
manipulated everyday
metal cutlery to create
stylish picture holders.

8

Iron Lamp
2002
Self-production
Iron, flexible tube, spotlight

A household iron proves
an unusual base for this
office lamp.

9–11

Spoon Chair
2001
Self-production
Stainless steel,
stainless-steel spoons

Exactly 114 spoons go
into creating the rippling
surface of this striking
chair's seat and back.

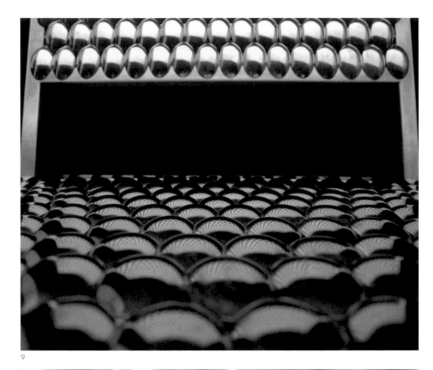

9

10

11

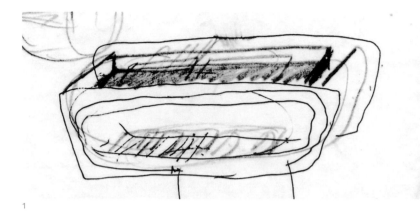

1

2

Jennifer Carpenter,
TRUCK Product Architecture,
New York

Jennifer Carpenter designs thoughtfully crafted furniture and accessories that bear a complex load of value. They are sturdy and practical yet have dashes of poetry. They are quiet enough to be fashion-resistant yet quirky enough to stand out from the herd. From snack bowls that nest to take the mess out of dipping to children's furniture that can be snapped together like a toy, they fit into contemporary lifestyles so snugly one may greet their arrival with a wistful, 'Ah, there you are!'

Trained as an architect, Carpenter worked for the New York firm Rogers Marvel designing Kate Spade store interiors before her employers founded TRUCK Product Architecture in 2000 as an extension of their practice, putting her in charge. She made an easy transition across scales: a virtuoso of biomorphic earthenware and etched glassware, she slid gracefully into improving the working life of taxi drivers (for an initiative to reinvent the taxi, she proposed a parkside rest station that allows drivers quick, bucolic breaks). Equally comfortable among urban art lovers, she recently teamed up with the legendary graphic designer Ivan Chermayeff to create the installation for a travelling outdoor photography exhibition of original commissions by Chuck Close, Mitch Epstein and Dayanita Singh.

Carpenter, who is the mother of two small boys, has made perhaps her most distinctive mark on children's products, designing an infant bed with a flip-down integral changing table that converts into a toddler bed once the child has grown up a little. 'Let's face it, baby products are not the sexiest arena,' she has said. 'Some designers would rather be working on the next iPod or the next running shoe than a breast pump ... But the breast pump is what really needs help.' **Julie Lasky**

1–2

Tambour Table
2004
TRUCK Product
Architecture
Solid wood,
PTD composite wood

*Impressive centrepiece
for a living room,
with storage area for
any unwanted clutter.*

3–5

Flip Down Crib
2006
Nurseryworks
Ptd (non toxic) wood, wood

*Adaptable crib with
flip-down changing table,
that can also be converted
into a comfortable toddler-
sized bed.*

3

4

5

Jennifer Carpenter,
TRUCK Product Architecture

6

STRETCH Fence
2005
Design Trust for Public
Space
Concept project
*Relief railing, integrated
with existing New York park
fencing, for taxi drivers to
rest against while they wait
for their next job.*

7–10

Clamshell Bowls
2004
Studio Nova
High-fired earthenware
*Sleek and stylish interlocking
set of three bowls.*

6

7

8

9

10

PIERRE CHARPIN, Ivry sur Seine

PIERRE CHARPIN's furniture and objects offer a unique combination of refined, almost abstract lines and bold colours, which bring cheerfulness and sensuality.

In his research into the fundamental traits of an object's purpose and shape, CHARPIN simplifies furniture down to its bare essentials, making an emblem of the object. His Stands (2002) have the purity of the representation of an idea – in this case the idea of props acting as supports for objects of domestic life. The issue of functionality is tackled with an intentional ambiguity: the elementary shapes of his realizations give the viewer freedom to interpret the possibilities of usage.

Articulation is another key notion behind CHARPIN's designs, both within individual objects and their environment, and also within grouped elements such as the Triplo Vase (2002), where individual tubes are gathered together to make a single vase, thereby creating a unified form. CHARPIN likes to create objects in series, to explore all the possibilities of an idea through different combinations of shapes and colours, as is the case in his Platform series (2006). Colour is a distinctive characteristic, both because of his use of a bright palette and because the plain colours interact with the volumes to influence our perception.

Working and teaching in France, CHARPIN also has very strong links with Milan, where he worked for George Sowden, a former member of the Memphis Group. Since then, he has been exhibiting in Paris and Milan, and collaborating with Alessi, Montina and Venini.

CHARPIN explores his vocabulary of shapes and colours in experimental projects such as Ceram X (2005) where his hide-and-seek erotic scenes made a new use of motifs on ceramics. CHARPIN is also suited to projects that veer more towards industrial design, which he demonstrated in 2004 by winning a competition organized by the Paris water company to design a glass carafe showcasing the city's tap water (Eau de Paris).
Didier Krzentowski

1

1

Platform Medium White
Wall mirror
2006
Kreo
Lacquered and polished
stainless steel.

2

Platform Coffee Table
2006
Kreo
Brushed and lacquered
aluminium

*These pieces were
exhibited at Galerie Kreo
in 2006 as part of Charpin's
solo exhibition, 'Platform'.*

3–4

Triplo Vase
2002
Venini
Blown multicoloured glass,
rubber band

*The Triplo Vase was
selected for the Compasso
d'Oro shortlist in 2004.*

2

3

4

5

6

7

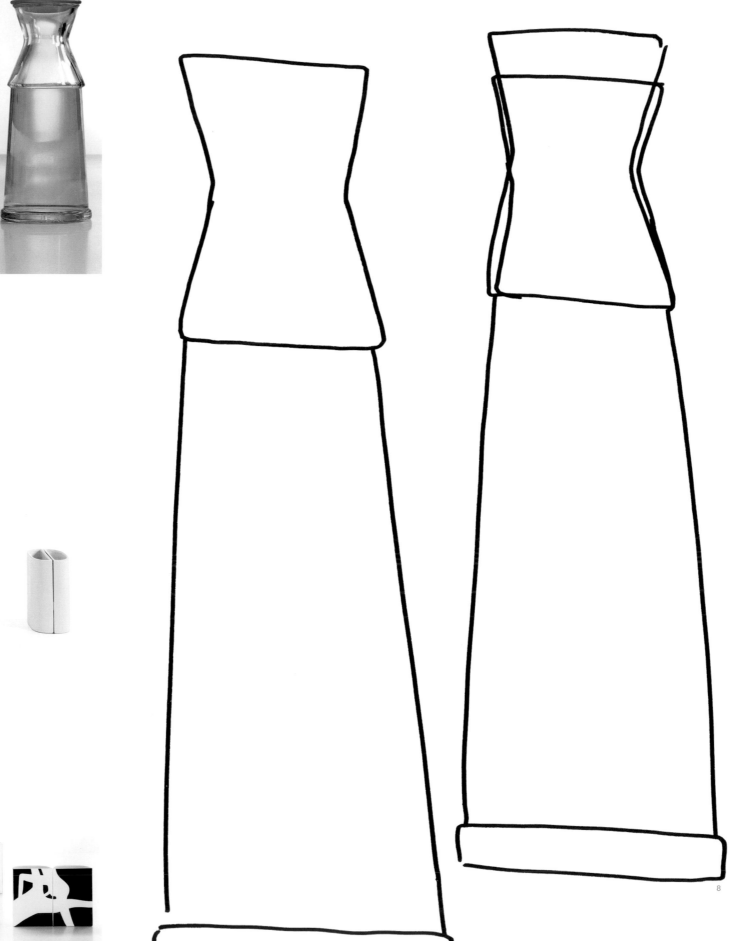

8

5 & 8

Carafe Eau de Paris
Glass carafe
2004
Eau de Paris
Moulded glass, plastic

*Winner of the Parisian
Water Service's 2004
competition to design
an enticing receptacle
for Parisian tap water.*

6–7

Ceram X Vases
2003/2005
Craft
Ceramic

*These ceramic vases have
graphic and colourful erotic
imagery on one side,
but can be turned around
when the neighbours
come over to become plain
white vases.*

9–10

Basket-Up
Coffee table
2004
Montina
Natural oakwood

*The coffee table has
a removable tray to
make this simple design
even more versatile.*

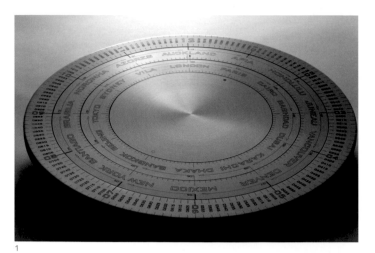

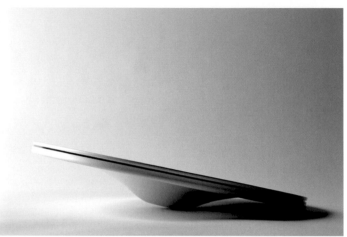

1–3

PI World Time Clock
2005
Lucas Chirnside studio
Hand-spun aluminium,
clock movement

Inspired by the elegant utopias of ancient horology, the limited edition PI simultaneously notes the time in 24 different zones of the world. PI shows that time is a shared condition that connects people around the world as much as it keeps them apart.

Lucas Chirnside, Melbourne

Lucas Chirnside completed a degree in architecture from RMIT University, Melbourne, in 2002. His foray into furniture and product design began during 2000 when he spent a year in Amsterdam studying at the Rietveld Academy and working with S333 architects. He has since maintained a commitment to simultaneous practice in architecture and product design. During the intense final months of his architecture degree, Chirnside was independently exhibiting prototypes of his Switch seating (2002) at Tokyo Designers Block 2002.

In late 2004 Chirnside took a five-month break from architecture work to undertake an Australia Council Residency in Milan, enabling him to foster connections with the European design industry and to focus on research and development of his own furniture and product ideas. Chirnside is fascinated with working at opposing scales: zooming out to work on large-scale urban design projects at 1:1,000 and zooming in to work on small products, often enlarging drawn components to a scale of 5:1 to resolve crucial details. This zooming in and out also informs Chirnside's conceptual interest in the interconnectivity between people, places and design around the world, a notion he calls 'universally local'.

These concerns are at the core of his provocatively elegant PI World Time Clock (2005), with which he won the Bombay Sapphire Design Discovery Award in Australia in 2005, and the more compact smlwrld Desk Clock (2006), which he exhibited at the Milan International Furniture Fair in 2006. Both of these remarkable timepieces remind us that the world's time zones connect us as much as they keep us apart. The clever use of a Moebius strip to organize localities from the 24 time zones on a 12-hour dial allows us to experience our own time relative to time everywhere else in the world. **Brian Parkes**

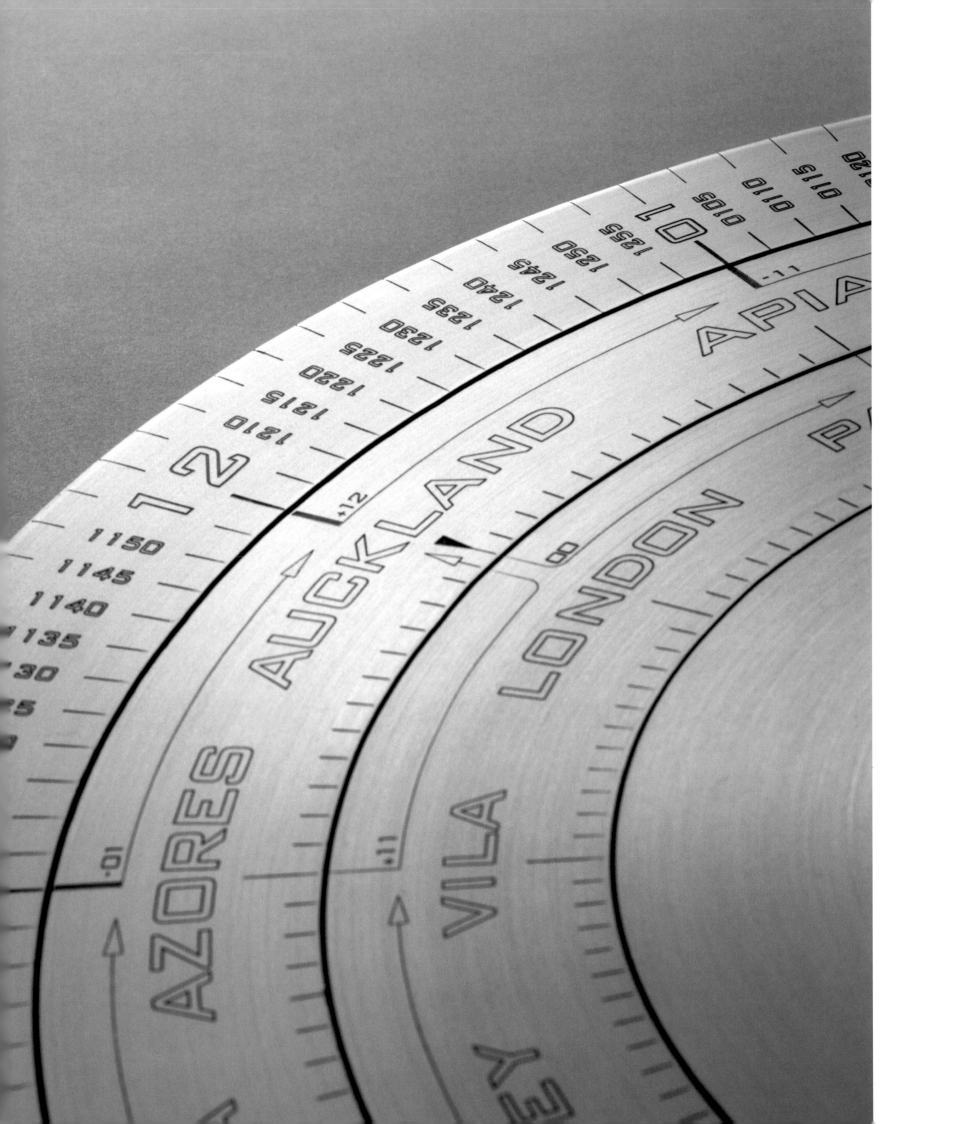

4

Lucas Chirnside

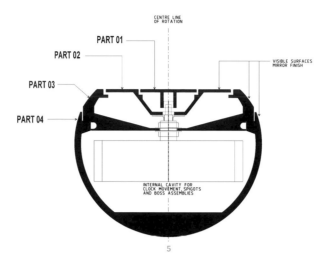

5

6

4–6

smlwrld Desk Clock
2006
Lucas Chirnside studio
Machined aluminium,
chrome electroplate,
clock movement

*The limited edition smlwrld
Desk Clock is inspired
by the image of quicksilver
beads, and represents
the impenetrable fluidity
of time in our lives.*

7–9

Polytope Seating
Landscape
2004
Edag Future/Lucas
Chirnside studio
CNC-milled
polyurethane foam

*Polytope is a seating
experiment proposing
4-D origami – folding
a cube. Interaction by
folding/unfolding is an
essential part of using
this furniture and gives it
an added dimension of
animation, a living quality.*

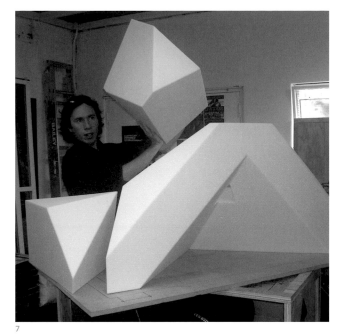

7

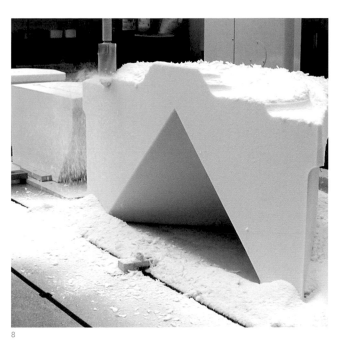

8

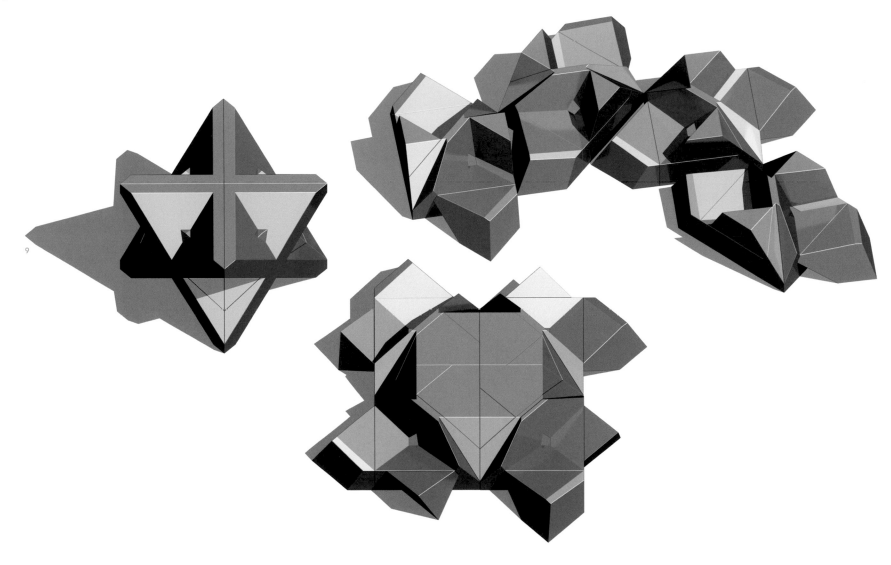

9

Rodrock – design drawing
Rocking chair
2005
Prosomedia
Lacquered beechwood

*Made completely of
beech, the Rodrock is
constructed through
binding solid wood
with numerous 5mm
wooden rods.*

2–3

Monza
Bar Stool
2007
CNC bent mild steel,
injection-moulded ABS

*The Monza is a stack-
ing bar stool that
combines functionality
with modern elegance.*

Clemens Weisshaar
& Reed Kram,
Munich and Stockholm

In 2001, having met while working as consultants on the information technology project for Koolhaas / AMO's Prada Epicentre stores, Clemens Weisshaar and Reed Kram combined their specific personal interests and founded a practice that works as a remote, multidisciplinary 'platform'. Their offices are in Stockholm, where Kram moved from his native USA, and Munich, where Weisshaar was born.

Coming from different design backgrounds – Kram from media and environmental design, and Weisshaar from product design – their strength is the ability to exploit the huge creative potential still latent in today's widespread technology. This gives them the freedom to look beyond standard mass production – the industrial model for the modern age. They challenge the stereotypical prototype by generating software that can control the birth and fabrication process, not of a single object, but of an entire generation of objects belonging to the same 'species'.

Kram and Weisshaar counteract the widespread pessimism prompted by the decline of manufacturing in Europe with a concrete, positive and pragmatic approach. Such optimistic projects include the Breeding Tables (2004), in which, for the first time, the computer is placed at the centre of the design, not only serving to check the process but forming an integral part of the project itself, from conception to production. Software is used as a virtual factory in which variations on a theme are continually produced, and from which the most convincing can be selected.

The Breeding Tables offer a solution to one of the key factors of design – that the basis of successful design is effective communication. Often all the components are there; it's a matter of constructing a network of relationships that allow the constituent parts to communicate.
Francesca Picchi

Component D
Bracket
Klammer

Component C_l
Hinge
Scharnier

Component C_r
Hinge
Scharnier

Component A_r
Frame Right
Rahmen r

Component A_l
Frame Right
Rahmen l

See Detail
Footrest
Fußstütze

Component B
Footrest
Fußstütze

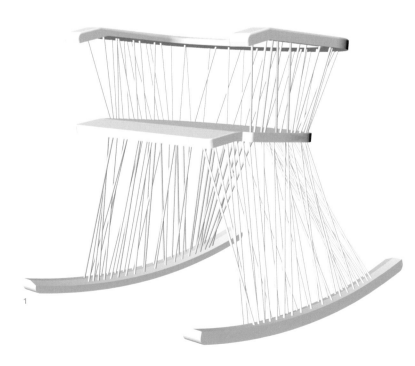

1

Component F
Seat
Sitz

See Detail Rubber Shock

Component E
Rubber Shock
Gummipuffer

MAG Weld here
Schweisspunkt

See Detail
Hinge
Scharnier

See Detail Frame Welds

MAG Weld here
Schweisspunkt

TIG Weld here
Schweissnaht

MAG Weld here
Schweisspunkt

Weld here

Detail Gliders TBD

2

3

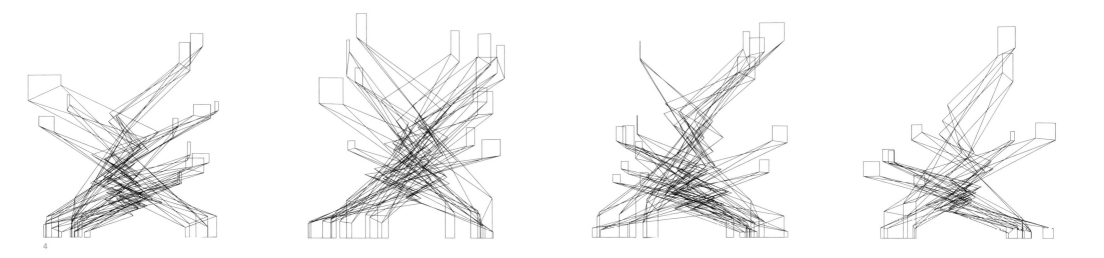

4

5

4–9

Breeding Tables
2004
Kram/Weisshaar AB
Laser-cut mild steel,
powder-coated and
brake-bent CNC

*Customized software was
used to generate slight
variations or 'mutations' so
that each table was unique.
The images show: script
output of table sequences
from the Breeding Tables
software, front elevation
(4); the finished product (5);
first series of Breeding
Tables (6); cutting patterns
from the initial two series of
tables (7); variables diagram
of the Breeding Tables
software, which define the
genetic code of each table
(8); second series of
Breeding Tables. (9).*

6

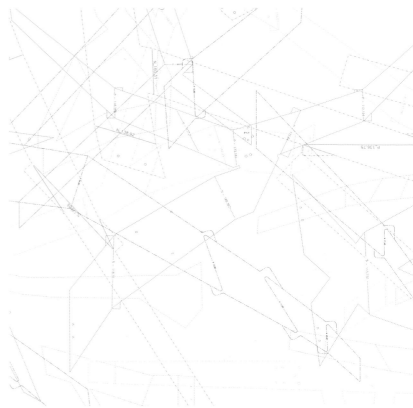

7

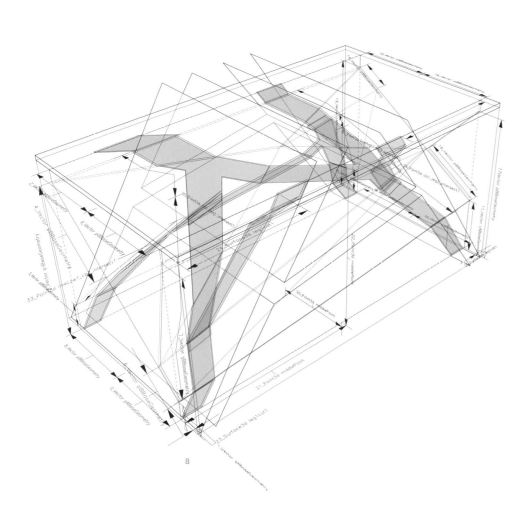

8

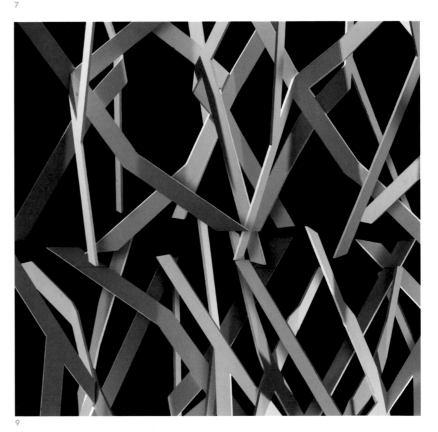

9

Kenneth Cobonpue, Cebu City

There is a generosity and boldness to Kenneth Cobonpue's designs that make them both accessible and aesthetically distinct. Space Age, organic, classically modernist, craft-based, sculptural – these are just a few thoughts that spring to mind on encountering his work. As a designer he utilizes a breadth of influences and ideas that perhaps come from his diverse design education – born in the Philippines, he studied at the Pratt Institute in New York and the Export-Akademie Baden-Württemberg in Reutlingen, Germany – as well as the rich tradition of design in his own family. His mother, Betty Cobonpue, started the design and manufacturing company that he now runs in the Philippines – Interior Crafts of the Islands, Inc. – and is herself famous for developing significant new techniques in working with rattan.

A key material for Cobonpue is rattan, which he uses with a mixture of industrial and natural materials such as steel, nylon wire, mahogany, cotton and abaca rope. His Yin & Yang armchair (1999), for example, uses rattan splits wrapped over a steel frame to create a form that is distinctly reminiscent of Le Corbusier but with a confident contemporary edge. And his Dragnet lounge chair (2005), inspired by fishnets moving through water, employs traditional Filipino weaving techniques but uses modern materials: polyester (indoor version) and Sunbrella fabric (outdoor version). With this variety of approach, Cobonpue's work pushes the boundaries of generic weaving into confidently industrial terrain.

Cobonpue has also pushed global boundaries: not only is he a designer, he is a manufacturer with a great understanding of the materials he is using, and an astute and charismatic businessman – qualities essential to get an Asian company noticed in a currently Western-orientated design world. Add to that Cobonpue's desire to promote Asian design internationally via Movement 8 – an alliance of Filipino designers who exhibit worldwide – and his role as a founding member of the Design Guild of the Philippines, and you have the complete package for a twenty-first-century global designer.
Tom Dixon

1

2

3

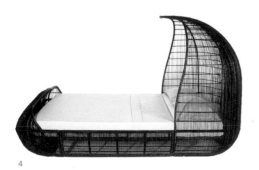

4

1–4

Voyage
Bed
2002
Interior Crafts of the
Islands, Inc.
Abaca or rattan vines,
steel, nylon filament

*The abaca or rattan vines
are skilfully woven over
a steel frame and tied with
nylon filament to create
a cosy cocoon for dreaming.*

5–9

YODA
Chair
2006
Interior Crafts of the
Islands, Inc.
Rattan vines, steel

*This chair cleverly utilizes
natural material tension
in its weaving, thereby
creating a very comfortable
and flexible seat.*

5

6

7

8

9

10

11

12

13

14

15

19

20

21

16

17

18

10–15

Yin & Yang
Seating
1999
Interior Crafts of the
Islands, Inc.
Natural and dyed rattan
splits, wicker, steel,
stainless steel

*The Yin & Yang design
heralded a new era for
Cobonpue's company –
Interior Crafts of the
Islands, Inc. Its structure
and craftsmanship
represent the aesthetic
the company is known
for today.*

16–18

Balou
Seating
2002
Interior Crafts of the
Islands, Inc.
Rattan vines, abaca rope,
cotton twine, steel

*Balou was inspired by the
idea of being ensconsed
in the warm embrace of
Balou – the bear in Kipling's
The Jungle Book.*

19–22

Dragnet
Lounge chair
2005
Interior Crafts of
the Islands, Inc.
Polycotton, powder-coated
steel, stainless steel

*The chair was inspired by
the movement of dragnets
in water.*

22

Paul Cocksedge, London

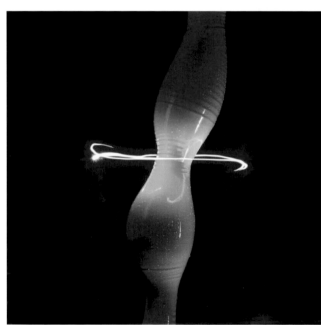

A rising star of lighting design, the British designer Paul Cocksedge creates visually spectacular, technically daring lamps. For example, Watt? (2003) rethinks the question of the switch: exploiting the conductive properties of graphite, Cocksedge's lamp incorporates an electrical circuit that is completed when a line is drawn with a pencil between two points. There is a similar approach with Bulb (2003): a flower, in a vase of water, acts as a conductor, turning on the light source concealed in the vase's base. The poetics are still more powerful: when the flower dies or is removed, the light goes out. When in use, the light is diffracted by the water, which, acting as a lens, projects a floral motif on to the room.

With NeON (2003), Cocksedge experiments with different aspects of light production. By enclosing natural gas in a bulb, radiance is generated via the passage of current through he gas. Some of these projects, owing to their experimental nature and volatility of materials, have been produced in limited editions. Light as Air (2006) is the most recent example. Seventy-seven unique lighting sculptures came out of Cocksedge's experiences with blowing glass: wondering whether other materials could be blown in this way, he discovered that PVC plastic could be treated in a similar manner. Instead of a glass-like texture and form, the objects look as though they have been created on a potter's wheel but with a high-gloss finish. An LED source is concealed within each sculpture, conveying the light as a 'breath' from within each piece.

This designer, who sculpts light and controls all the parameters of its production, is already established as a key figure in this particular field. Cocksedge himself sums up his aim 'to add the passion and emotional side to light, something that is not driven by function at all'. **Pierre Keller**

1–2
Light As Air
Lighting sculptures
2006
Paul Cocksedge Studio
PVC, LEDs

Cocksedge experimented by using glass-blowing techniques on other suitable materials. In Light As Air he concentrated his efforts on PVC, and the results are astonishing.

3
Bulb
Light
2003
Paul Cocksedge Studio
Glass, steel

Caption: A vase of water that miraculously lights up when a flower is inserted.

4–6
Watt?
Light
2003
Paul Cocksedge Studio
Steel, paper, pencil, graphite

Watt? utilizes the natural conductive properties of graphite. Connecting two points by drawing a pencil line completes the circuit and the light is switched on; when the line is rubbed out, the light is switched off.

4

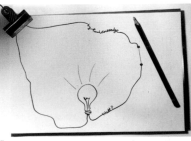

5

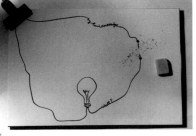

6

Paul Cocksedge

7

9

8

10

12

7–10

**Styrene
Lampshades
2002–3
Paul Cocksedge Studio
Polystyrene**

*By applying heat to
polystyrene, the form of
the shades changes,
and with it the properties
of the polystyrene. As
a result, the limited edition
lampshades are organic,
durable and unique forms.*

11–12

**NeON
Lights
2003
Paul Cocksedge Studio
Glass vessels filled with gas**

*The handmade vessels are
filled with natural gas and
then charged with current.
By day they are translucent
with a hint of natural
colour, by night they have
a beautiful aesthetic all
of their own with strikingly
vivid colours.*

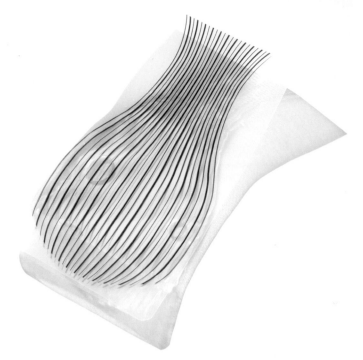

1

D-BROS, Tokyo

One of D-BROS's main strengths as a design group is its graphics expertise, and knowing the group's background immediately explains why it is so proficient in this area. D-BROS is actually a department handling product design within Draft, an advertising company run by Satoru Miyata. It started out in 1995 as a project seeking a design style different from run-of-the-mill commissioned advertising work. Rather than just designing separate products such as interior goods and stationery, D-BROS also provides a full commercial service including production, PR, advertising, retail negotiations and inventory management. The main designers are Ryosuke Uehara and Yoshie Watanabe.

In addition to the impact of its graphics, D-BROS's designs also have a story. Its bestselling design overseas – Hope Forever Blossoming (2003–6) – was inspired by shampoo refill packs. The product seems to be simply a flat plastic bag, but when you pour water into it, it expands and both looks and behaves like a glass flower vase. Other popular designs include cup and saucer sets, Waltz (2005) and Rivulet of the Heart (2003). The cup is coated in palladium so that the saucer's decorative pattern is reflected in its shiny mirrored surface, creating a visually pleasing relationship between the two.

D-BROS's policy is to focus first on new categories and systems, and then to develop a design concept. Starting the design before taking on commissions gives the designers freedom to express themselves. One of their latest designs, Hotel Butterfly (2006), uses a butterfly motif from previous work but adds a world view. Uehara explains how this fascination with butterflies lies in the 2-D flatness of the living creature as well as in its patterns: 'This is like our design work [in that it] moves between two and three dimensions.' **Chieko Yoshiie**

1–3

**Hope Forever Blossoming
Vase
2003–6
D-BROS
Polyamide, polyethylene,
PET**

The vases were inspired by shampoo refill packs. The vase is flat, but when you pour water into it, it looks like glass and can hold water and flowers.

2

3

4

5

6

7

4

Hotel Butterfly
Light
2006
D-BROS
Polypropylene

*The butterflies appear
to be drawn to the light
and flutter round the lamp.
This effect was created
by affixing the butterflies
to a lampshade made
of a polypropylene sheet.*

5

Hotel Butterfly
Bookmark
2006
D-BROS
Paper

*D-BROS created a fictitious
hotel, 'Hotel Butterfly',
to introduce a line of new
products, of which this
bookmark is one. When
the bookmark is placed
inside a book it seems as
if a butterfly is perching
on top of it.*

6–7

Rivulet of the Heart
Cup and saucer
2003
D-BROS
China

Waltz
Cup and saucer
2005
D-BROS
China

*In these designs, the
cup is coated in a mirror
finish so that the pattern
on the saucer is reflected
onto the cup.*

8

Seconds Tick Away
Wall clock
2004
D-BROS
Glass, wood, quartz
movement

*The clock is made from
a patterned glass plate,
which makes the movement
of the hands fascinating
to watch; it looks different
depending on which angle
the clock is viewed from.*

8

1

DEMAKERSVAN, Rotterdam

Dutch twins Jeroen and Joep Verhoeven set up the DEMAKERSVAN workshop with Judith de Graauw in Rotterdam after meeting at Eindhoven's Design Academy. The design trio is part of a recent Dutch phenomenon – that of young creative people becoming involved in the entrepreneurial aspects of design. DEMAKERSVAN is an independent enterprise that is devoted to the manufacture of the studio's own projects, whilst also putting pressure on industry to break from its limitations in favour of aesthetic ends. Upsetting traditional balances and accepted trends, the studio's work is a mix of handicraft and industrial processes, digital techniques and the production of one-off pieces. Involving other practitioners from different disciplines (designers, writers, filmmakers, artists, scientists), they set out to overcome the idea that creative work is the expression of an individual alone.

A project that illustrates DEMAKERSVAN's work is How To Plant a Fence (2005), made out of a standard mesh of industrially manufactured wire, which the three designers 'embroidered' by hand as if it were a delicate piece of linen. The metal fence is a clear example of how even ordinary objects that fill everyday life with a silent, neutral presence can reveal unexpected 'poetic' qualities. In 2005, as a part of his graduation project, Jeroen Verhoeven created the Cinderella Table (2005). His design was based on his research of CNC industrialized wood processes. In designing the table he used seventeenth-century drawings of furniture, which he manipulated with morphing software. Working side by side with skilled boat makers, Verhoeven then created a table cut from 57 layers of birch plywood. In 2006 the Cinderella Table was celebrated with its induction into the permanent collection of New York's Museum of Modern Art.

DEMAKERSVAN's mission is to reveal the unexpected 'poetic' qualities of objects that seemingly have nothing new to say; to show, for example, the innate potential in functional forms that owe their *raison d'être* to the logic of mechanical production (related to the concept of standardization), now being called into question by new digital technology that can mass-produce complex or ornamental forms with incredible ease. **Francesca Picchi**

1

Lost&Found
Stool
2005
DEMAKERSVAN
Leather, wood

This carefully made leather and wood stool makes a virtue out of its rather curious sense of symmetry, and is an unusual and comfortable domestic seat.

2

Minimizing Einstein
Radio
2005
DEMAKERSVAN
Porcelain, wire, Bakelite

You search for different channels on this ingeniously basic radio by moving the tuner up and down the pole.

3–6

Cinderella Table
2005
DEMAKERSVAN
Finnish birch

This beautiful wooden table has an appealingly organic shape, but is actually made by a 5-axis CNC robot.

2

3

4

5

6

7

7

Waiting For You
Chairs
2005
DEMAKERSVAN
Rubber, RVS metal

*These handmade chairs
brilliantly evoke the
idea of a person waiting
for somebody in the
way their structure apes
human gestures of
fidgety impatience.*

8–10

How to Plant a Fence
Wire fence
2005
IDfence
Galvanized metal, coated
PVC

*The urban utility of an
inner-city fence is
here transformed into
an object of beauty.*

8

9

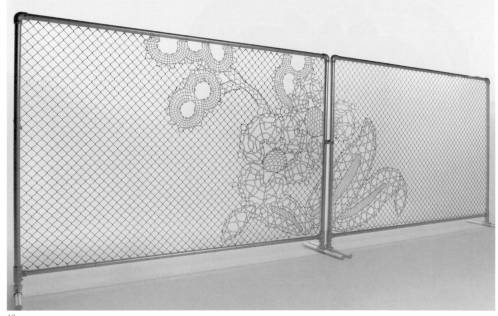

10

Stefan Diez, Munich

Stefan Diez epitomizes the designer-maker. For him the creative process is based on an unfailing faith in manual work as a process for creativity. During his research Diez puts immense energy into studying objects, which triggers a process of experimentation and refinement that does not end until he reaches his final solution.

The basic idea of his Instant Lounge project (1999/2003) was a single sheet simply folded into the shape of a chair, which he developed together with Christophe de la Fontaine. Experimenting with an infinite number of different layers of materials, they came to the solution of a composite with an expanded polypropylene core covered on both sides with fabric.

Each project taken over by Diez entails a large number of prototypes, tests, sketches, models and experiments: no stage in the creative process has any initial certainties, except that a result will be reached at the end of the search. The process involves not only his Munich studio – the hub of his creative activity – but also the company commissioning the work and the factory producing it.

The Shuttle project (2006) for Rosenthal, one of Europe's longest established porcelain manufacturers, is significant in this sense. Diez has produced a project for them that is exemplary in terms of the worldwide geographic scale of design today: the porcelain is manufactured in Germany, the glass in Hong Kong, and the plastic connection that holds them together is made in China, using German technology.

Diez embodies the concept of the designer as being at the centre of a network of relationships that he constructs and expands himself, mediating between different realities and setting in motion a circuit of inter-relationships that can best exploit the characteristics of each other. **Francesca Picchi**

1

2

3

1–2
BENT
Tables and chairs
2006
Moroso
Aluminium
In collaboration with
Christophe de la Fontaine

The BENT tables and chairs, are the result of a project which Diez began working on in 1999, and are made by bending 3-mm-thick sheets of aluminium.

3–7
UPON Series
Prototypes
2006
Various materials

Diez has experimented with a concertina form to create tables, benches, coat hangers and floor pieces.

8
UPON Wall
Coat hanger
2006
Schönbuch
Sheet metal

Here one of the prototypes is realized as a functional coat hanger.

4

5

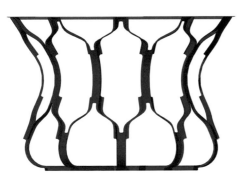

6

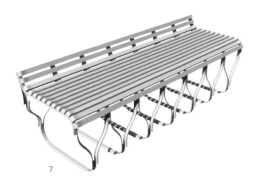

7

8

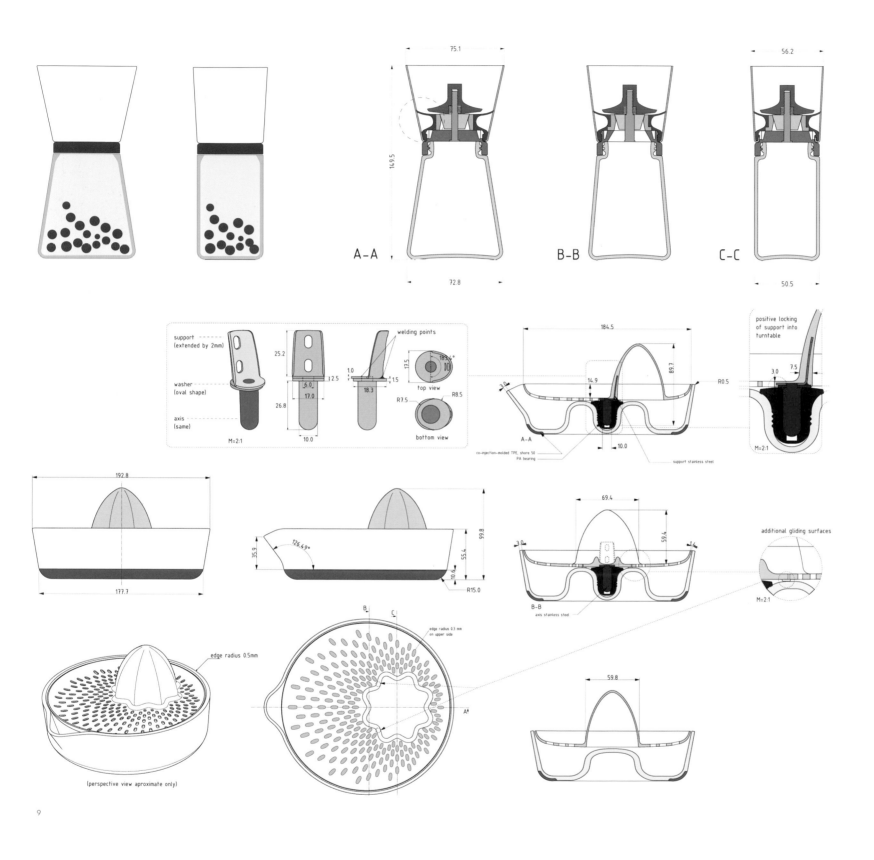

9

GENIO
Cookware
2005
Design sketches

*Detailed drawings
of the GENIO range.*

10

GENIO
Cookware
2005
Rosenthal-Thomas
Cast iron, porcelain

*The fully realized, stylish
cookware set.*

1–3
Vague à l'Âme
Carpet seat
2004
Fabrication Parsua pour
l'espace Chevallier
Hand-knotted wool carpet,
inflatable PVC ball, feather
cushions

*Vague à l'Âme is a
place of rest: the PVC
ball is inserted under
the carpet, which
generates comfortable
folds and creases.*

Florence Doléac, Paris

'Phlogistic' is a 17th-century chemical term, which describes an ancient principle relating to the combustion of materials. Florence Doléac practises 'flogistique', as the title of her first exhibition cleverly reminded us, marking her determination to transform the way things are, to 'set fire' to the ordinary and re-create a space of pleasure and freedom.

A 1994 graduate of the École Nationale Supérieure de Création Industrielle, Les Ateliers in Paris and a founding member of the Radi Designers group in 1997, since 2003 Doléac has practised on her own setting out to 're-label the everyday', as she herself puts it. She systematically scrutinizes domestic space, overturning values, changing habits, shifting the relationships we have with objects and the familiar world in general. She has invented a survival box for the person who can no longer bear their partner's snoring. And she offers to transform the housewife into an artist with her Fée du Logis kit (2004). With her Princesse (2004), she drapes chairs with a bride's train. Essentially, in everything she produces, Doléac defunctionalizes and restores meaning, destabilizes and restores a sense of playfulness.

Rather than adding value to objects, Doléac encourages us to question our existing relationships with everyday objects, counteracting our tendency to take them for granted as mere slaves, bereft of life and quality. Imagine just for a moment an apartment that has had the 'flogistique' treatment: opening a door fitted with a soft handle that changes shape to fit your hand – Poigneé Molle (2002); seeing the television rather than watching it –Télélumine (2004); or feeling a water-rug – Piétinoires (2004) – beneath your feet. Projects filled with subtle sensations and archetypes that are then magnified, distorted or mutated; this is what Doléac's wonderful world is composed of. **Pierre Keller**

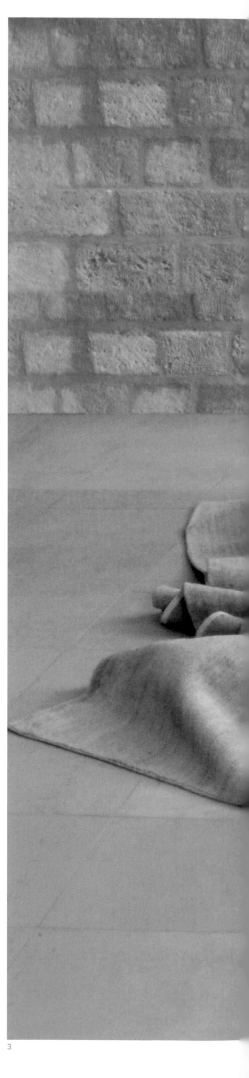

4

5

6

Florence Doléac

7

4–7

Patapouf
Seating
2005
Fermob
Polystyrene balls, epoxy
painted curved metal,
impermeable bag with
washable cover

*Patapouf offers two very
comfortable positions:
sitting or lying flat-out. It
is designed especially to
sink into after a big meal.*

8–10

Biscuit en Couvercle
2004
Sébastien Gaudard,
Délicabar, Le Bon Marché
Rive Gauche
Coffee cup and saucer, cake

*The hazelnut cake with
a cube of biscuit on top
keeps the drink warm,
whilst being simultaneously
dunked into the coffee.*

11–12

Point de Suspension
Peg
2004
Limited edition for
Toolsgalerie
Resined rope

*This coat peg is made from
just one knot.*

8

9

10

11

12

1

2

3

4

Doshi Levien, London

You can spot Doshi Levien's designs from a distance. Working from a tiny Brick Lane studio in East London, the husband-and-wife design duo brings a unique persona to an already crowded market-place. Jonathan Levien and Nipa Doshi combine meticulous design techniques with a striking graphic and fashion edge, and merge European style with strong Indian influences. It is rare for designers to develop such a distinctive aesthetic style from their own cultural experiences, but it is exactly this quality that makes Doshi Levien's work so distinctive and successful. While Levien is technically precise, Doshi's design style is behind the team's visual signature. She brings warm, more generous lines, bold colours and patterns to their products.

Doshi Levien's award-winning Mosaic range for Tefal (2003) demonstrates how they meld their individual skills to great effect. The beautifully formed pots and pans have wooden handles, are non-stick and easy to clean, and encourage world cooking via the Tajine pan, the Paella pan, the Wok pan, the Plancha and the Karhai pan. Aside from their functionality, the pan shapes are appealing and fun, and their colour motifs are more extravagantly explored on their bases, which look very striking when lined up, hanging from kitchen hooks.

In 2005 they were invited to create the imaginary 'Wellbeing Centre' installation for the Wellcome Trust – a series of window displays that communicated the history and work of the Trust. They responded with great imagination to the commission, exploring ideas of health through self-reflection and knowledge. Their first colourful and almost fairy-tale-like installation presented a doctor surrounded by the 'tools' of his trade, including a fantastical stethoscope through which you can listen to both your heart and your soul, a bejewelled doctor's bag and the elixir of life.

Doshi Levien is powerful and proficient as a design team and you can only describe its work as rich. It brings a richness to what it does, which makes its products almost impossible to resist. **Tom Dixon**

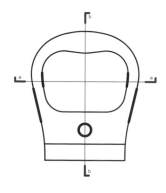

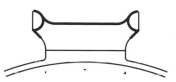

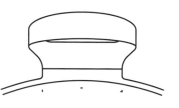

5

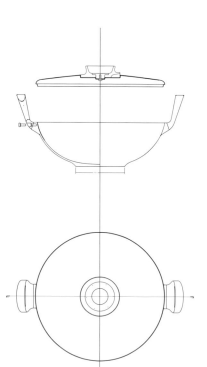

6

1–4

Mosaic range for Tefal –
base patterns
2003
Tefal
Screen-printed designs

*Each pot in the Mosaic
range expresses strong
cultural identity through
material, colour, and the
varying translation of
the screen-printed Tefal
marque on the base.*

5–10

Mosaic range for Tefal
Cooking pans
2003
Tefal
Aluminium, Bakelite,
stainless steel, terracotta

*Doshi Levien translated
authentic characteristics
of Asian, Moroccan
and Latin American
cookware into accessible
cookware products.*

7

8

9

10

11

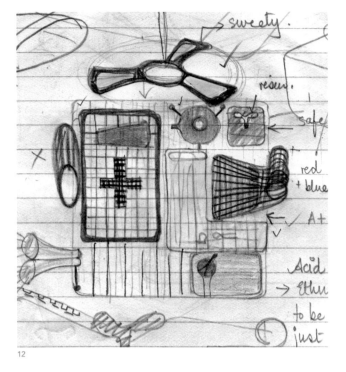

12

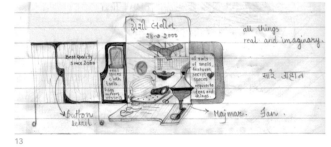

13

14

15

16

11

Jonathan Levien and Nipa Doshi, 2005.

12–13

My World
Installation
2005
Design sketches

The installation for My World is partly inspired by the shops and workshops of ancient but still functioning markets in India. Doshi Levien have created a liminal space between two worlds: Indian and European; imaginary and real.

14

Matlo
Drinking vessel
2006
Self-production
Slip-cast terracotta, diatomite water filters, brass

The Matlo is a slip-cast version of a traditional Indian water cooler, which has evolved to incorporate filtration and could be batch-produced from a mould. Doshi Levien propose it as an environmentally sound alternative to bottled water and electric coolers.

15

Fan
2006
Prototype
Three-part moulded resin

Inspired by old colonial fans whose generous blades create a soothing sound, this fan is designed as an antidote to very efficient but soulless contemporary fans.

16

Marble Table
2006
Texxus
CNC-machined marble

This table combines the archetypal form of a work desk with sensual materials and opulent form, thereby expressing the idea of work as a source of pleasure.

1

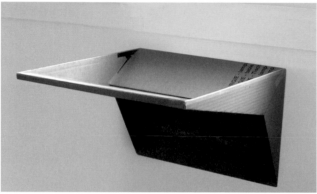

2

3

david dubois, Paris

david dubois' products stem from the creative gaps that exist between different design disciplines such as fashion, art and gardening. His objects act as devices that graft on to other existing objects, giving them new functions. The result is a disturbing hybridization of elements from our everyday lives, which seem to lead a life of their own.

Having studied art in the 1990s and design since 2000, one of dubois' favourite design combinations is a blend of garments and household objects, a freedom from categories that he was seeking when he collaborated with BLESS from 2001 to 2005. For example, his Vase Coton (2005) are tubes of fabric that become vases when sheathing a bottle; in the same way his leather-zipped pocket (2004) becomes a *vide poche* (an empty pocket) when fitted on to a door handle. Another significant collaboration is with Martin Szekely, with whom dubois worked as an assistant.

Another of dubois' favourite themes is the infiltration of nature, as demonstrated in 'Tutoring', his 2005 solo show in a chapel in the South of France. The garden plants were colonized by geometrical structures, which acted as tutors, disrupting the plants' organic shapes with vertical and horizontal lines. Pieces of furniture were also presented: an outdoor bench with its feet in the pond, bench-like furniture incorporating an indoor fountain at one end, and a patio pot with plants that grew into a tutor.

A commission in 2006 by the Musée d'Art Moderne Grand-Duc Jean, Luxembourg, provided an opportunity for dubois to create an ensemble of public space furniture that playfully interpenetrated the museum's garden and inner spaces. Benches echo the lapping of water and plants give the impression of offering a seat to visitors, while bicycle racks are carved out of tree trunks. In a situation where the use of lockers is compulsory for security reasons, his System Bag (2006) becomes a particularly appropriate makeshift bag: the strap, with a hook at both ends, clips into the handles of a museum plastic bag allowing visitors to carry around their valuables during the visit.

dubois' interest in cross-disciplinary collaborative projects continues with his current project designing new stores for Christian Lacroix. **Didier Krzentowski**

4

5

6

1–3

Sink
2005
DMA
Nickel steel

*The sink's simple form
is built on the principles
of origami.*

4–6

Chêne-à-vélo
Bicycle rack
2006
MUDAM
Oak, buckle metal

*These oak trunks are
simply notched to hold
bicycles securely. They
were produced by MUDAM
for the garden of the
Luxembourg Museum.*

david dubois

7

8

7–8

Glass Stool
2006
Prototype
Glass

*The stool is made up
of three superimposed
containers and a lid.*

9–10

Tutoring
Flowerpot holder
2005
Wood, plastic vat

*This flowerpot holder
comes with integrated
tutors to help the plants
grow more easily.*

11–12

Water-Bench
2006
MUDAM
Varnished cedar, zinc vats,
electric water pump

*Also created for the
Luxembourg Museum,
these benches integrate
an invisible fountain,
with only the sound of
the water perceptible
to the user.*

13–14

System Bag
2006
MUDAM
Leather, plastic loops, white
serigraphy

*This project provides
visitors to the Luxembourg
Museum with a strap to
which they can attach a
clear plastic museum bag.
This allows them to carry
their belongings around
during their visit without
breaching security.*

9

10

11

12

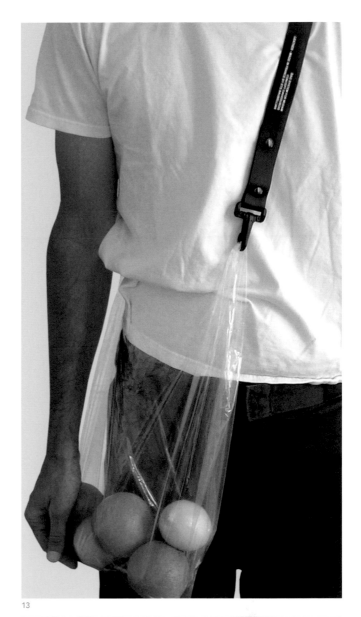

13

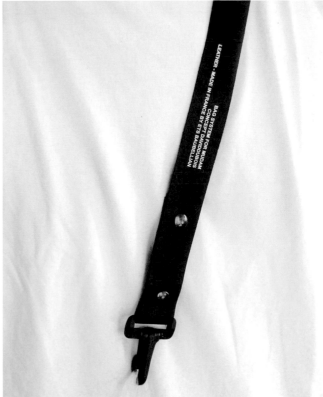

14

1

2

Piet Hein Eek, Geldrop

People weary of the same old standard designs and those who are searching for unusual furniture appreciate how Piet Hein Eek's unique aesthetic makes his designs distinctive. Many of his clients say that they are seduced in just one glance, after which they never forget his work, and keep coming back to buy again and again.

Eek uses everyday and discarded materials – wall and flooring materials, industrial fencing, broken windows and doors – in his iconic designs, such as his series of tables and chairs. His graduation work was an expression of his central idea: 'Because everyone's work is perfectly beautiful, it all becomes the same, so I aimed to create a design that was attractive through its lack of perfection.' He bases his work on the concept that design is the product of materials and processes. For example, his Scrap Wood Piling Stool (1999) was formed from a pile of discarded materials. Rather than use a measuring rule, he relied on his craftsman's eye to organize the pile, and then applied nine coats of lacquer to bring out the glossy lustre and emphasize its unique presence.

The combination of materials and technologies is different for each piece but Eek's constant theme is the use of commonplace materials and out-moded methods and tools, such as the CNC router, in time-consuming labours of love. He also devotes attention to details such as metal fixtures and trims. His studio employs 15 artisans, but as he is also an artisan, his conceptualizations guide the designs.

Eek is currently working on a scrap wood house for the Kröller Müller Museum in Holland as well as commissions for the Museum of Modern Art in New York and a Dutch bakery chain, which has led to a series of interior designs, including lighting, all in his very characteristic style. Whatever the world trend in furniture design, we can be sure his works will continue to reflect his one-off artistic style. **Chieko Yoshiie**

1–3

Ceramic Jugs
2005
Eek en Ruijgrok
Ceramics

These jugs are made of flat ceramics, a technique best described as being like using a pasta roller to create the ceramics instead of plaster moulds.

Piet Hein Eek

5

4

6

4–6

Scrap Wood Cabinet
2004
Eek en Ruijgrok
Leftover scrap wood

This project uses worth-less materials but involves an enormous amount of labour – exactly the opposite of the normal production routine.

7–8

Football Lamp
2006
Eek en Ruijgrok
Stainless steel

This dazzling lamp was inspired by the construction of a football.

9–11

Desk
Crisis Furniture Collection
2004
Eek en Ruijgrok
Plywood

The parts of this desk are made with a computerized router and fixed with a very simple nut and bolt system.

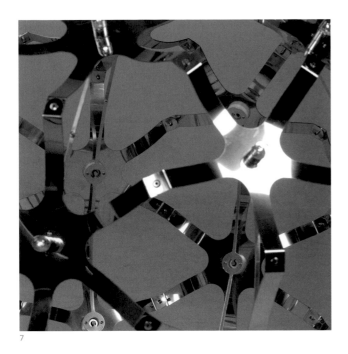

7

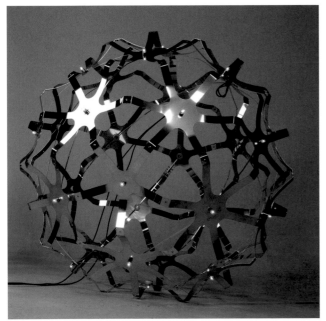

8

9

10

11

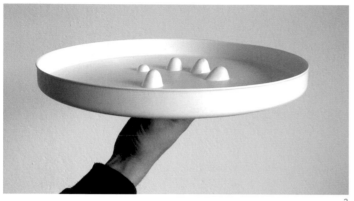

1–2
Drink'Tray
2003
Ligne Roset
Polypropylene

*This moulded tray allows
the carrier to transport
drinks effortlessly whilst
holding the tray securely
using the protruding
finger spaces.*

3
Earthquake Table
Prototype
2001
ECAL
Steel

4
Earthquake Table/
Adult version

5
Earthquake Table/
Nostalgic version

6
Earthquake Table/
Cheese Fondue version

*A simple utility table
that its user can shelter
underneath during an
earthquake. The underside
of the tabletop is stocked
with differently themed
supplies to sustain
and amuse the person
it shelters.*

Martino d'Esposito, Lausanne

A graduate of ECAL – École cantonale d'art de Lausanne, Martino d'Esposito declares that he wants to be a 'megaphone for objects': that is, to act as an intermediary between the object and its user. Interested more by symbolism, figures of style and the human factor than by industrial production or limited edition furniture, d'Esposito belongs to the school of design that believes that the object must first and foremost communicate with its user.

The moulded form of d'Esposito's Drink'Tray (2003) gives the server a helping hand by adapting to his or her movements; his Smoker Corner ashtray (2001) keeps smoke away from non-smokers by attaching to the corner of a table; the Grow Bag (2003), a flexible polyurethane container, expands according to the needs of the growing plant; the Twins chairs (2002), for public waiting areas, displays its wish to group people together; the Earthquake Table (2001), usually a domestic occasional table, is designed to save lives in times of crisis.

These are more than just objects; in the way that they are used, as part of our everyday lives, they offer us 'soul' – encouraging us to play or dream – without sacrificing any part of their original function. While the majority of these projects were developed as prototypes, firms such as Ligne Roset and Cinna are today distributing several.

D'Esposito is not afraid of tackling a wide range of design projects, as his sewer grid (Vortex, 2006), designed for the city of Lausanne, in collaboration with a local manufacturer of cast iron, shows. He has recently established a design studio with fellow ECAL graduate Alexandre Gaillard.

As a designer d'Esposito hopes to continue to 'generate objects that "speak" a little to the user and/or manufacturer'. Of his work he says, 'There is always a hidden message in the products. My design is a form of mass-communication.
Pierre Keller

3

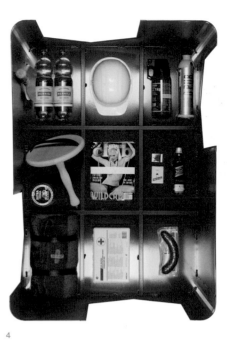

4

5

6

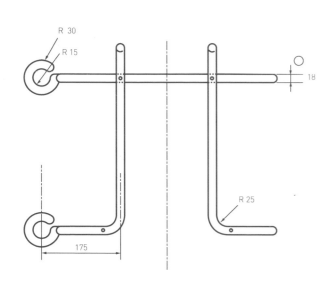

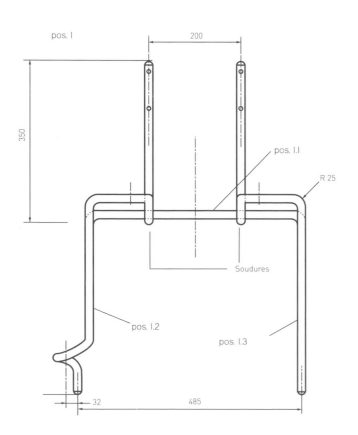

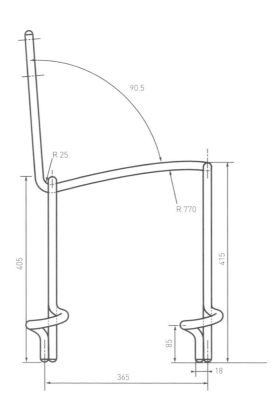

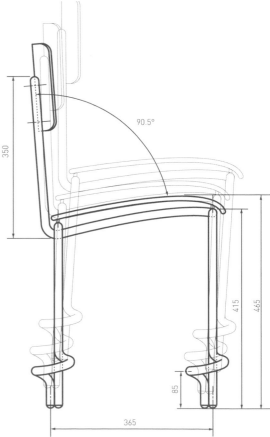

7

Twins
Chairs
2002
Chromed steel, plywood

These chairs encourage their users to sit closer together by connecting to one another; they can be arranged into a more sociable circular formation

8–9

Grow Bag
2003
Ligne Roset
PVC foam

The Grow Bag expands as the plant grows and lays down deeper roots so that re-potting is unnecessary.

10–11

Vortex
Sewer grid
2006
FMG.SA
Spheroid cast iron

Featured in 'INOUT' (2006), an exhibition of outdoor design presented by a group of young Swiss designers in Lausanne, this sewer cover is both urban decoration and an aid to a practical drainage system.

8

9

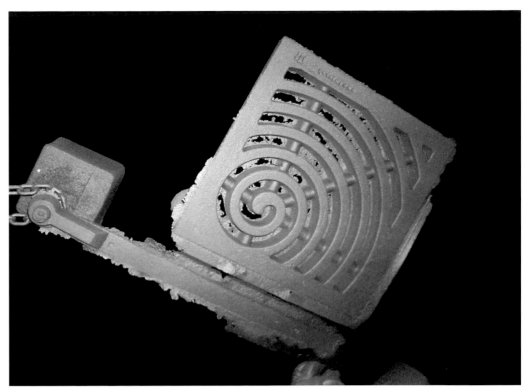

10

11

FRONT, Stockholm

The Swedish group FRONT is a four-woman collective formed in Stockholm in 2003. Sofia Lagerkvist, Charlotte von der Lancken, Anna Lindgren and Katja Sävström, met at Konstfack, the University College of Arts, Crafts and Design based in Stockholm. One of their first collective works was called Design By, in which the four young Swedes proposed a sort of mapping of spontaneous forms, exploring the morphological implications of instinctive animal behaviour. The Design By project was the starting point for FRONT's research in this field and allowed it to question the conventions of the designer's role as creator presiding over the formal process. They sought to counteract the designer's creative control with an impersonal point of view in which random elements intervene to decide the formal result. As well as promoting an aesthetic based on spontaneity, the group reminds us that nature around us is full of signs by highlighting and intercepting the slightest, barely perceptible indication.

The Animals project (2003), which features Wallpaper by Rats (2003) may resemble a, perhaps naive, catalogue of 'natural' and spontaneous forms but it also seems to be an attempt to initiate a programmed destruction process. This theme becomes explicit when the crater left by an explosion serves as a model for a seat – Chair By Explosion (2003) – and the form of a vase – Shoot Vases in Glass, Representation of Things (2005) – incorporates within it the image of its potential fate of falling and being smashed to pieces. Animal Thing (2006) also shows the group's playful side by taking realistic glassfibre animal forms and turning them into useful household objects.

On the one hand, FRONT steps aside to avoid dominating the formal design process but, on the other, it seems to do everything in its power to entrust the identity of an object to its missing parts. This highlighting of random zones, holes, cavities and fissures may be a way of questioning the design of contrived and self-referential objects, in which nothing can be added or removed. With the same spirit, FRONT uses 3-D modelling and numerical control milling machines to survey 'found forms' that can then be made into everyday objects.

Francesca Picchi

1

2

1–2
Loudspeakers
2003
FRONT
Hand blown glass

The use of hand blown glass makes these beautiful loudspeakers an unusual way to listen to music.

3–6

Sketch Furniture
2005
Barry Friedman Gallery
Rapid prototyping

Invisible pen strokes in the air, recorded with Motion capture, become real pieces of furniture through laser sintering.

3

4

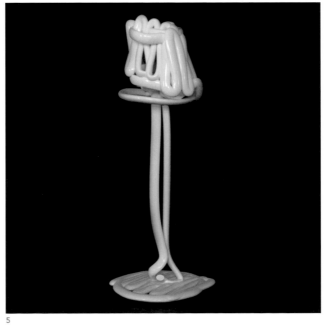

5

6

7

7–8

Chainsaw Chair
2005
FRONT
Wood

FRONT asked a chainsaw sculptor to create a chair out of a solid piece of wood. The result is a dense and organic furniture piece.

9–11

Wallpaper by Rats
2003
FRONT
Wallpaper gnawed by rats

The holes gnawed by the rats in the paper reveals the old wallpaper that has been covered up, creating a delicate effect.

12–14

Animal Thing: Horse Lamp; Pig Tray; Rabbit Lamp
2006
Moooi
Glassfibre

These Animal Things give traditional-style domestic furniture an ironic modern twist.

8

9

10

11

12

13

14

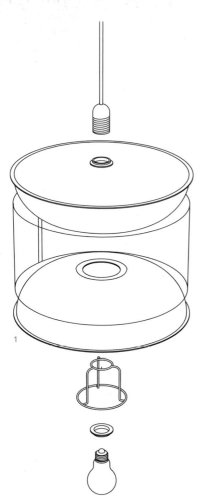

1

1–2

**Waist
Lamp
2003
Habitat
Polycarbonate, plastic
colour filter, wire**

*Two hemispherical shades
divide the light of one
bulb into direct, indirect
and atmospheric light.*

FUCHS + FUNKE, Berlin

Wilm Fuchs and Kai Funke set up their design office in Berlin in 2001 after meeting at the Institute for Product and Process Design at the University of Arts in Berlin. They like to experiment with already familiar objects and ideas, and by doing so find 'gaps, leaps and missing links' that other designers have side-stepped or overlooked. They seek to create practical, everyday objects that fit into our increasingly cluttered and restricted environments.

The Knock Down Lamp prototype (2002) is a perfect example. It's a beautiful object in both its aesthetic and simplicity of function. It uses magnetic connections to hold the component parts together, which when required can be pulled apart to store the lamp neatly away. The magnets and the use of a counterweight also mean that the lamp can be easily adjusted on its base axis into any position required while maintaining a clean and unfettered line, free from external cables.

Similarly, the Waist lamp for Habitat (2003) is a lesson in refined design and is very cheap to produce. Two half globes, placed together in an egg timer-like shape, are enclosed by a tinted cylindrical shade that houses just one standard light bulb. However, the structure of the lamp enables a complex flow of light to issue and creates an unusual light quality that merges down, indirect and excess atmospheric light.

One of FUCHS + FUNKE's latest prototypes, the Mono chair (2006), marks a new move into more organic forms. The basis of the design still lies in disciplined construction, but the resultant contours are highly sculptural.

FUCHS + FUNKE's ideas are clever and original and they balance their imaginative use of industrial components and elegant simple lines with a subtle humour that encourages us to take an interest in the world of objects and to share their pleasure in effective design. **Tom Dixon**

2

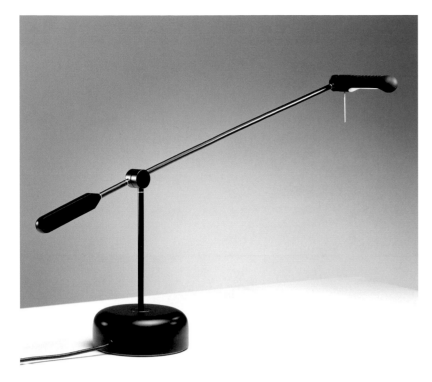

5

6

3–4

Papton
Chair
2004
FUCHS + FUNKE
Honeycomb cardboard

*A few folds transform
a composite panel into
a real lightweight chair,
weighing just 2kg (4.5lbs).*

5–6

Knock Down Lamp
2002
Prototype
Metal,
electric components,
magnets, low voltage
halogen bulb
In collaboration with Felix
Almstadt

*Magnetic connections
and a counterweight
ensure smooth
adjustment unlimited
by any cable contortion.*

7

Mono
Chair
2006
Prototype
Polyurethane foam

*An organic design
emerges from
constructional
requirements and
seamless transitions.*

7

1

2

Martino Gamper, London

Martino Gamper's early work revealed a near-obsessional interest in corners, spaces generally neglected and rarely considered by other designers who prefer to design furniture for the centre of the domestic space. During this period his designs included a corner bookshelf, a corner light, a corner sofa, and a corner bench. All these special units were part of his degree show for the Royal College of Art in 2000.

Alongside this concern with underused spaces, Gamper nurtured an interest in unwanted objects. He focused on bringing new life to objects made from materials found whilst wandering around rubbish dumps in the proposal for a 'community art' event entitled '100 Chairs in 100 Days' (2006). It would be easy to conclude, then, that Gamper harbours a special interest for things everyone else overlooks, and that he prefers to use what others deem to be of no further interest, both places and objects, as materials from which to draw new experiences. In the case of the chairs project, Gamper used a collection of discarded items to make 100 new designs. The process of creating this mismatched family of objects involved both fusing chairs or bits of them together, and adding new materials. But to ensure that the work remained spontaneous, each new chair had to be completed within the space of a single day.

The mere fact that these objects have escaped their fate as waste gives them a strange power. In some ways, like people, the objects bear signs of their history. 'Some are beautiful and some pathetic, but all of it, even the most abject of pieces, has a spring in its step, a euphoria inspired by the escape of destiny.' This deliberately marginal choice of the material of expression underpins Gamper's interest in what he himself has described as the 'psycho-social aspects of the furnishing project' and design in general. Training himself to consider uncommon perspectives, Gamper has instigated a new way of looking at everyday objects.
Francesca Picchi

1–2
Tree Log
Seating
2004
Martino Gamper,
Rainer Spehl & åbäke
Leather

Upholstered in leather, the Tree Log is a visually playful piece; it is also surprisingly comfortable.

3
Chair
2006
Martino Gamper
Mixed media

Part of the '100 Chairs in 100 Days' project, in which Gamper sourced discarded chairs and redesigned and reconstructed them. He saw the work as a 3-D design sketchbook with the best results becoming models for possible mass production.

4–5
Arnold Circus Stool
2006
Martino Gamper
Rotation-moulded plastic

The stool is the official seat for events at Arnold Circus in Shoreditch: a circular park currently undergoing regeneration that was built on the rubble of old East End slums.

3

4

6–7
Chair
2006
Martino Gamper
Mixed media

*Pieces from the
'100 Chairs in 100 Days'
project, these models
have been made
from discarded chairs
to make new and
innovative designs.*

8
Martino Gamper's studio
2006

*A photo of Gamper's studio
showing various pieces
of furniture he designed
between 2004 and 2006.*

6

7

1

Chairs – D/side Series
2002
Limited edition – Attitudes
Art Gallery
Plywood

*A visual play between
two dimensions is
explored in this series
of wooden furniture.*

2–3

Inflatable Bottle Cooler
2000
Editions ECAL
PCV, rope

*Keep your drinks cool
with this innovative
floating bottle cooler.*

4

Newton Fruit Bowl/
Green Apples
2004
Self-production
Ceramic

*This playful fruit bowl
mimics the moment
before the apple
dropped on Newton.*

Alexis Georgacopoulos, Lausanne

The name and designs of Alexis Georgacopoulos are closely associated with ECAL – École cantonale d'art de Lausanne, where he has been head of the industrial design department since 2000. He set up ECAL's Botte-Culs (milking stools, 2002) and Conductor's Batons (2004) projects, and has been involved in many workshops in conjunction with companies such as Serralunga, B&B Italia, Coca-Cola or Christofle and designers such as Ronan Bouroullec, Richard Hutten and the Campana brothers.

In his own work, Georgacopoulos epitomizes design that is both functional and amusing, with humour as the counterweight to technology. 'An object must induce sympathy,' he asserts, and it has to be said that his objects succeed. Whether handling his Inflatable Bottle Cooler (2000) for aperitifs on Lake Geneva or his Newton Fruit Bowl (2004) with a circular side opening from which fruit can protrude or sometimes even fall, using them is designed to give the user extra pleasure. The Stando vase (2005), with its long transparent glass stem, holds tall flowers upright, while the water is contained in a black and relatively small blown-glass bowl. The contrast between these two elements creates an optical illusion – as if the flowers were floating.

Georgacopoulos' CMYK trays (prototypes, 2006) are emblematic of his quest to incorporate pleasure. A series of trays for odds and ends made of thermo-lacquered steel and representing both the four basic colours and the four geometric forms (Cyan Square, Magenta Circle, Yellow Triangle, Black Rectangle), they operate on the basis of a simultaneous return to fundamentals and the world of nostalgic playful memories.

'Pleasure, play and classicism': this could be Georgacopoulos' typically Hellenic credo that inspires his designs – conceptual lightness with simple functions and straight-to-the-point effectiveness. **Pierre Keller**

2

3

1

4

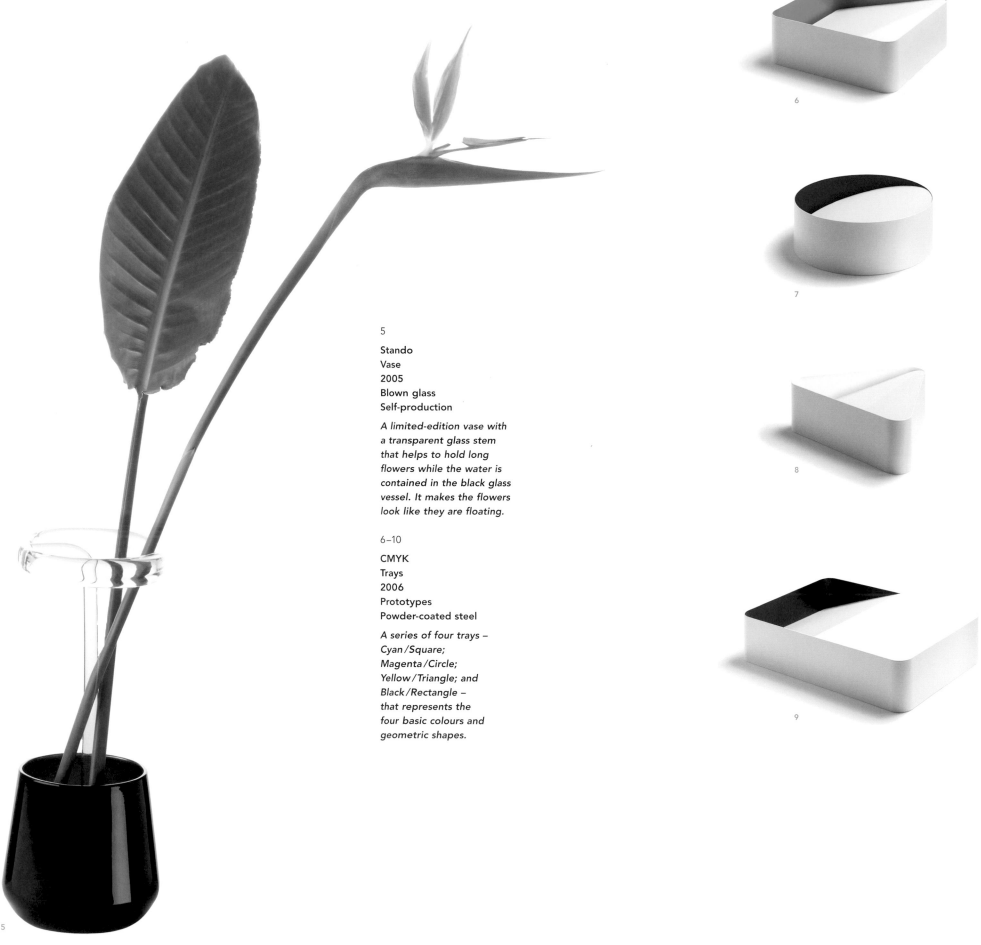

5

Stando
Vase
2005
Blown glass
Self-production

A limited-edition vase with a transparent glass stem that helps to hold long flowers while the water is contained in the black glass vessel. It makes the flowers look like they are floating.

6–10

CMYK
Trays
2006
Prototypes
Powder-coated steel

A series of four trays – Cyan/Square; Magenta/Circle; Yellow/Triangle; and Black/Rectangle – that represents the four basic colours and geometric shapes.

6

7

8

9

5

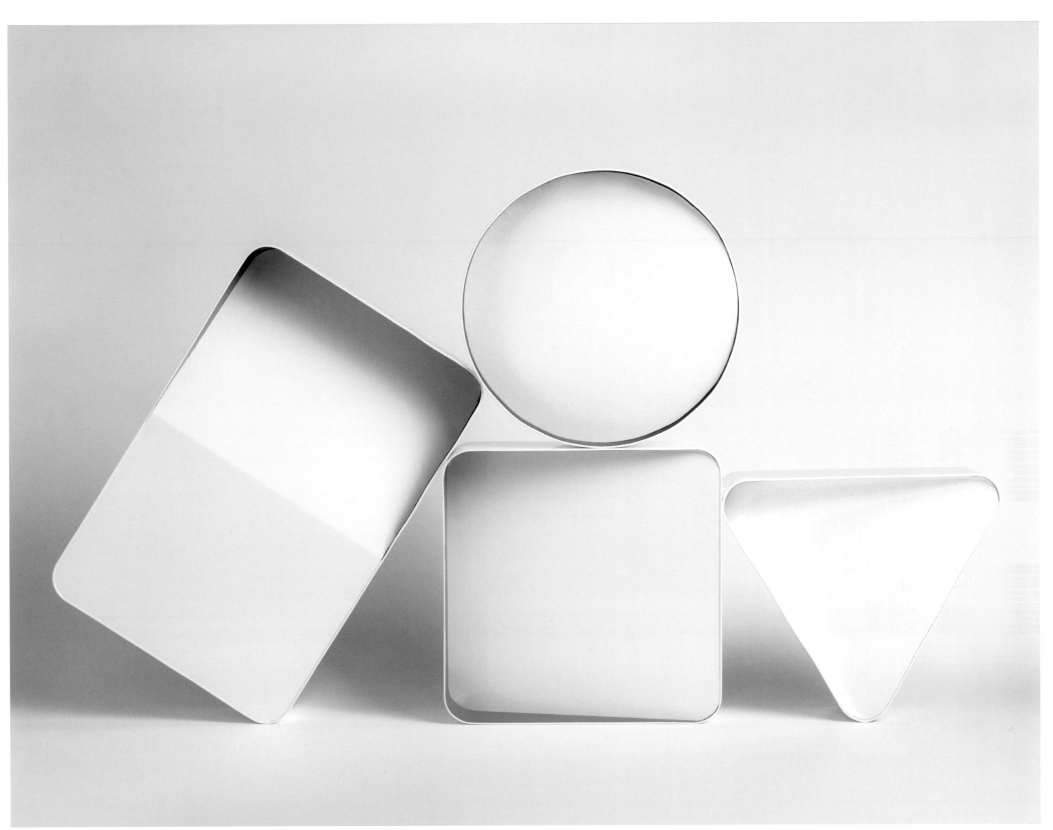

1

2

Adam Goodrum, Sydney

Adam Goodrum, who graduated from Sydney's University of Technology in 1993, returned to Australia after working in London for two and a half years to establish his design practice at the beginning of 2003. He has since emerged as one of the rising stars of Australian design. In 2004 Goodrum won the Bombay Sapphire Design Discovery Award, a prestigious annual prize for Australian product designers. Winning the award enabled him to exhibit his Eve Ottoman (2004) at the Milan International Furniture Fair in 2005. This sculpturally compelling piece cradles the seated figure and illustrates Goodrum's overriding concern to merge functionality with aesthetic beauty. A concern more recently demonstrated in his stunning glass Boab Light (2006).

During the last decade, Goodrum has assembled an impressive portfolio of self-initiated work, with projects ranging from an innovative series of figurative plastic pegs to a moulded plywood screen system, most of which have been realized as immaculately finished prototypes. Goodrum's skills and resourcefulness have aroused the interest of several prominent global manufacturers.

Goodrum has a particular fascination for folding systems and designed his first folding chair in 1992, while still studying. It is a theme he has revisited many times, perhaps most elegantly with his Stitch Chair (2004), which folds piano-accordion style along its vertical axis to a width of only 15 mm and was selected as one of the Best Designs in 2004 by *I.D.* International Design Review. The following year Goodrum extended this concept further to develop his extraordinary Folding House. Having long had an interest in socially and ethically responsible design, he wanted to use his folding system ideas to create a lightweight emergency shelter that could be easily transported and rapidly assembled for use in refugee and disaster relief situations. Folding House is made from heavy-duty cardboard coated with an inert waterproof sealer. In a matter of seconds and with one smooth movement, it transforms from a flat-pack sheet to a temporary home. And continuing his interest in easily stored utility furniture, Goodrum's Out Door Table (2006) is a prototype for a large outdoor table that can be packed away when not in use. **Brian Parkes**

1–3

Boab Light
2006
Prototype
Glass

*These richly coloured glass
lights are hand blown to
create unusual organic and
sensual forms.*

3

Adam Goodrum

4–5

Out Door Table
2006
Prototype
Timber, stainless steel

Streamlined and sparse in design, this table has a removable top so that the product can be easily dismantled for storage.

6–7

Lissajous
Chair
2004
Prototype
Bright mild steel, stainless-steel cable

Sixty metres of continuous coiled metal go into making this strikingly architectural chair.

8–9

Armour Screen
2006
Prototype
Plywood, aluminium, acrylic

This complex overlapping form creates a stunning screen that can be used as a screen, room divider or wall enhancer.

4

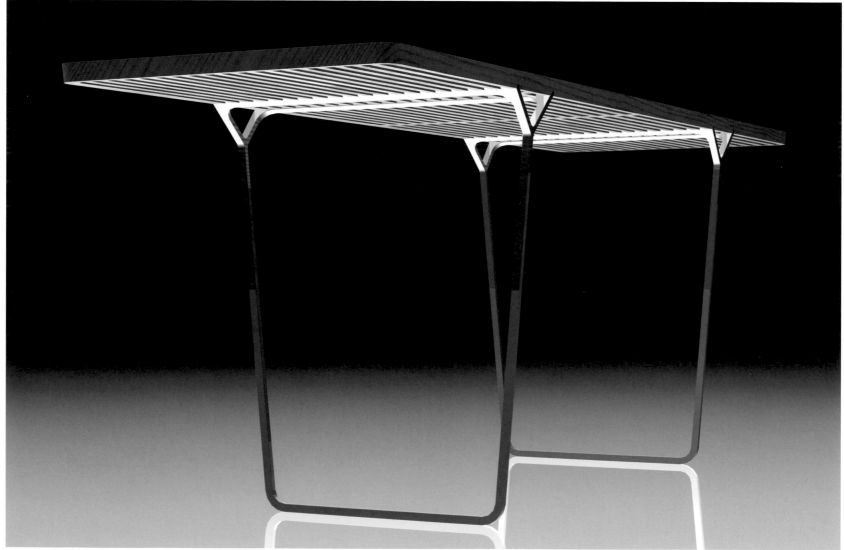

5

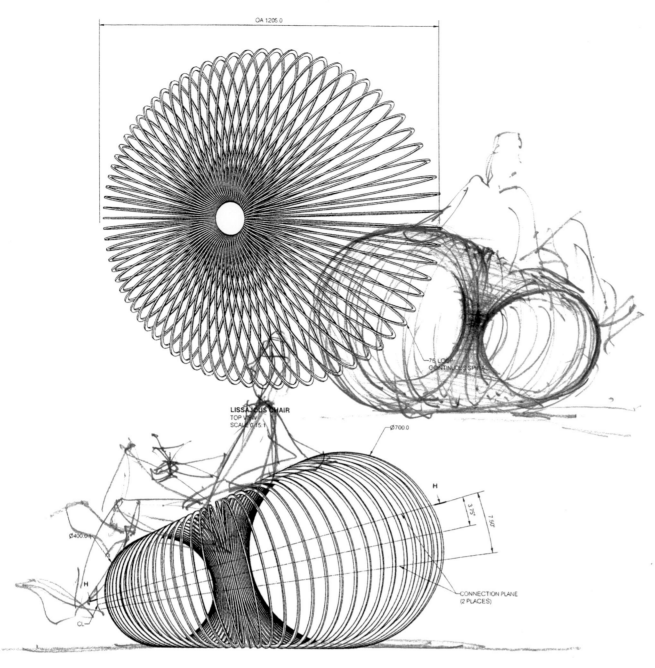

OA 1205.0

-75 LO.
CONTINUOUS SPIRAL

LISSAJOUS CHAIR
TOP VIEW
SCALE 0.15:1

Ø700.0

H

3.75

7.50

H

Ø400.0

CL

CONNECTION PLANE
(2 PLACES)

LISSAJOUS CHAIR
RIGHT SIDE VIEW
SCALE 0.15:1

6

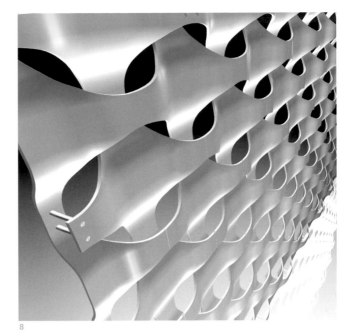

8

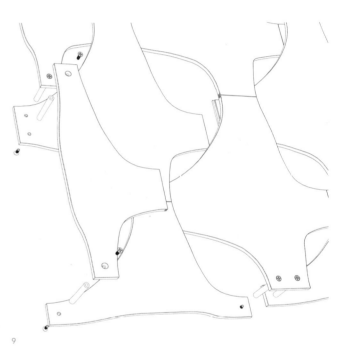

9

7

graf:decorative mode no.3, Osaka

Based in its own special graf building in Osaka, this Japanese design team also has offices in London and Tokyo. Its trademark, slightly nostalgic, wooden furniture is produced in a workshop at its headquarters, which also features a product showroom, café restaurant, gallery and design studio.

This diversity is typical of graf but could perhaps derive from the fact that the company was not initially a product-design business. In 1993, a designer, a product designer, a carpenter, a cabinet-maker, a chef and an artist teamed up to form a design group working in various fields such as furniture, space design, house design and graphics. This wide range of disciplines and skills has enabled graf to find solutions to architectural problems while maintaining an artistic perspective, and then to utilize its manual skills to realize its designs. Thus it chooses to expand beyond pure design, and embrace both the manufacture and sale of its own items.

One of the founders, designer Shigeki Hattori says, 'graf is not just a design office – we want it to be an environment where ideas that improve life can be realized. Therefore, we need the stimulation of life by the restaurant and gallery.' The name 'graf' originates from the word 'oscillograph', which the group chose to symbolize the importance of measuring, from its own standpoint, the times in which we are living. Hattori continues: 'Our final future target is to create an operating system not unlike those found on personal computers, with various "software" modules to solve different design problems. At the moment, with 40 employees, graf has been working on collaborative projects with various artists, attracting worldwide attention and moving graf into new territory.' However, although graf's work is becoming more diverse, the tradition of furniture-making still continues at its Osaka workshop. **Chieko Yoshiie**

1

1

3/6 Series
Furniture
1999
graf:decorative mode no.3
Wood, wool

*Original furniture line
made-to-order from the
Osaka studio in Japan.*

2

Nauplius
Stool
2001
SOGO furniture Co., Ltd.
Beech, cotton rope

*An elegant and graceful
chair and stand, which
combines light wood with
high quality cotton rope,
making for a simple yet
effective design.*

2

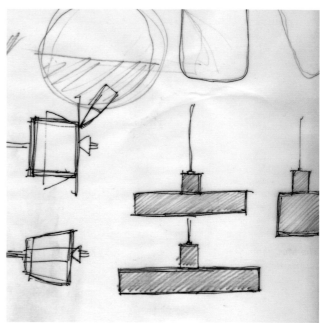

3

4

5

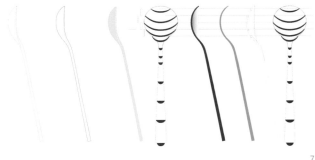

3

Mushroom Stools
2002
FRP

The Mushroom Stool tries to bring art into everyday life by incorporating colour and nature into design.

4–5

Deco Lamp
2005
Maxray Inc.
Steel

Cylindrical hanging lamps with accompanying first sketches.

6-8

SUNAO series
Cutlery
2003
Tsubame Shinko Co., Ltd.
Stainless steel

Described as 'tools for living' this set of cutlery still manages to retain traces of the Japanese style promoted by graf.

Tal Gur, Tel Aviv and Jerusalem

Tal Gur, an inhabitant of the Gilgal kibbutz, is one of Israel's leading designers. He has challenged established industrial technologies by relocating, elevating and transforming the plastic material used in rotation modelling into domestic objects with a new sophisticated appeal.

Gur makes plastic products with an atypical shape: the thickness is uneven and the line flows as if the material were fluid. This unevenness creates the attractive texture of light which is distinctive to his designs. His lighting objects, Yam Yabasha (1999), Shlonsky (2002) and Luba (2002), excellently represent these characteristics. Although these products are reproduced over and over again, each one looks a little bit different; just as all human faces are different, each of his works possesses individual character.

Gur's craft-like approach is very much influenced by the conditions of the Israeli economy where the infrastructure of the manufacturing industry is vulnerable and craft traditions are still relatively strong. This may have worked in his favour, making his products appealingly different from the glossy surfaces of mass-produced products.

Left Turn Vase (2004) and Flo Flow Flowers (2005) demonstrate the aesthetics of combination. They achieve both complete functionality and exuberance of form by combining the same modules. The Sturdy Straws Chair (2003) employs this 'combination' and 'repetition' more rigorously by using gathered straws to create both a specific form and decorative patterns.

As Nirith Nelson, the independent curator comments, Gur has found 'a way that will enable a change in an industrial process from one product to another and thus will grant it with an individual look'. Gur also takes local manufacturing facilities into consideration when designing a product. His designs, which embody an individual and specific identity, at the same time symbolize the community region that he cares so much about. **Sang-kyu Kim**

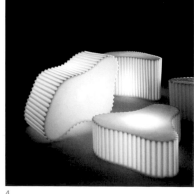

1–2

Gwanju Chair
2005
Tal GuR design
Metal wires, rattan

*The organic shape of
the chair is intended
to prioritize and draw
attention to a sense
of touch, as the rattan
surfaces of this chair are
woven by blind workers.*

3–4

Yam Yabasha
Light
1999
Hofit
Polyethylene

*The name of the piece is
derived from a children's
game, literally meaning
Sea (Yam), Land (Yabasha).*

5–8

Eash
Light
2000
Hofit
Polyethylene

*The Eash was designed
to portray the tumult at
the turn of the millennium
when everyone was
anxious and worried.
After it passed smoothly,
Gur created a smiling
twenty-first century Eash.*

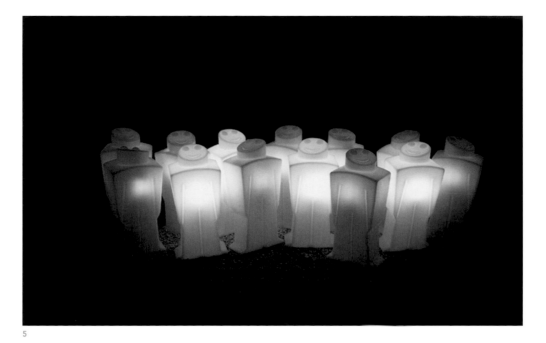

5

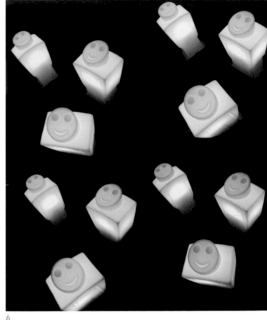

6

7

8

Tal Gur

9

9
Serve-Bowl
Fruit bowl
2004
Tal GuR design
Porcelain, plastic string

*Gur wove a coloured plastic
string with a random
pattern to create the base
of this attractive fruit bowl.*

10
Sturdy Straws Green Chair
2003
Tal GuR design
Plastic drinking straws

11
Sturdy Straws Orange Stool
2003
Tal GuR design
Plastic drinking straws

12
Sturdy Straws Chair –
Hokusai version
2003
Tal GuR design
Plastic drinking straws

*Fifteen thousand plastic
straws are used to make
just one colourful chair,
transforming the fragile
straws into a cheerful
solid chair.*

13–14
A Matter of Taste
Table
2005
Tal GuR design
Plywood
Fireproof nylon,
polyester stuffing

*This modular honeycomb
table is comprised of
several small tables to
enable a wide range of
sizes and compositions.*

13

14

Ineke Hans, Arnhem

Ineke Hans' work is straight, efficient and different. While Hans questions people's responses towards objects, she explores the latter's expected and unexpected functions with a self-imposed style. Archetypes, pictograms, childhood references and folklore codes are played with and twisted into clever products and robust furniture.

Born in The Netherlands, Hans spent some years in London while studying at the Royal College of Art and working for Habitat as a freelance designer. Today, having worked continuously since founding her studio in Arnhem in 1998, her oeuvre is made up of numerous and diverse products.

Each project is nourished by an analytic and intuitive approach based on meticulous observation of the user. Hans reflects on the essential function of the object as well as the psychological relationship between user and object.

Thus, the stainless-steel Garlic Crusher, designed for Royal VKB in 2005 and winner of the Red Dot Design Award, allies the primary function of crushing with one of the most traditionally enjoyable gestures of home cooking. Another example is the Jolly Jubilee easy chair, made of MDF and designed for Arco in 2005. Described by Hans as an 'easy chair', it combines the sobriety of a useful seat with universally comfortable proportions and a toy-like appearance in accordance with childhood archetypal codes. She responded to the challenge of designing children's furniture with her self-initiated Black Beauties collection (2000). Eluding adult prejudices, the colour black is chosen in order to highlight shapes and to suggest multiple imaginary uses to children.

Research into new materials and fabrication processes has led Hans to remarkable innovations. For example, the Black Gold modular porcelain ceramic collection (2002) proposes an infinite number of strange yet useful objects by assembling a limited series of shapes (tubes in three sizes, a corner and a plunger). A step further in her exploration is her outdoor furniture collection made from recycled plastic traditionally found along Dutch canals. Mischievously moulded in the shape of wooden planks, it is unexpectedly resistant to wind, water, salt, acid and UV rays.

In her consideration of human behaviour and our expectations of objects, at the same time exploring new materials, Hans always finds ways to surprise us with a definitively different approach. **Didier Krzentowski**

3

2

1

1

Black Gold
Five-armed candleholder
2002
Erik Jan Kwakkel
Black porcelain

As part of her 'modular porcelain' project, Ineke Hans produced only five shapes/moulds, and used them to make vases, coffee pots, candleholders and chandeliers in graphically striking black porcelain.

2–3

Pixelated Peacock
Light
2006
Swarovski
Crystal beads

The images of the peacocks are made using 104,328 Swarovski crystal beads, creating the effect of an embroidered pattern.

4

Forest for the Trees
Coat stand
2005
Lensvelt BV
Epoxy-coated metal

These stands are ideal for reception areas. Here they are seen at Club 11, Amsterdam with Hans' Relax Set.

5

Relax Set
Table and chairs
2005
INEKEHANS/ARNHEM
Recycled plastic

Although wood-like in appearance, the sets are made from 100% recycled plastic that is found along Dutch canals.

6

Beer Set
Beer table set
2005
INEKEHANS/ARNHEM
Recycled plastic

Beer-garden furniture, also made of 100% recycled materials.

4

6

5

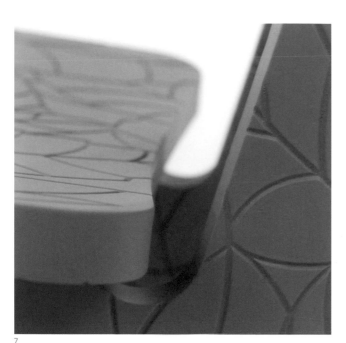

7

7–8

Jolly Jubilee
Easy chair
2005
Arco
MDF

This easy chair, made from one sheet of MDF indented with a leaf pattern, was designed as part of Arco's 100th birthday celebrations.

9

Bowls & Spoons
2005
Royal VKB
Porcelain, stainless steel

The appealing design of Bowl & Spoons is based on the traditional Chinese noodle bowl; it can be used for eating soup or serving sauce. It won the Red Dot Award in 2006.

10

Garlic Crusher
2005
Royal VKB
Stainless steel

This tool pulverizes the garlic for the most intense flavour, and when rinsing the crusher, the special stainless steel it is made from removes the odour of garlic from your hands.

8

9

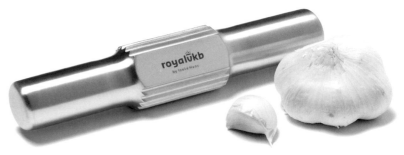

10

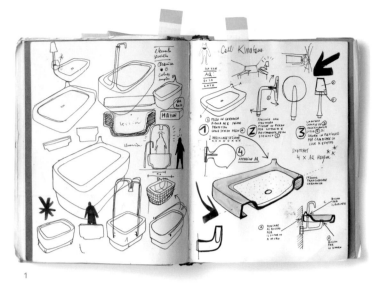

1

2

3

4

Jaime Hayon, Barcelona

Jaime Hayon's video shows himself up close, repeatedly being slapped in the face. As his head is knocked from side to side, Hayon remains motionless and unfazed. In 1997, this work, which he conceived and directed, earned the young designer, fresh out of graduate school, a job at Fabrica, the Benetton Group's research centre in Treviso, mentored by Oliviero Toscani. Only a year later Hayon was head of Fabrica's design department, responsible for a score of projects in industrial design, exhibition design, communication and curatorship.

Hayon invents his own world, keenly absorbing the eclectic reality that surrounds him and responding with new creations. He is a compulsive draughtsman and his work as a whole is infused with artistic processes. Although production systems interest him, they do so only in as much as they represent the means to achieve his creative goals. Nevertheless, he does have a special interest in handmade processes and their potential in an age of mass industry – notions that can be recognized in the projects Mediterranean Digital Baroque and Mon Cirque. In a similar way, his bathroom collection (2004–6), adapted for production by Artquitect, highlights Hayon's fondness for identifiable references combined with a gamble that overturns conservative limits and standardized, dictated uses of objects.

Designers such as Hayon prove that contemporary society has been able to devise tools that allow individuals to control and operate a whole production system and market that, until recently, had been exclusively dominated by major companies and corporations. In today's emerging design scene, he represents a generation that has given unparalleled visibility to issues of identity, individual production, networking, communication and entrepreneurship, coupled with a considerable success in penetrating the market. Hayon's incredible drive and unique perspective significantly contribute to opening up new dimensions in our everyday lives. **Guta Moura Guedes**

1

Technical research sketches
for AQHayon Collection
2004

2–4

AQHayon Sink and Mirror
2004
Artquitect Editions
Ceramic sink, mirror

5–8

AQHayon Bathtub
2006
Artquitect Editions
Fibreglass, lacquered and
polished wood, ceramic

*The AQHayon Collection
combines organic shapes
with minimal colours,
and brings a new versatility
to bathroom furniture.
The collection won the
Elle Deco International
Design Award for best
bathroom in 2005.*

9–10

Poltrona with cover
Chair
2006
BD Ediciones de Diseño
Lacquered rotation-
moulded polyethylene,
handmade leather interior

*The Showtime Collection,
of which the Poltrona
is part, was inspired by
old MGM musicals. The
Poltrona combines highly
contrasting materials –
glossy plastic and leather –
to forge a new type
of elegance.*

5

9

6

7

8

10

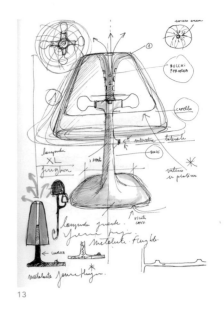

13

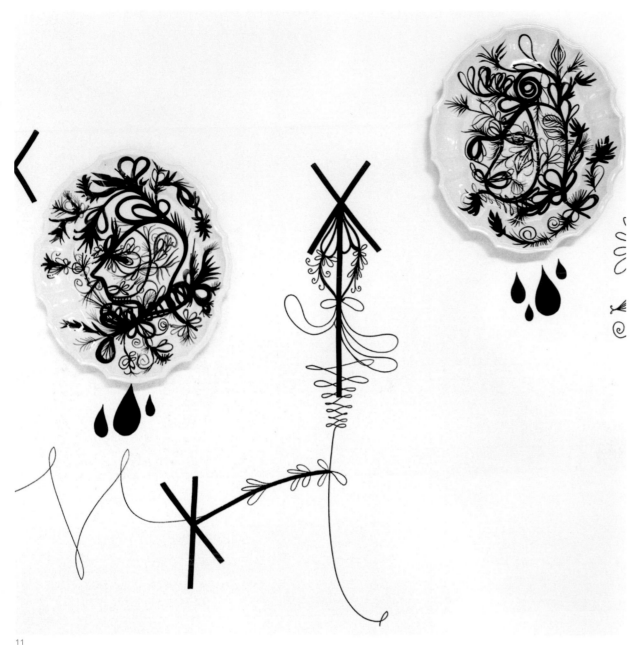

11

14

15

16

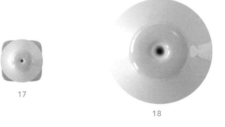

17

18

19

12

20

11–12

Mon Cirque Ceramic Plates
Decorative plates
2005
Bosa/Jaime Hayon
Ceramic and ink

*Each piece was hand
painted by Hayon with
different motifs.*

13–20

Funghi
Lamp
2006
Metalarte
Ceramic

*The light from the Funghi
lamp creates a subtle
and mysterious aura, as if
you were walking under
the leaves of forest trees.*

21–22

Multileg Cabinet and Vases
2006
BD Ediciones de Diseño
Cabinet: Lacquered wood
DM, glass
Vase: Glazed ceramic

*The Multileg Cabinet is
a modular system with
12 legs that can be chosen
and combined in different
ways. The Vases are based
on ancient funerary urns,
and their quirky shapes and
random openings allow
flowers to be displayed in
unusual ways.*

21

22

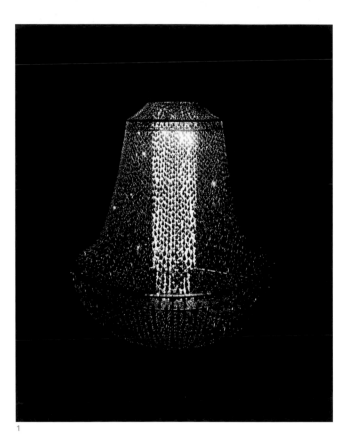

1

Simon Heijdens, Rotterdam

Simon Heijdens' work references nature as a model of transformation and change. He sees this model as a constant process that governs the transformation of things over time, a concept usually denied to objects, which are normally considered unchanging. Time may well be abstract but Heijdens does all he can to materialize it. 'Inserting a story dimension' into objects is one of the recurrent themes of his work, a propensity that stems from the filmmaking studies in which he trained before studying at the Design Academy, Eindoven.

His fascination for the way things evolve over time and with use inspired his Broken White project (2004), for which he designed a set of ceramics that are immaculate and white, when new. As the bowls and plates are used, delicate cracks appear in the glazing and with time slowly spread to form a floral decoration. As this artificial growth is directly correlated with use of the object, the most used items – the favourites – stand out: a sign that portrays the intimate relationship between the user and the object. Heijdens sees objects and their environment acting as collectors of images, sensations and memories.
In his designs he observes and plays with what is already inherent in their character, as with Clean Carpets (2004), where the pattern is derived from a slight manipulation of the material itself. Here with a high-pressure sprayer and a template, grime is removed from dirty city pavements in the patterned form of a carpet.

Nature's continuous metamorphosis is an indication of the endless stories that can be told, alluding to the mysterious energy that runs through things and subtends their use. As Wittgenstein said, 'use is the life-breath of objects'. **Francesca Picchi**

1

Rising Slowly
Chandelier
2006
Swarovski
Swarovski crystals, medical
wire, wind sensor

The chandelier's crystals are threaded with contracting wires that react to wind sensors and reproduce the ripples of the real wind outside the building.

2

Lightweeds
Illustrations
2006

3

Lightweeds
Light installation
2006
Computers, projectors,
various weather
and motion sensors

White silhouettes of plants are projected onto the walls of a space. They are 'alive': they grow with rainfall and sunshine that is measured from outside, move with the real wind, and turn towards the position of the sun.

4–6

Moving Wallpaper
Interactive wallpaper
2002
Self-production/Droog
Design
Paper, colour-changing
pigment

Wallpaper with animated and changeable patterns inside the paper. Due to the pigment used there is no light projection involved, and you can turn the design on and off, or select another one.

2

3

4

5

6

8

9

silhouette of a tree grows on façade

tree moves to actual measured wind

tree looses a leaf when some passes, and shedding tree reveals human activity

leaves break of the tree and fall on the floor nearby

leaves roll out when someone walks through

7

10

7–9

Tree
Location-sensitive
outdoor light installation
2004
Self-production
Computers, projectors,
wind-, sound- and
motion-sensors

10

Tree
Location-sensitive outdoor
light installation
2004
Illustrations

*A life-sized tree projected
onto the facade of a
building that moves with
the wind as it is measured
around the actual building.
Each time someone passes
the tree, one of its leaves
breaks off and falls onto
the pavement.*

11

Clean Carpets
2004
Self-production
High-pressure cleaner,
template

*The piece is made by
removing grime from the
pavement with a template
and high-pressure cleaner.*

12–13

Broken White
Ceramics
2004
Artecnica
Ceramic

*These ceramics evolve
with their user. When new,
the cups and dishes are
white and undecorated; the
more the object is used,
however, the more hairline
cracks appear in the glaze,
slowly forming a floral
decoration that grows as
a flower would.*

11

12

13

1

Iron Chair X
Prototype
2006
Jackson Hong Design
ABS plastic, aluminium,
stainless steel

*Iron Chair X is designed for
people who love silence.
If a talkative person tries
to sit on it, eight spikes
are released so that they
can't sit down!*

2–4

Lovely Polisher X
Cleaning Machine
2005
Jackson Hong Design
Mixed media

*The Lovely Polisher X
allows the wearer to show
their loyalty and affection
for a person without
words: they remove dust
on the shoes or body of
their chosen subject by
brushing and vacuuming.*

Jackson Hong, Seoul

Jackson Hong is a designer with an artistic imagination who often crosses the border between art and design. There have been artists such as Donald Judd who have designed products, and product designers who have held exhibitions in galleries. However, Hong's work takes a somewhat different approach.

In his career as a product designer at Samsung Motors and at ECCO Design, Hong has always treated objects as commercial goods. Yet he constantly explores the relationship between the human being and the object, endlessly reassessing his design processes to find mechanical solutions to create thoroughly industrial products. His work's formal perfection, however, means that his designs are elegant enough to be displayed in exhibitions, even though they were designed to be manufactured on a production line.

In the past Hong has struggled to get his designs into production. This is due to the conditions of Korea's consumer product dominated market that means manufacturers are unwilling to produce objects with a strong individual style. Thus far he has displayed his designs in galleries, but he is finally preparing for his products to be manufactured.

Hong feels that the products he designed when he worked for commercial design studios do not represent his ideas accurately. That is why he has produced works such as External Intestine X (2005). In a delicate way, this design engages with the fear of obesity and the desire to overeat at the same time, actually proposing to solve the conflict. Masked Citizen X version 2.0 (2006) is a mask that outwardly displays an artificial expression and is designed to be a defensive device in city life. The mask demonstrates the unique characteristics of Hong's work.

A designer often designs products for the public in general without a specific user in mind. Or sometimes they design for a specific group such as physically disabled people or product design enthusiasts, in order to help them or satisfy them. In that mode of thinking, Hong is currently developing a line of design products for people who are emotionally unstable or nervous. **Sang-kyu Kim**

2

3

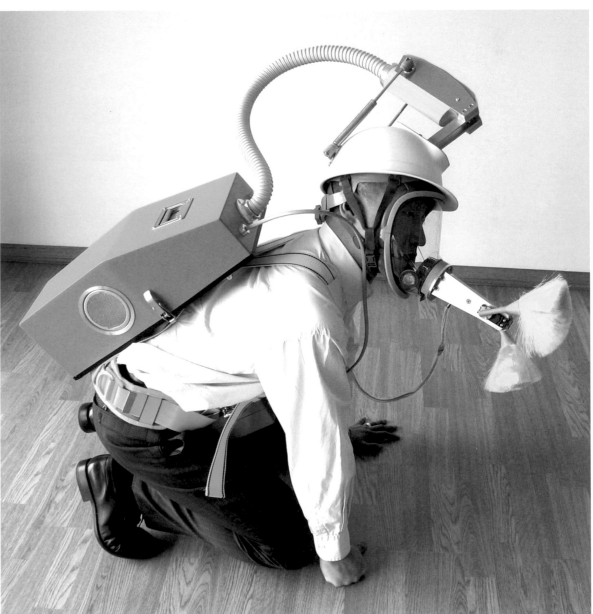

4

5

5–6

Reactive Hair X
Prototype
2006
Jackson Hong Design
ABS plastic, aluminium,
stainless steel

*If the wearer becomes
afraid that someone is
going to attack him/her,
the headgear releases
blades to warn them away.*

7–11

Masked Citizen X version
2.0
Prototype
2006
Jackson Hong Design
ABS plastic, aluminium

*Masked Citizen provides
a refuge for people
who are tired of social
communication. It displays
an artificial expression
through LCD panels on
the eyes and mouth.*

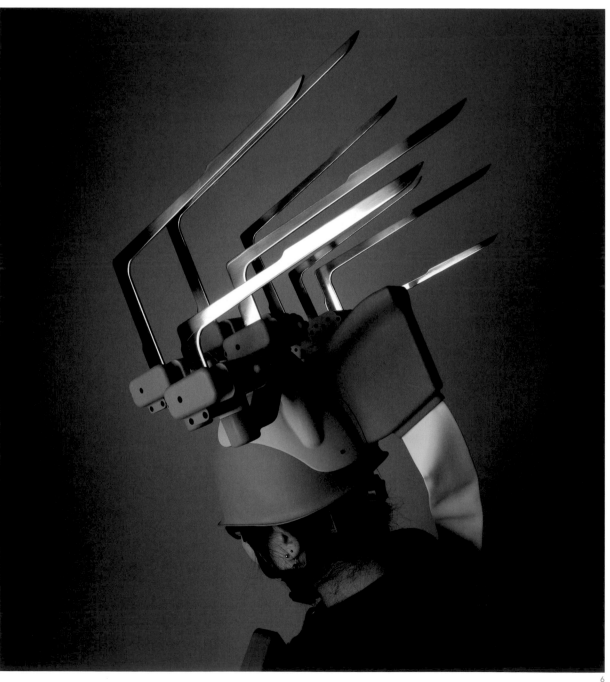

6

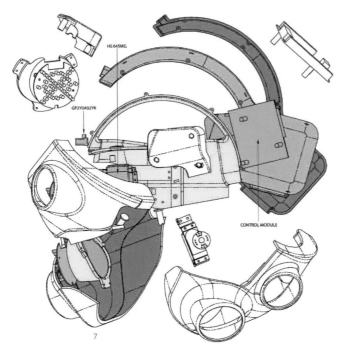

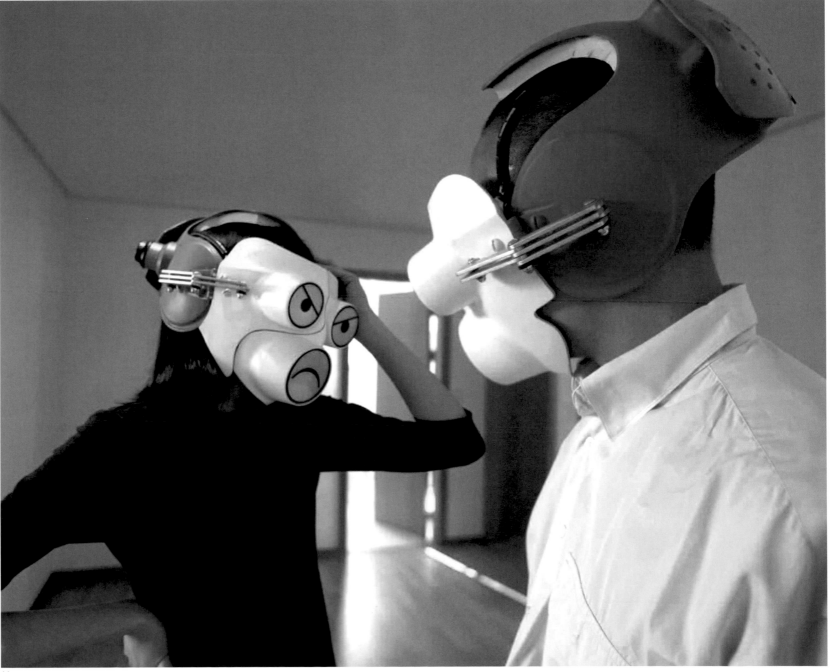

Junya Ishigami, Tokyo

Junya Ishigami graduated from Tokyo University of Fine Art and Music in 2000, and worked for Kazuyo Sejima & Associates before opening his own architectural practice in Tokyo in 2004. Examining Ishigami's work in the fields of both architecture and design reveals a common thread running throughout: whatever the function or scale of the project, his strong sensitivity encourages us to question our perception of space.

Ishigami uses the slightest changes in our perception of the world around us to express a concept of beauty that contains something of the mysterious. He creates a subtle sense of alienation that stems from precise form, material and execution, combined with calculated rigour. Perfect objects, excellent manufacturing standards, careful structural conception and sophisticated materials all help to create an abstract and subtly surreal dimension in his work.

One of Ishigami's most significant works is Table (2005). The initial aspiration – to achieve a table surface as thin as a sheet of paper (in fact, it is 4.5 mm / 0.2 inchs thick) – was followed by technical research. Structural analysis prompted him to pre-compress sheet steel in the opposite direction to that of the natural flexion stress: the pre-tension produces a concave form that when rested on the floor returns to the customary flat form we expect of a table top. Ishigami reduces the table to the bare minimum so that even the thin finishing layer of veneer performs a function by controlling horizontal oscillations, thereby stabilizing the overall form. Despite having all the usual requisites (four legs and a top), the table almost ceases to be a recognizable object because it is so remarkably slender. Like a suspended plane, it runs parallel to the floor and ceiling as if creating a new dimension, and objects placed on it seem to float in mid-air. **Francesca Picchi**

1

1–2

Low Chair
2005
Mituisi
Polystyrene foam, race cloth

This immensely light chair looks like a prototype model but is, in fact, a fully functioning chair. The density of the polystyrene foam varies: the leg ends have less air and are therefore denser for stability, whereas the seat is less dense and is therefore softer and more comfortable for the user.

2

3–5
Table
2005
Akahoshi
Steel, aluminium, wood

Table seems to pare the idea of a table down to the bare minimum, and yet it is also a luxurious item that is the result of progressive technology and brilliant structural design. The table top has a thickness of only 4.5 mm so that the items placed on it almost seem to be floating in mid-air.

4

5

Ichiro Iwasaki, Tokyo

Ichiro Iwasaki joined the Sony Design Center in 1986, taking responsibility for the company's TV and car audio designs. He moved to Italy in 1991 and, after working as a designer in Milan, returned to Japan in 1995 to form his own design office. Making use of his early knowledge and experience, his designs were centred on electronics. He received wide recognition after being involved with the Korean audio brand, MUTECH, of which he was design director for five years.

Iwasaki's audio and telephone designs are based on the concepts of ease of use and simplicity. By clearly defining the product's functions, he aims to make them easy to understand at a glance, and familiarize intuitively. His designs for mobile phones, for example, are highly acclaimed, and he is gradually expanding this field of work. At present, he is also designing espresso machines, tableware, furniture and even fishing reels. Iwasaki's Stainless Steel Kettle for PYPEX (2006) combines his creative principles in its elegant poise and practical, usable design.

Whatever the product, Iwasaki's first inspiration comes from the basic function of the object, followed by how the design can be made to match its various uses. By focusing on how the product actually works, when and where it will be used, under what conditions and how it feels to the user, he searches for the 'what it is' that exemplifies them all. Only then does he mould those ideas into shapes.

Consequently, Iwasaki's designs do not all conform to a single style; he may make a number of models for any one product, based on his beliefs that 'design cannot be rushed' and 'goods in daily use require more time'. As a result, he aims to arrive at a design which is 'not over-designed but not cold and inorganic either'. Iwasaki declares, 'I can design only by coming face-to-face individually with each project', a worthy mission that places him within a unique group of artisans working today. **Chieko Yoshiie**

1–2

Takata 04-neo
Child car seat
2003
Takata
Polypropylene, urethane foam, fabric

The seat was selected to be displayed at MoMA's 'SAFE: Design Takes on Risk' exhibition in 2005.

3

Telephone 510
2004
MUTECH
ABS resin, acryl

This cordless phone looks as if it is floating on air, and goes some way to redefine the home phone.

4

Stainless Steel Kettle
2006
PYPEX
Stainless steel, silicon

The kettle's upright poise and tapered grip ensure safe and easy pouring.

5

6

7

8

9

5

Drop
Plates
2004
Contrast
Porcelain

*The free and imperfect
shapes of the plates
were designed to
accommodate and assist
all types of cooking.*

6

Cameo
Cutlery
2005
Contrast
Stainless steel,
Acrylonite-Styrene
copolymer

*Iwasaki imagined the
relaxed mood of
a picnic as inspiration
for this design.*

7

Harp
Teapot
2004
Contrast
Porcelain, plywood

*Suitable for Japanese or
English tea, this teapot
is a formal blending of
the two cultures.*

8

O-Bon
Trays
2004
Contrast
Porcelain, maplewood

*Either use O-Bon as two
separate trays, or slot them
together to form one unique
and delicate tray.*

9

Aroma
Espresso maker
2004
Contrast
Aluminium, polyamide

*The unaffected form and
tender texture of the
Aroma creates a familiar
and homely mood.*

10–12

Stainless Steel Kettle
2006
PYPEX
Design drawings and models.

10

11

12

13

13

Meteo
Side table
2005
AIDEC
MDF, steel

*This asymmetrical design
subtly changes its
expression depending on
the direction in which it is
placed.*

ixi, Okinawa

ixi is a design network established by Xavier Moulin and Izumi Kohama. After working in France and Italy, they settled in Okinawa, Japan, sharing ideas with Japanese and foreign designers, and developing new designs.

New design usually replaces one object with another, thereby initiating change. However, the design of Moulin and Kohama never challenges the existing order of objects. Rather, they aim to fill empty spaces other designs have missed, or make small modifications to suit a specific purpose.

The Stool Pants (2002), introduced in the 2002 'SKIN' exhibition at the Cooper-Hewitt Design Museum, demonstrates the originality of ixi's design. This garment extends the function of trousers, providing a readymade rug, so that the wearer can sit down on the ground. The Pady Bag (2001) serves as a rug and a bag at the same time. This wearable furniture is part of the HomeWear series, designed to provide users with in-built comfort wherever they go.

The Mangastrap (2001) is a response to the daily behaviour of Japanese people on overcrowded underground trains, standing and reading their manga (comics); the strap allows them to hold a comic and read in comfort. High Tension (2004) is a suit with a system of hooks and straps that lock the body into various positions when connected, thus transforming the entire body into a piece of furniture. This design displays the characteristics of *chindogu* – the transformation of an object to emphasize its function rather than its form.

In contrast with the HomeWear series, Up (1998, 2006), part of the Smart Home Fitness series, deliberately makes people uncomfortable. It forces them to exercise: objects are placed in the container at the top of the ladder, out of arm's reach, so that in order to obtain them you must climb up and down.

Moulin and Kohama want objects to be assimilated into their environments rather than to stand out from them. Therefore, they don't stick to a certain style, shape or space; instead, they describe their designs as 'shapeless' and 'all-terrain'. Their office and home, which they call 'ixi land', is also absorbed by the surrounding nature of Okinawa island, the building itself being an extension of their design.
Sang-kyu Kim

1

2

3

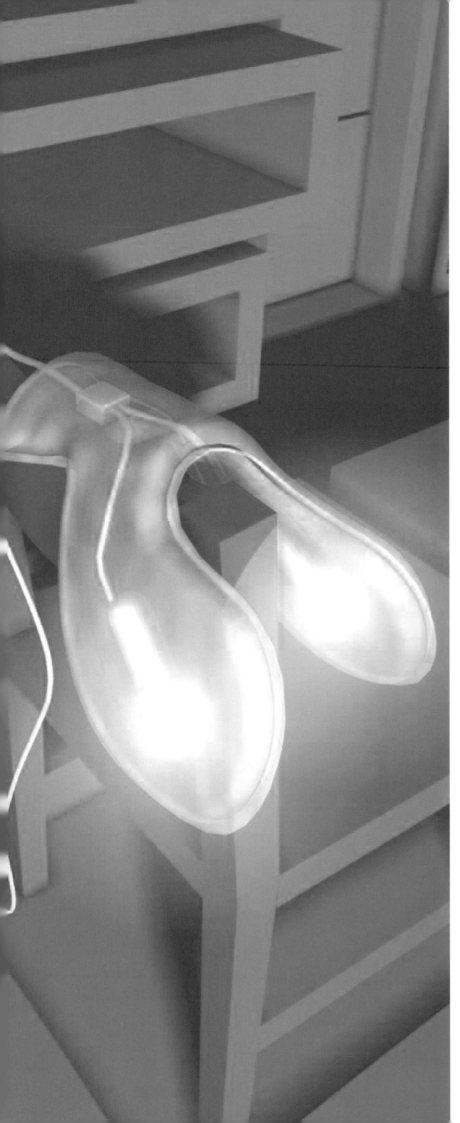

4

5

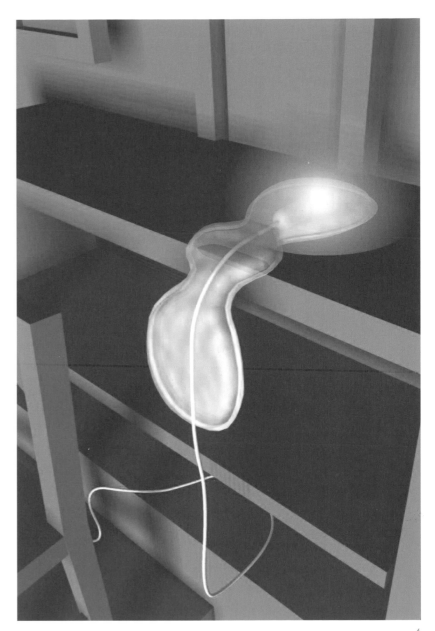

6

1–2

Mangastrap
2001
ixi network
Fabric

*Part of ixi's HomeWear
range of wearable objects
and furniture, this jacket
is designed to allow manga
fans to read comics in
comfort no matter where
they are.*

3–6

Balladeuse
Light
2000
ixi network
Levagel, P.U. film, fluo
compact light

*The Balladeuses are
fluid and flexible, they
crawl around all terrains
bringing light to the
most inaccessible nooks.*

7

8

9

7–9

Pady Bag
2001
ixi network
Fabric

*Part of the HomeWear
range, this multifunctional
and stylish bag can be
adapted to become a
padded cushion for any
standard chair.*

10–12

Up
Storage unit
1998, 2006
Kenji Nakachi
Iron

*This ladder storage
unit is designed to make
people to do easy and
natural exercises in the
home every day.*

11

12

10

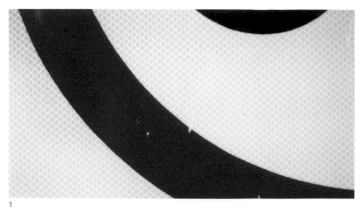

1–6

Sign Stool
2004
Trent Jansen Studio
Re-used road signs

The road signs used to create the Sign Stool come complete with all the characteristics of their original purpose, including colouful vinyl labels and the odd bit of evidence from their life by the roadside.

Trent Jansen, Sydney

Trent Jansen is a graduate of Sydney's University of New South Wales. Its Design School is notable for its encouragement of artistic enquiry and its requirement that students work across multiple design disciplines. This training clearly suited Jansen's probing mind and varied interests as he has developed a practice that incorporates graphic, web and product design. After completing his studies in 2003, he undertook a three-month internship with Marcel Wanders Studio in Amsterdam and received widespread attention in Australia after winning Object Gallery's annual New Design national graduate award in 2004.

The award-winning work was Jansen's Sign Stool (2004), which incorporates the striking re-use of metal road signs, complete with the reflective luminosity and other characteristics of their original purpose. The Sign Stool led Jansen to develop the Tokyo Sign Project for Tokyo Design Tide 2005, in which he assembled, sold and exhibited a simpler, more archetypal stool made from old road signs in a shopfront on the main street of the Harajuku district. He encouraged passers-by to consider the simplicity and importance of sustainable design. The project also allowed Jansen to collaborate with Japanese fashion designer Masahiro Nakagawa – renowned for his Tokyo Recycle Project, transforming old clothes into one-off couture pieces. Jansen's stools have been upholstered with Nakagawa's recycled clothing. Most recently, Jansen has reconsidered the properties of the road sign material and has created a series of attachable bicycle reflectors that help make cyclists more visible when riding at night.

Archetype and irony are regular features of Jansen's work. His Topple Lamp (2006), for ISM Objects, explores the idea of a precious object being lovingly looked after for decades by a grandparent, only to end up in a garage sale or charity bin once they have passed away. With an internal mercury switch, the traditional-looking lamp has to be treated with disrespect – toppled – to turn it on. **Brian Parkes**

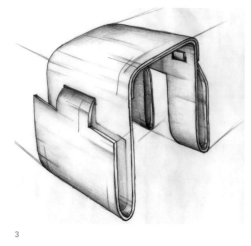

3

4

5

7

8

9

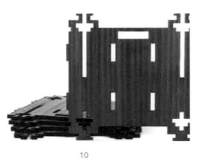

10

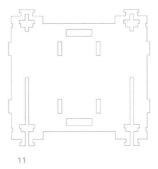

11

12

7–9

Topple Lamp
2006
ISM Objects
Ceramic, cotton

*The lamp takes an
old-fashioned form, but
the twist is in the function:
the lamp has to 'toppled'
to be switched on.*

10–12

Red x
Storage system
2005
Prototype
Plywood

*Red x is a bookcase
designed to take up as
little space as possible
when dismantled,
whllst providing as much
storage space as possible
when assembled.*

Patrick Jouin, Paris

A slim, white cylinder, ribbed like a parasol, spreads open in one effortless motion to become a seat. Nothing in the complexity of Patrick Jouin's Oneshot stool (2006) for Materialise.MGX hints that the product emerged from a vat of resin as a single, resolved piece, untouched by any hand. Jouin, a Paris-based industrial and interior designer, is at the forefront of investigations into stereolithography, in which a computer-controlled laser congeals liquid plastic. He's been pushing the limits of this rapid-prototyping technology ever since 2004, when he launched Solid, a furniture collection that includes chairs that look like masses of hardened strings and a table resembling a hollow cube with perforations.

Such experiments represent only one facet of Jouin's prismatic career. Since leaving Philippe Starck's office in 1999 to go out on his own, he has designed dozens of projects, including armchairs for Cassina, earthenware dishes for the 400-year-old French company Gian, and interiors for Alain Ducasse's Paris restaurant and of the Manhattan eatery Gilt. 'The direction I've followed from the beginning is no direction,' he says. Striving to be 'in the present, but never in fashion', Jouin went so far as to abort work on an unproduced Kartell chair because it was already looking derivative. (The chair was based on Optic, a hinged plastic box, introduced at the 2006 Milan International Furniture Fair. Optic's irregularly faceted surface, however, had caught the eye of too many imitators, and Jouin wanted nothing more to do with it.)

Jouin claims that his personal ennui motivates his innovation, but he is equally compelled by cultural demands. His work with rapid prototyping addresses not just a thirst for new aesthetic languages but also the modern obsession with speed. Collaborating with experts in brain chemistry, he is also developing an alarm clock that lulls users to sleep with one hue of light and wakes them gently with another. He sees the revival of ornament in design as 'human', but warns that it can turn into a prison when it fails to break away from familiar motifs. 'It's not easy to invent new decoration,' he muses. 'I think it's really the goal of design now: to reinvent.'
Julie Lasky

1–2

Ether
2006
Murano Due
Polished steel, blown glass

The light streams down between the delicate, transparent blown-glass balls to create a stunning light effect.

2

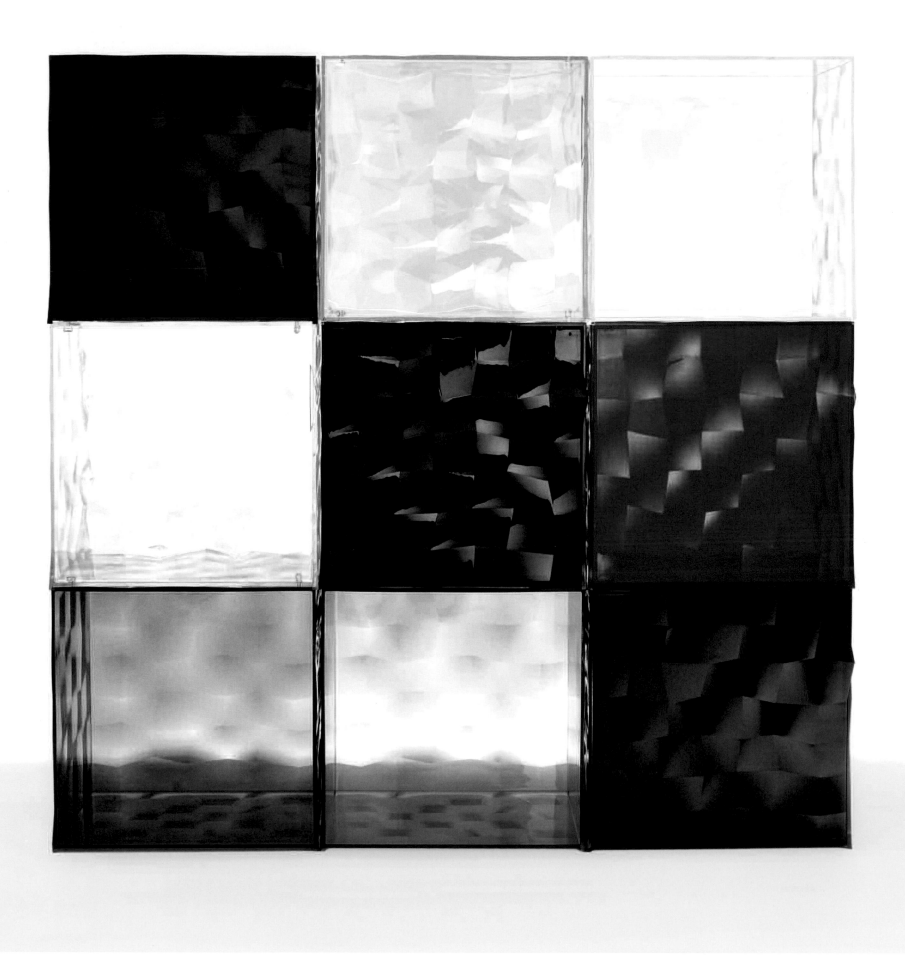

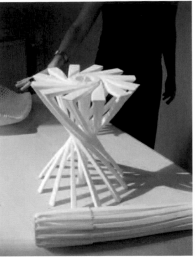

3

Optic Cube
2006
Kartell
Transparent or smooth
batch-dyed PMMA

*These faceted cubes can
be arranged or stacked
to create unusual,
flexible arrangements.*

4

Solid C1
Chair
2004
Materialise.MGX
Epoxy resin

*The form of this chair
is reminiscent of blades
of grass or ribbons
waving in the wind and
weaving together.*

5–7

Oneshot
Foldable stool
2006
Materialise.MGX
Polyamide

*Made using polyamide
treated with Selective
Laser Sintering (SLS), this
portable foldable stool is
both durable and practical.*

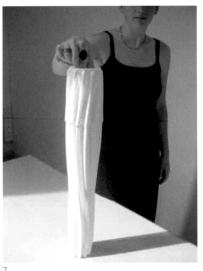

1

2

Chris Kabel, Rotterdam

Chris Kabel describes his designs as 'poetic reassemblages'. They result from an analysis of objects, which enables him to differentiate function from decoration and shape from material.

Now based in Rotterdam, Kabel studied at the Design Academy, Eindhoven, where he caught the attention of Droog Design. His diploma project Sticky Lamp (2001) was included in their catalogue in 2002, marking the beginning of an ongoing collaboration including commissions and exhibitions. moooi and Royal VKB followed suit, and some of Kabel's products have been acquired by the Museum Boijmans van Beuningen, the Stedelijk in Amsterdam and the Fonds National d'Art Contemporain. Kabel was also nominated in 2003 for the Rotterdam Design Prize with his 1TOTREE lamp.

Kabel's products achieve a balance between sensitive evocations and statements about function. He likes to figure out how decoration can add new use to an object. For instance, cut-out lace, a purely ornamental material, creates a pattern of foliage when the sun projects through his parasol Shadylace (2004). He values his industrial design education, which explains why issues of production and use are at the heart of his design objectives.

Kabel's lamp Flames (2003) is representative of a process where the inner and more functional structure of an object merges with the decorative outer layer. The raw industrial gas pipes and tank are painted white, their slender shape conjured into a romantic chandelier casting a flickering light. A similar process led to the Mesh Chair (2006). While he was working on an upholstered chair, Kabel discovered that if he treated the metal as if it were a textile, using patterned metal pieces and replacing sewing with welding, he could 'tailor' it in the form of a chair that looked like it was upholstered.

Kabel playfully revealed his questioning of some aspects of the design world with his Black Money Vase (2005). A vase that appears as a heap of coins made of porcelain covered with the imprint of coins, it highlights the tendency of design to be, at times, driven more by profit than by creation. Here the decorative surface reflects and creates the value of the vase.

Having taken part in many significant group shows, in 2006 Kabel was rewarded with a solo show in Paris, another sign that he is being recognized as a promising independent designer. **Didier Krzentowski**

1–3
Black Money Vase (BMV)
2005
Limited edition
Black porcelain, coins

4
Platinum Money Vase (PMV)
2005
Limited edition –
Toolsgalerie Ltd
Platinum-lustred porcelain,
coins

Adorned with euro coins,
these vases are a mild
and very ornamental critique
of capitalism. Price has
become decoration, which
in turn adds value to
the object.

3

4

5

6

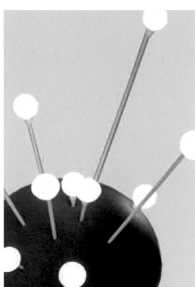

7

5–7

Pinlight
2006
Limited edition –
Toolsgalerie Ltd
Flocked foam, electrical
components, LED

*Since LED lamps resemble
needles, the designer
used this aspect to create
a lamp. When you stick the
Pinlights into the cushion,
they light up.*

8–11

Shadylace
Parasol
2004
Droog Design
Lace, parasol frame

*Commissioned by Lille2004
with Droog Design, the
parasol was developed with
local lace industries in Lille.*

12–13

Flames
Candle
2003
moooi
Powder-coated gas
pipes and fittings,
gas-cartridge holder

*A refillable gas-fuelled
candle.*

14–17

Tailor Chair
2005
Limited edition
Powder-coated
expanded metal

*The construction parts
of the Tailor Chair are
assembled as a piece
of clothing would be,
although they are welded
instead of stitched.*

Chris Kabel

8

9   11

10

12

13

14

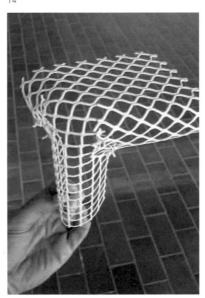

15

16

17

Lambert Kamps, Groningen

1–3

Air Bridge
2001
Lambert Kamps
Soft PVC

*The bridge was
successfully used at
the Blue Moon Festival
in Groningen, The
Netherlands in 2001.
It allowed visitors to
cross a river in an
inflatable tunnel.*

Confined to a waterlogged fragment of turf, the Dutch
are geniuses at making room. They have built spindly
houses and reclaimed land from the sea, and now Lambert
Kamps is dreaming of more radical solutions. A resident
of Groningen, where he was born and educated,
Kamps envisions architecture that expands when occupied
and collapses when empty. His 15-metre-wide (50-foot)
Spacemaker tent (2002) registers visitors' presences not
just by billowing and sinking but also by illuminating – floor
sensors trigger lights with every footstep. Like much
of Kamps' work, Spacemaker seems alive with use, as if
reprimanding the immobility of traditional architecture.

Kinetic constructions have preoccupied Kamps from the
time he was a student at Groningen's Minerva Art Academy
and made a suite of pneumatic furniture threaded together
by a hose and inflated with an air pump. In 2001 he won
a citywide competition to build, for an arts festival, an inflatable
bridge – a 15-metre-long (50-foot) span that supported
the crossings of over 500 people each day for a month. He
assembles such pieces himself (early training in fashion has
made him handy with a sewing machine), scrounging chunks
of PVC from a ruined building for Spacemaker's floor tiles,
or squirrelling away the cast-off skins of hot-air balloons for
a kilometre-long inflatable sphere that he plans to construct –
a 'new planet' that will float, sinking in the evening sky as
ambient temperatures cool its internal gasses.

Kamps begins with an art concept and spins off design
applications. Having mastered the science of programmable
LEDs for Spacemaker, for instance, he found a commercial
use in illuminated polyester roof tiles that work as architectural
signage. Similarly, after studying hydraulics for a public art
commission that involved replanting a fallen tree, he
transferred the technology to a lamp containing a rod that
rises and descends to adjust brightness.

Kamps' preoccupation with the vitality of inanimate forms
has taken darker turns as well: his installation at the 2005
Milan International Furniture Fair showed motorized furniture
pieces crashing together and gradually destroying each
other as each seemingly tried to secure the best position in
the room. **Julie Lasky**

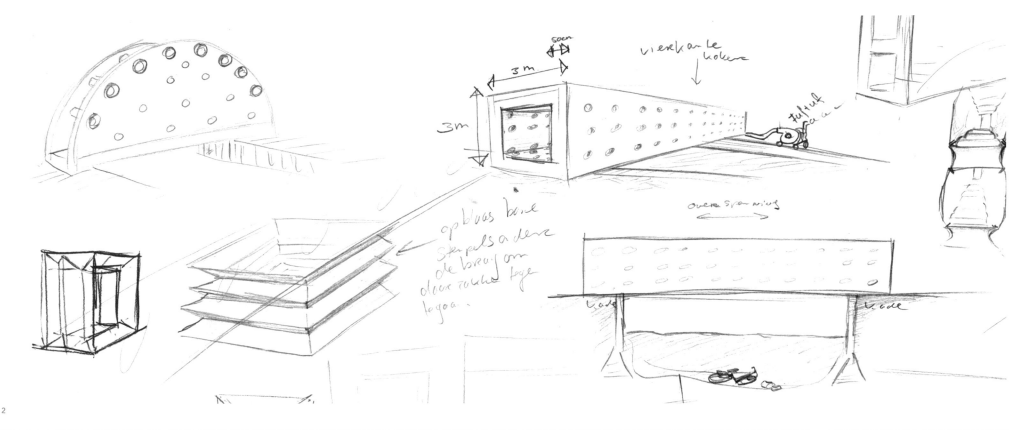

2

3

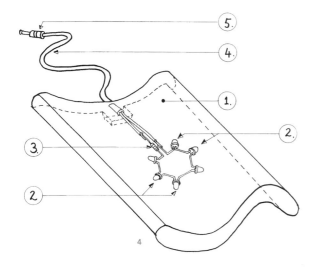

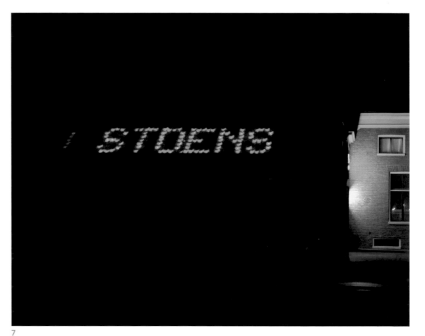

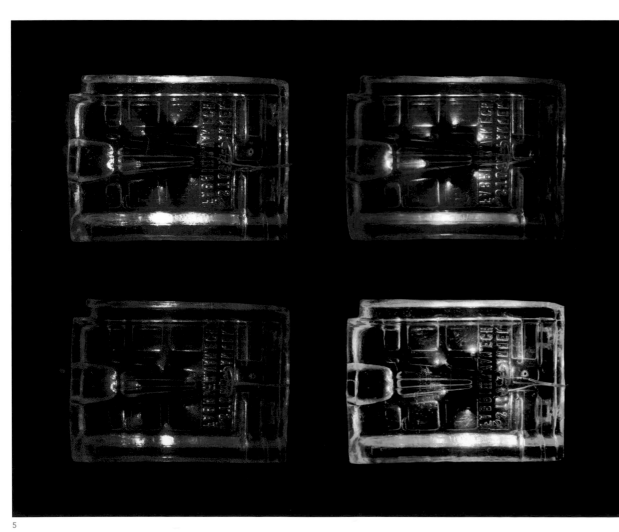

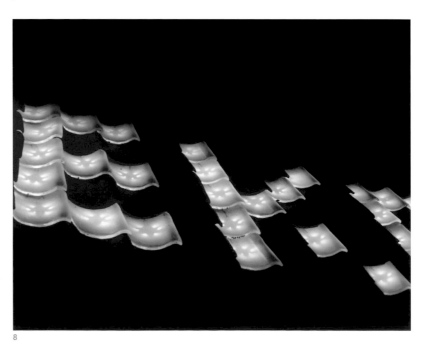

4–8

Light-Emitting Roof Tiles
2004–5
Lambert Kamps
Epoxy, electrics

The tiles, which display huge logos, numbers or words on rooftops, can be used for commercial purposes ... or just for fun.

9–11

Magnet Clock
2005
Lambert Kamps
Steel, wood, electrics

The clock is operated by a matrix of electronic magnets. These hold the iron powder that forms the time display onto the clock's surface.

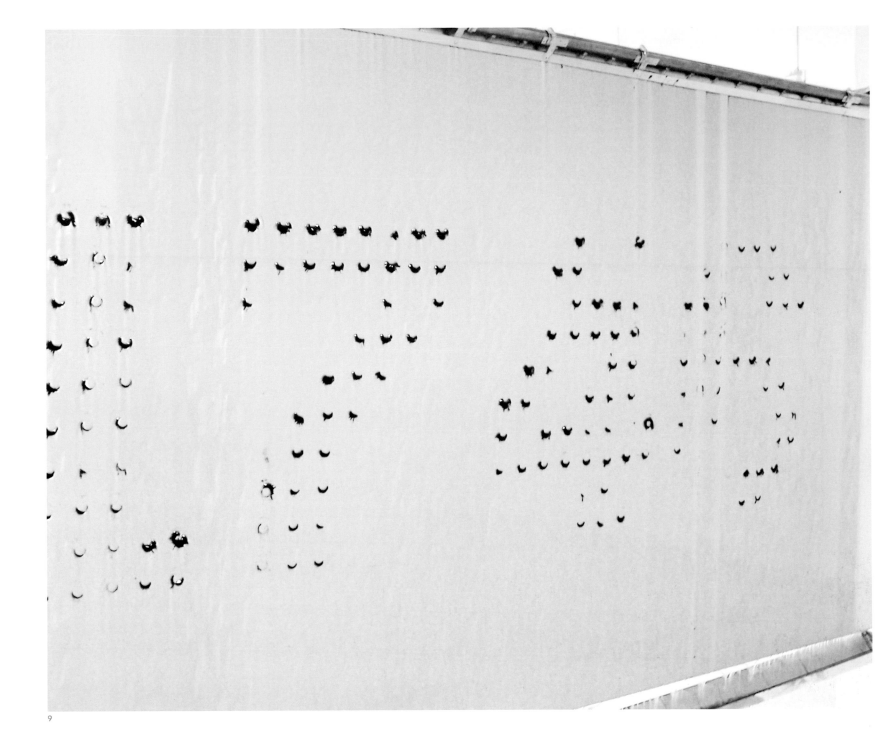

9

10

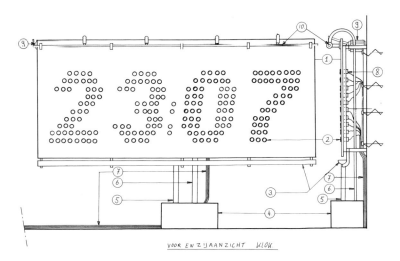

11

VOOR EN ZIJAANZICHT KLOK

Yun-je Kang, Samsung Electronics Co., LTD., Seoul

1

2

Yun-je Kang is creative director for Samsung Electronics' AV cluster design team. His 12 years designing electronic products as an in-house designer is a typical career path for a Korean designer. For many Korean companies the brand name is considered of paramount importance and therefore design is entrusted to an in-house team rather than an individual designer. However, recently there has been a trend to reveal individual designers' names and acknowledge the positive impact of individual creativity on company image and competitiveness. Kang is one of the current electronics designers enjoying this recognition.

Kang creates highly marketable products endowed with new values. But it's hard to achieve a unique design style with products targeting a global market and where the focus is more on appealing to universal tastes and anticipating or satisfying worldwide trends. This is especially true in the LCD market, which demands ever slimmer and wider panels at reasonable prices. It's difficult to succeed here with innovative designs.

In the AV industry technology develops at a similar speed to that of mobile devices, and the market share can change dramatically. This means a smaller permissible range of design and bigger financial risks. This tension between design and commercial requirements is well represented in Kang's L7 Project (2005). Shying away from boxy forms, Kang makes use of verticals to create an elegant and slim home theatre system, in an attempt to strike a balance between his own expectations (for a good/ideal form) and maximizing market sales. In the recent Bordeaux Project (2005), Kang employs poetic imagery to break from the stereotypes of home theatres. And in the case of the Pocket Imager (2005), Kang successfully works with the tensions that exist between the designer's expectation and market requirements to create the world's smallest LDP projector (laserdisc player) with an LED light source. Elegant, minimal and compact, the Pocket Imager (2005) can be used both for private entertainment or made portable for business purposes.

To balance creativity with the universal needs of a global company such as Samsung requires experience and skill – both of which Yun-je Kang confidently possesses. **Sang-kyu Kim**

3

4

5

1–7

Bordeaux Project – LCD TV
and Home Theater System
2005
Samsung Electronics
ABS, Steam mould
(high glossy moulding)
In collaboration with
Seung-Ho Lee and
Jae-hyung Kim

Bordeaux LCD TV presents
strong colour contrasts
and a pleasing soft image.

6

7

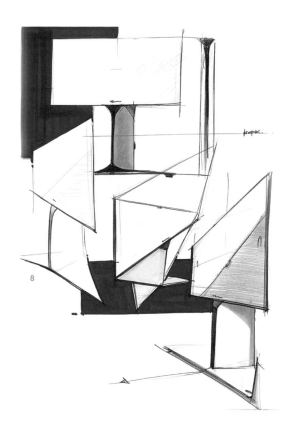

8

Yun-je Kang, Samsung
Electronics Co., LTD.

11

9

10

224

8–11

**L7 Project (50" DLP TV
and Home Theater System)**
2005
**Samsung Electronics ABS,
aluminium, glass**

*A new concept DLP-
projection TV, which is built
and arranged vertically
so it is as slim as LCD and
PDP televisions.*

12–14

Pocket Imager
2005
**Samsung Electronics
ABS, aluminium**
**In collaboration with
Kyung-hoon Kim**

*A minimal design concept
emphasizing simplicity
for a clean and compact
appearance, the Pocket
Imager is the world's
smallest DLP projector that
utilizes an LED light source.*

13

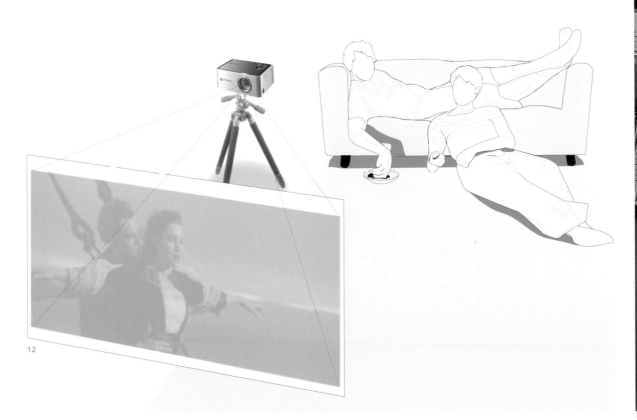

12

14

1

Sugarcubes
Prototype
2005
Corian

Part of the B-Sides collection, this spoon's cubed recess cleverly enables the user to scoop the shape and quantity of a sugar cube each time.

2–5

Digit
Table piece
2005
Paola C./Fabrica
Ceramic

The Digit's witty design, inspired by both birthday cake candles and digital timepieces, can be used to celebrate all ages up to your 99th birthday.

Meriç Kara, Istanbul

At the centre of Meriç Kara's creations lie a finely tuned curiosity, sense of irony and restlessness. Kara wants to raise questions with her projects. She wants to toy with our desires and habits and discover our truths and lies. She wants to know whether we are resigned to conformity, dominated by a mass-produced and overbearing consumer society, or rather ready and willing to take on a more creative role in our material reality, aware that there is room for a more active attitude that will benefit everyone.

Born in Izmir, Turkey, in 1977, Kara studied industrial design in Metu, one of Turkey's most prestigious universities, and gained her master's degree at the Domus Academy. Two years later she was taken on by Fabrica, the Benetton Group's research centre in Treviso. Her experience in this creative lab honed her analytical discourse on reality, as well as furnishing her with new resources to understand and handle it. One of Kara's most exciting projects, B-Sides (2005), is a collection of image-products that she has been working on since 2002: 'Some have a connection with people through habits or accidents, or they only have a sense of humour. Their common point is the "underlying idea". Some are experimental and you wouldn't know what it was if you didn't know the name.' Cotton in Cotton is a good example. While her image-products are shaped by semiotics, communication and the notion of interactivity, her adapted, more marketable products – such as Digit (2005) distributed by Paola C. – also retain these distinguishing features.

Kara has recently moved to Istanbul, back to her native, culturally rich and stimulating country. She works as a freelance designer, collaborating with other designers, and her client list includes Harvey Nichols, Istanbul_cosmetics, Benetton and Sisley. She has also designed a line for the museum shop of Santralistanbul, a contemporary art museum opening later this year. Broadening her scope, she has undertaken concept design in various fields, including advertising, event design and installation. Still committed to design ideas as much as to practice, Kara is scheduled to participate in a number of international exhibitions in the near future. **Guta Moura Guedes**

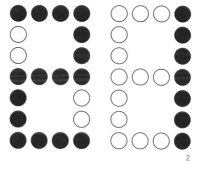

2

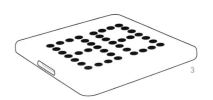

3

4

6

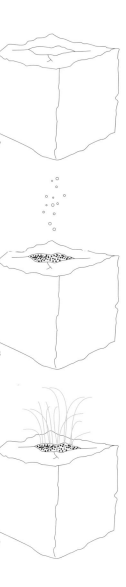

7

8

9

6–9

Blockcrack
Prototype
2005
Cement

The Blockcrack evokes the romance of plants growing through cracks in the pavement, and contrasts nature with a practical man-made material.

10

Shirtvase
Prototype
2005
Polypropylene

The small plastic vase allows the corsage flower to flourish throughout the day, rather than quickly wilting.

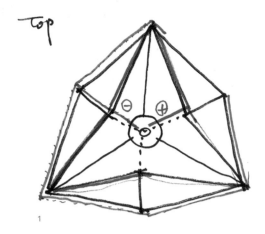

Top

⊖ ⊕

1

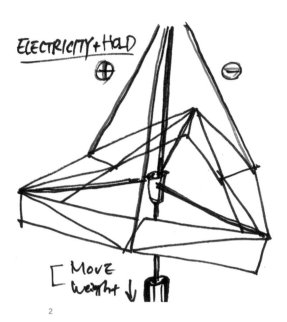

ELECTRICITY + HOLD

⊕ ⊖

[MOVE weight ↓]

2

1–11

Metamorphosis Chandelier
2005
Self-production
Aluminium, metal wire,
PC, fabric, 12V bulbs

*By pulling down the
string, the Metamorphosis
lives up to its name by
dramatically rotating
200 degrees. The user can
also adjust the direction
of the light and transform
the mood of the space.*

Bosung Kim, California

In his designs Bosung Kim attempts to establish new relationships with products. He creates scenarios that seek to amalgamate the relationship between objects with the relationship between objects and human beings.

Kim states that he is particularly interested in sequence, metamorphosis and other related subjects, as is evident in works such as Metamorphosis Chandelier (2005) and Burn the Candle at Both Ends (2004). Not only is his key concept visible in such designs but it is also presented in his modular personal computer, Lluon (2002), which is manufactured by TG. He explains such achievements: 'A designer must go beyond being an innovator of product and service ... a designer is closer to an inventor, as he essentially needs to have the urge to create something new to be put into practice. In other words, a designer must always retain the attitude of a "genuine creator".'

It is only natural for a product designer to be attracted to mechanical objects, but Kim is more interested in biological matters. The two prototypes for GM Love (2004) were based on DNA and genetic issues, and investigate how biotechnology could affect human love and relationships. For example, part of this project, Bedside Table (2004), allows a potential mother to investigate her partner's DNA and generate images of what their baby might look like.

When Kim worked for KODAS, a Korean design office, he ran a large project where he focused on architecture and interior design as inspiration for his product design. Metamorphosis Chandelier and Nomadic Eye Chandelier (2005) were part of this, and connected object and space simultaneously with the user. The user can change the shape and quality of light using a pull rope that actually changes the chandelier's physical form. The extent of the user's participation goes beyond simply pulling the rope as it allows the physical operation to become a dramatic experience.

Although Kim's designs are based on ideas such as mutation or heterosis – notions that are essentially conceptual – his works are also incredibly practical. **Sang-kyu Kim**

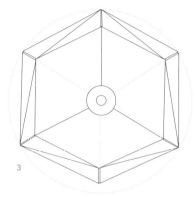

3

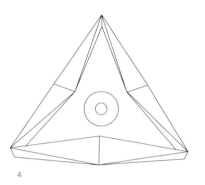

4

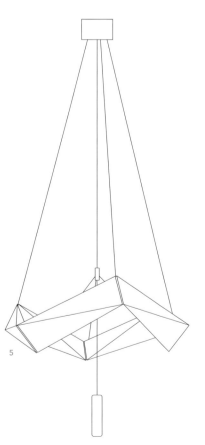

5

6

7

8

9

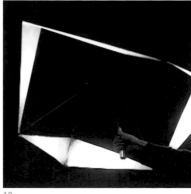

10

11

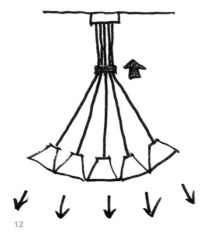

12

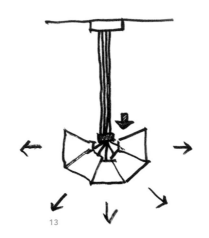

13

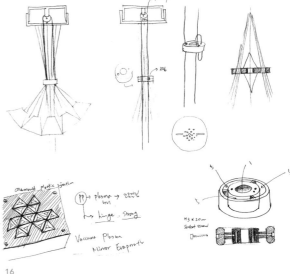

16

14

15

17

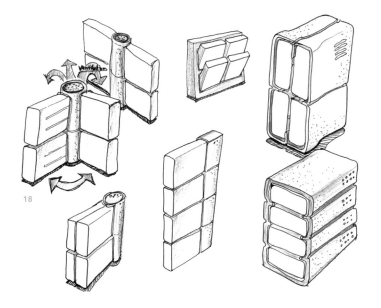

18

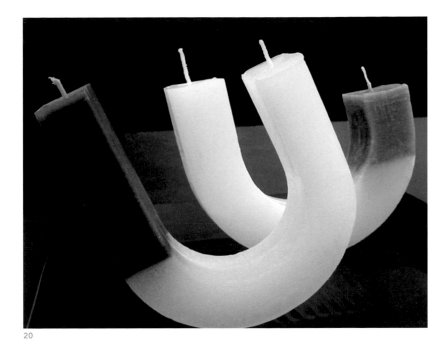

20

19

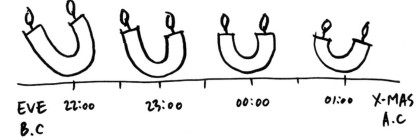

EVE 22:00   23:00   00:00   01:00 X-MAS
B.C   A.C

21

12–17

Nomadic Eye Chandelier
2005
Self-production
Mirror, pygmy bulbs, fabric,
metal wire, pulley

*This lighting piece uses an
icosahedron structure
and mirrors to control the
direction and the ambience
of the room.*

18–19

Lluon Modular PC
2002
TG
ABS, steel, LED

*The Lluon is for those
unfamiliar with PCs and
hardware. It is easier to
manipulate and upgrade
due to its modular
system of upgrading.*

20–21

Burn the Candle at Both
Ends
Prototype
2004
Wax, vaper

*Using wax that melts at
different speeds, this candle
is designed to be lit 2 hours
before Christmas Day.
By midnight it will have
become perfectly even.*

solitaire ©thrink

Before play commences, 32 olives are placed in the small dips, except for the one in the centre.

The objective of the game is to remove (and eat) the olives until only one remains.

An olive can be removed only when an adjacent olive jumps over it. It can only jump into an empty dip and only over one olive at a time. Diagonal moves are not allowed.

The pips are discarded in the four large dips provided. Olives can be replaced with nuts or any similar snack. Bon Appétit.

1

André Klauser, London

If one were to sum up André Klauser's work in only a couple of words, one would probably say: 'Playing games.' Although Klauser's designs involve more than just his obvious pleasure in engaging users, of playing with them and challenging them to take part in a game, this approach is a leitmotif that can be traced through his entire production.

Born in Hanover in 1972, Klauser graduated from Fachhochschule, Münster in 2000. A stint at Jasper Morrison's studio in 1999 shaped and coloured the course of his work. During his stay in London, Klauser decided to attend a postgraduate course at the Royal College of Art, which he completed in 2002. Another strong connection, which led to several joint projects, was with Barnaby Barford, a London-based artist and designer whose work crosses and blurs the boundaries between art, craft and design.

In 2002 Klauser set up his own studio and since then has been developing and producing several projects for a range of companies. Two good examples of his passion for, literally, playing games are Solitaire Olive Dish (2003), which invites users to play with their olives, and O's and X's Ashtray (2003), both produced by Thorsten van Elten. On a more subtle note, Flip Glass Series (2004), manufactured by Moeve, is Klauser's take on the eternal question 'Is the glass half full or half empty?' and Conversation Cushion (2004), manufactured by Thorsten van Elten, toys with the idea of boring small talk and uncomfortable silences and how to avoid them.

While the elements of communication and interaction are paramount in many of his designs, creating complex constructions with rich semiotic undertones, Klauser is equally at home with more minimalist approaches where acute perceptiveness is key. The Dolomite Series (2004), produced by Moeve, the Mauro chair (designed for the Promosedia International Design Competion) and the e27 Light (2001) are examples of his startling sense of balance, understated elegance and stream-lined visual effect. **Guta Moura Guedes**

1–2
Solitaire Olive Dish
2003
Thorsten van Elten
Ceramic
In collaboration with
Barnaby Barford

A snack dish that actually encourages you to play with your food.

3

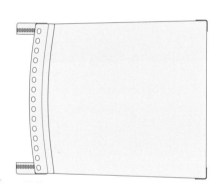

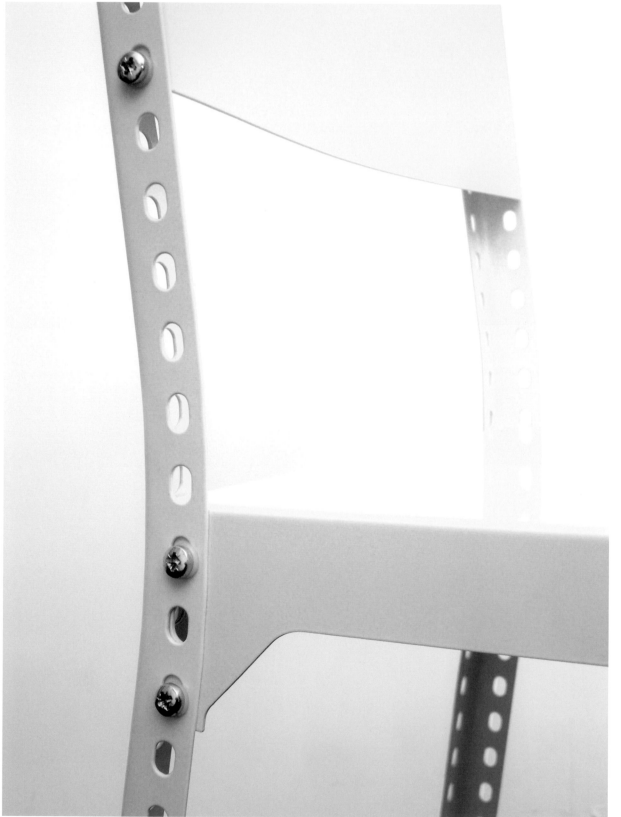

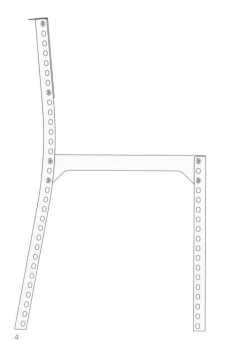

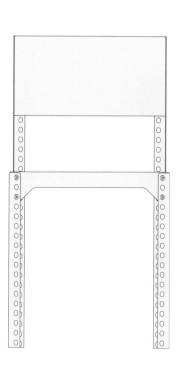

4

5

6

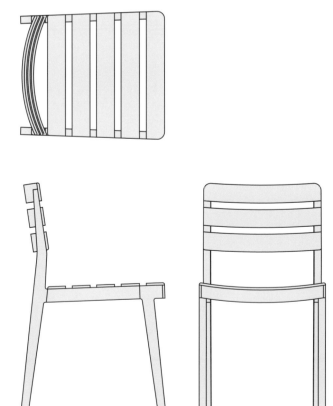

7

3–5

Meccano Chair
2005
Self-production
Powder-coated mild steel,
bolt fixings

*Inspired by industrial
metal shelving, and named
after the popular metal
toy of the same name.*

6–7

Mandi Chair
2006
Adal
Solid beechwood

*This functional chair was
originally designed for the
2005 Promosedia Design
Competition in Udine, Italy.*

8–9

Cone Tables
2007
Elmar Flötotto
Metal, MDF
In collaboration with Ed
Carpenter

*A system that allows for
a large range of tables
with the use of very few
different components.*

8

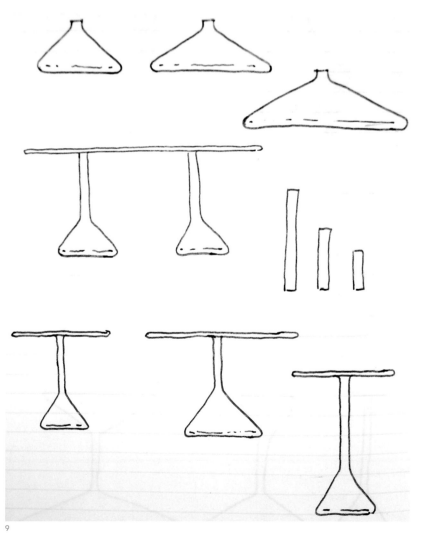

9

Tetê Knecht, Lausanne

The complexity and seductiveness of materials have always proved highly stimulating for designers and creative practitioners in general. The work of Brazilian Tetê Knecht often springs from direct contact with materials, a tendency she herself acknowledges when talking about the production of her objects. A profound immersion in production processes and an experimentation that borders on the physical are ubiquitous throughout her creative practice and close to the techniques and environments found in crafts.

Born in São Paulo, where she graduated in design at the city's Faculdade de Belas Artes, Knecht gained her master's degree in industrial design at the École cantonale d'art de Lausanne, where she currently lives and works. Her experiments with materials started at the beginning of her career, as part of what she called a 'three-dimensional kitchen' where she researched possible combinations of fixed and flexible materials, striving to solve some of the issues her projects threw up.

Whether all things begin with matter, or whether everything is hinged on the project, is not a fundamental concern in Knecht's work: in her case both aspects feed into each other. The Sabot shoes (2005), for instance, arose during the course of manufacturing the Grand Pouf de Paille (2005). She explains: 'While I was applying layer upon layer of straw and latex to make the mould for the pouffe, my naked feet were also getting coated in this mix, the soles thickened, the feet shaped it and it started to feel comfortable.'

Knecht's work for the Notech Design group also speaks volumes about her inclination towards sustainable design and practices connected to recycling: such as Luxe? (2002), a dress made out of inner tubes from lorry tyres, or the Frufru lamp (2000) – a good example of her poetic recycling essays, rich with powerful symbolic connotations. In the field of research into materials and design practices that are more intimately connected to manual techniques and limited series, Knecht asserts her unique outlook and sensitivity, epitomizing the cross between Brazilian culture and European context.
Guta Moura Guedes

1–2

Grand Pouf de Paille
2005
Tetê Knecht
Straw, latex

The combination of straw and latex creates a smooth and very comfortable surface that is also durable.

3–4

Oil
Vase
2005
Tetê Knecht
Coal, silicone

The Oil vase is a combination of contrasts: the texture and blackness of the coal with the transparency and flexibility of the silicone.

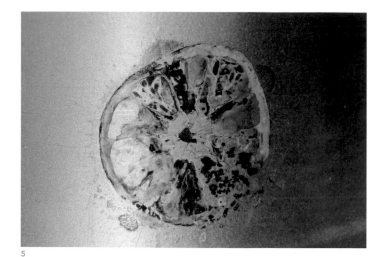

5

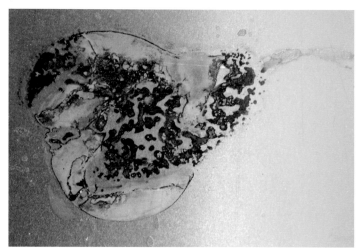

6

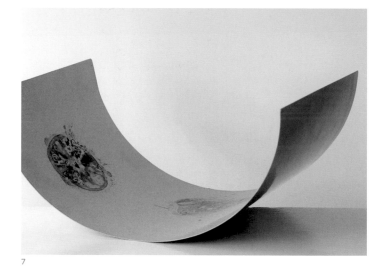

7

Tetê Knecht

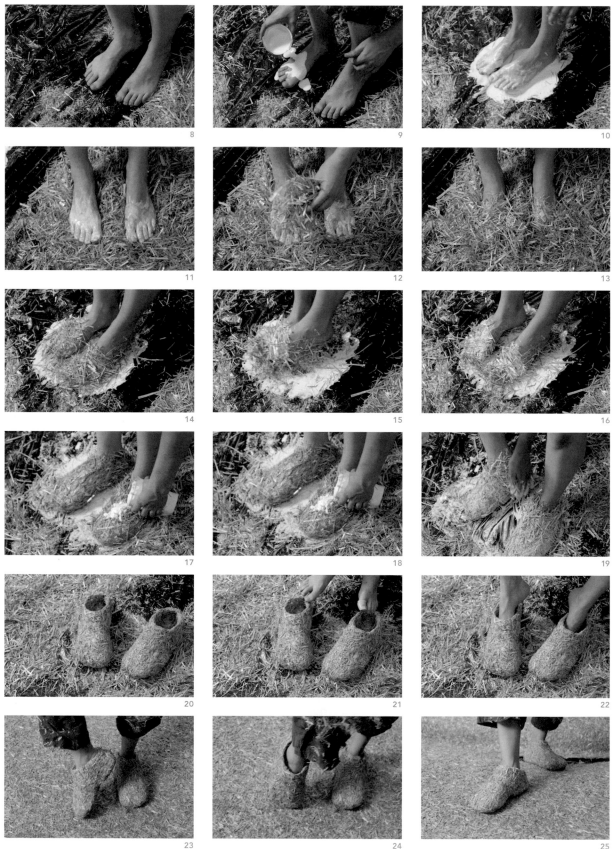

5–7
Rust
Fruit bowl
2004
Tetê Knecht
Iron

Slices of fruit have printed their shape onto the bowl using oxidation, leaving on the object the memory of use.

8–9
Sabot
Shoe
2005
Tetê Knecht
Straw, latex

While working on her Grand Pouf de Paille Knecht discovered that the straw and latex mix, which got stuck to her feet, was actually very comfortable to walk on. Thus the Sabot was born.

1
Burstlight
2006

*Inspiration for the Burstlight
came from fireworks
exploding in the night sky.*

2
Burstlight
2006
korban/flaubert
**Anodized aluminium,
fluorescent light source**

*The Burstlight is
a wall light realized as
an explosion, but it is
also based on organic
spiral growth patterns
seen in nature. The
concealed light source
lends a delicacy and
variation to the dramatic
exploding spines.*

korban/flaubert, Sydney

korban/flaubert is a design and production partnership between metal specialist Janos Korban and architect Stefanie Flaubert. After the couple left Australia in 1990, Korban (who had previously studied psychology and genetics) developing sophisticated metalworking skills within the innovative architectural metal workshop K. M. Hardwork and Flaubert working for the German architect Günter Behnisch, they established korban/flaubert in Stuttgart in 1993. Their practice was based on their collaborative research, including the exploration of experimental form-finding ideas. The pair returned to Australia in 1995 and re-established the business in Sydney, continuing to produce experimental work as well as simpler, more functional furniture and lighting products suited to more pragmatic needs, such as the Block Table (1999) and the Bongo Stool (1999).

Since establishing its current studio-cum-showroom in 2001, korban/flaubert has gained significant international exposure. It has exhibited at Designers Block in London and Tokyo, and at the Milan International Furniture Fair. It has also sold work to, and undertaken commissions for, a wide range of clients in Europe, North America, Asia and Australia. korban/flaubert describes its studio as a laboratory for form in which it playfully explores certain themes such as 'material and action' or 'rhythm/repetition/sequence' to create both functional and purely sculptural forms. Function is rarely predetermined. Instead, potential functions emerge after intensive processes of analysis, experiments based on mathematical sequences, model-making and prototyping.

korban/flaubert's critically acclaimed Cellscreen (2003) is a salient example of this. The honeycomb-like structure began as a series of simple geometric experiments with a single line-length and a repeated five-way joint. The resulting object's potential use as a decorative screen was only revealed once it was scaled up in size. korban/flaubert continues its preoccupation with geometry and mathematics. Its dramatic Burstlight (2006) follows this trajectory, echoing the radial growth and spiralling sequences so commonly found in nature. **Brian Parkes**

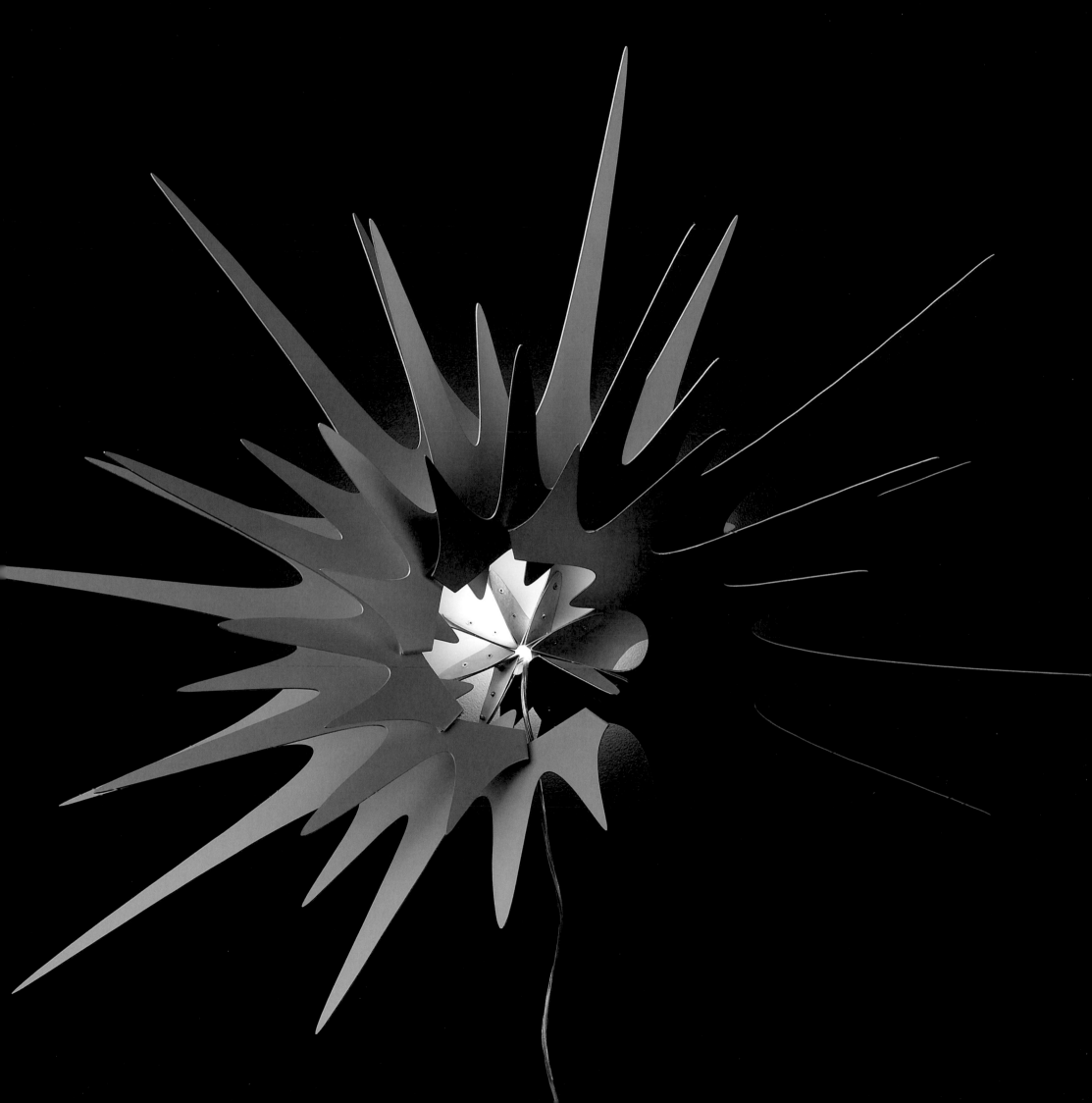

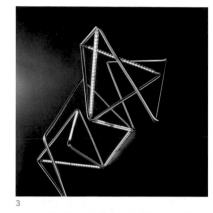

3

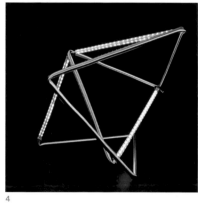

4

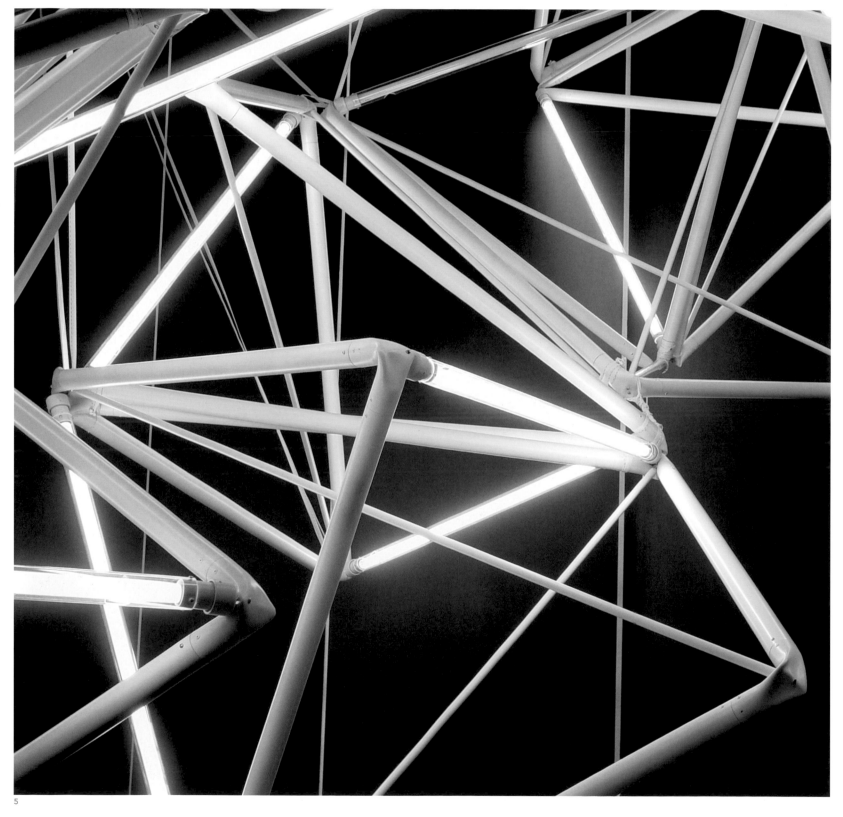

5

6

3–4
Weblight
2005
korban/flaubert
Stainless steel,
polycarbonate, LEDs

*Weblight is a single zigzag
line of light. A flippable
3-D sculptural model, it
connects to itself to form
increasingly complex
forms in light.*

5

Adventure
Installation at Object
gallery, Sydney
2005
korban/flaubert
PVC, aluminium,
fluorescent tubes

*The installation emerges as
a growing structure which
begins to give substance
to the space it is located
in by picking out lines in
light and metal: these are
fixed-length modules which
move against each other as
the visitor moves through
the space.*

6–7

Cellscreen
2003
korban/flaubert
Anodized aluminium

*The Cellscreen achieves
visual density and
decorative effect by the
simplest means. The
geometry was inspired
by the decorative economy
of patterns in Islamic
architecture, mathematical
sequences and natural
growth patterns.*

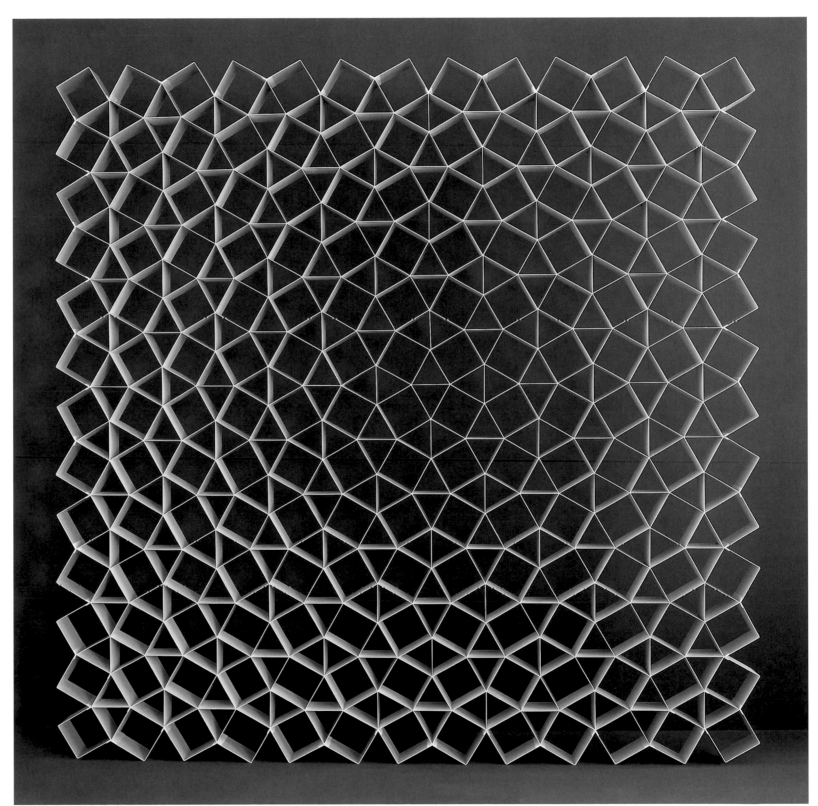

7

Joris Laarman, Rotterdam

When Joris Laarman presented Heatwave (2003), a radiator that resembled lingerie embroidery, as his graduation project at Eindhoven's Design Academy, he was trying to demonstrate that, functionally speaking, 'simple form is not always better than decoration'.

The radiator is usually considered to be a generic functional object, associated with austere and simple geometric forms. But Laarman's floral, baroque version actually emerged as a true formal rendition of the technological essence of the radiator by fully responding to the technical requirements of its functional mission – i.e. heat production and thermal exchange. Laarman's version was based on the simple observation that in terms of heat distribution it is far more effective for a radiator to develop its surface area to the maximum, stretching out, for instance, along walls like a rambling plant, rather than remaining confined and hidden in some remote corner of the house.

In his project for an Ideal House, realized for the Cologne Fair 2006, he designed a contemporary model of an urban tree house made of standard steel tubes of different sizes. Claiming 'My surroundings are my house', Laarman refused to make his house a fixed and closed entity, instead creating a totally open space with a special feeling of intimacy. Furnished with projects both by himself and by his friends, he describes the house as a place dealing with the luxury of uncertainty and imagination to set against the luxury of comfortable living. On entering the house in order to go upstairs you can either choose to use the staircase or to climb up a wall decoration that also functions as a climbing wall, Ivy (2003).

What I like about Laarman's work is its remarkable capacity to question what is usually taken for granted. He introduces small shifts of meaning to transform an everyday object into something that reveals the surprising beauty concealed within. **Francesca Picchi**

1–2

Ivy
Wall Decoration
2004
JL Laboratory
Polyconcrete

Ivy is like lingerie for the wall. An adventurous decoration that works as a free-climbing wall to see your 'over-designed' interior from a different perspective, or find alternative ways of going upstairs.

3–5

Heatwave
Radiator
2003
Droog Design/Jaga
Concrete

A highly decorative and ornate radiator that makes a stunning feature out of what is traditionally a forgotten and mundane appliance.

3

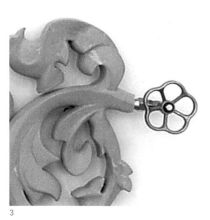

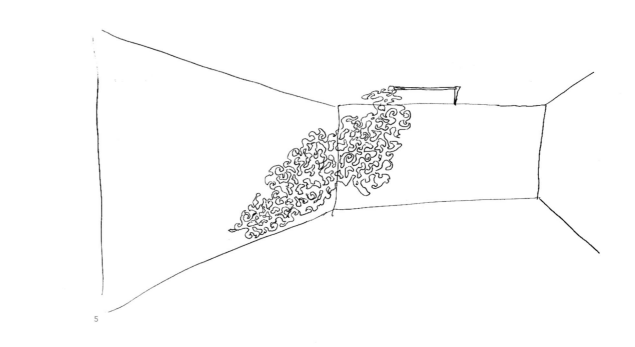

5

4

6

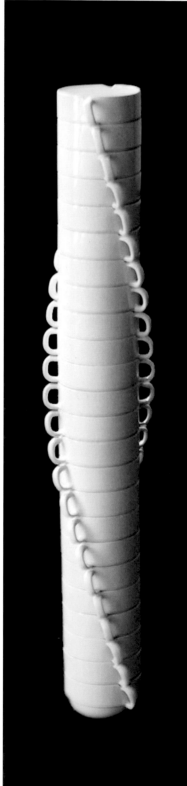

7

8

9

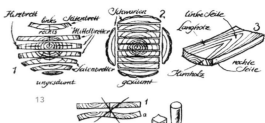

13

6–8
Stakhanov
2004
Droog B.V.
Ceramics

*A range of ceramics
that is able to be
stacked easily and
safety; the stacks create
interesting sculptures
in their own right.*

9–14
Scheluw
2004
JL Laboratory
Fresh white beechwood

*By using a material that
is highly sensitive to
temperature and moisture,
Laarman ensures that each
chair develops its own
individual character through
the alterations that occur
to the material over time.*

10

11

14

12

1

Cigar Case
2006
Christofle
Silver

*Cigar holder that allows
you to slide the cigar out
easily with your thumb.*

2

Fountain Mirror
2006
Prototype
Mirror steel

*By adding a mirror this
fountain is transformed into
an 'outdoor bathroom'.*

Nicolas Le Moigne, Lausanne

Aged only 26, Nicolas Le Moigne, a Fribourg resident of
French origin, already has one commercial success to his
credit. His Verso Diverso watering can (2006), developed
during his bachelor's degree in industrial design at
ECAL – École cantonale d'art de Lausanne, was not only
awarded first prize at the MACEF International Home
Fair in Milan, but is also a useful device, distributed by
Viceversa, that is selling well!

This is proof that thinking about everyday objects and
how we use them can take on the aspect of a re-design
that is functional rather than technological. By adding
a spout to an empty bottle, Le Moigne has transformed
a piece of rubbish into a watering can.

For his Pot-au-Mur (2003), a rotationally moulded
flowerpot for Serralunga, Le Moigne's considerations went
towards functionality. The flowerpot, designed during
a workshop at ECAL, can be directly attached to the wall,
thus liberating the ground surface whilst looking as if it
were fully integrated into the building's architecture.

Le Moigne's scheme to adapt public fountains into
'outdoor bathrooms' with his Fountain Mirror (2006), which
includes a system of hooks, is more poetic. This is a simple
proposal that modifies our everyday surroundings, just like
his Public Clock (2005) for 'reading' the time, which is
written out in words. Developed with Omega Electronics,
this was tried out on one of Geneva's main squares.

If the simplicity of Le Moigne's suggestions has tended
to provoke controversy about urban issues, it is to the
credit of this young designer who says he is not looking
for the 'revolutionary' in design, but for 'little changes, the
aesthetic or functional detail that helps in everyday life'.
When it comes down to it, in all its modesty, this could well
be a fitting manifesto for present-day design. **Pierre Keller**

3–4

Verso Diverso
Watering can
2006
Viceversa
Polypropylene

*Screw on spout that fits
a variety of plastic or
glass bottles. Comes in
a range of colours.*

5–6

Trompe l'Oeil
Vase
2005
BanalExtra
Enamelled steel

*The Trompe l'Oeil
transforms any drinking
glass into a flower vase.*

7–9

Public Clock for the
City of Geneva
2005
Omega Electronics
Aluminium, Plexiglas

*Suspended public clock
with rotating flaps and
the time written out in
word form.*

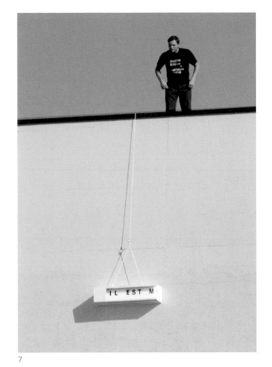

7

8

9

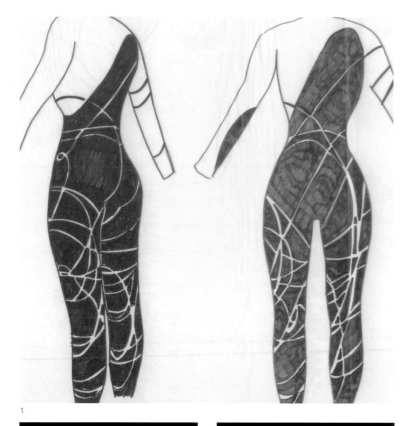

1

Simone LeAmon, Melbourne and Milan

Simone LeAmon, who originally trained in sculpture at Melbourne's Victorian College of the Arts, has increasingly engaged with the notions of function, consumption and desire in her interdisciplinary work, leading her to complete a master's degree in industrial design at RMIT University in 2004. Conceptually based exhibitions and performances remain important strategies in her pursuit of unorthodox and often transgressive design methodologies. LeAmon continues to work as an artist across a variety of media, including drawing, sculpture, performance and digital media, and she often initiates collaborations in order to develop design projects.

LeAmon has designed for projects in areas as varied as jewellery, lighting, graphics and public art, and was recently contracted by the Italian lighting company Oluce. She has even developed a concept for a women's motorcycle racing-suit: Bodywork (2003). This virtual one-piece leather suit is provocative in a subculture dominated by men and extends from LeAmon's abiding interest in the politics of a consumer society.

LeAmon was awarded a studio residency in Milan in 2001. This opportunity enabled her to engage with the Italian design and manufacturing industry, and to participate in the Salone del Mobile during the Milan International Furniture Fair. Her performance-based project took the form of a game where visitors were invited to play with a series of conceptual models with interlocking parts that could be pulled apart and reconfigured in unlimited ways.

LeAmon has since exhibited in Milan in 2003 and 2005. Her 2005 presentation, A Supersystem Concept for 5 Products: Post Object Confessions, consisted of digitally rendered images of five new lifestyle products derived from the volume and timeless aesthetic appeal of the Ducati Superbike motorcycle. Each product was accompanied by a narrative text and edited video footage of a related performance, including one in which LeAmon is seen kissing motorcycles at busy road junctions in Melbourne and in Milan. **Brian Parkes**

2

3

4

1

Bodywork
360 arterial study
of racing bodysuit
2003
Ink on paper

2–4

Bodywork
Computerized racing suit
2003
3-D computer model

*Bodywork is a 3-D
computer-generated model
of LeAmon's body in a
custom-designed motorbike
racing suit. A translation
of tyre marks retrieved from
the Phillip Island racetrack
is traced onto the body of
the suit.*

5

A Supersystem Concept
for 5 Products
2003– 6
Design sketch

*A design methodology
that reduces parent objects
and spaces to pure volume.
The concept informs five
products: Cuff; Little Kettle;
Rest; Head Light; and Purse.*

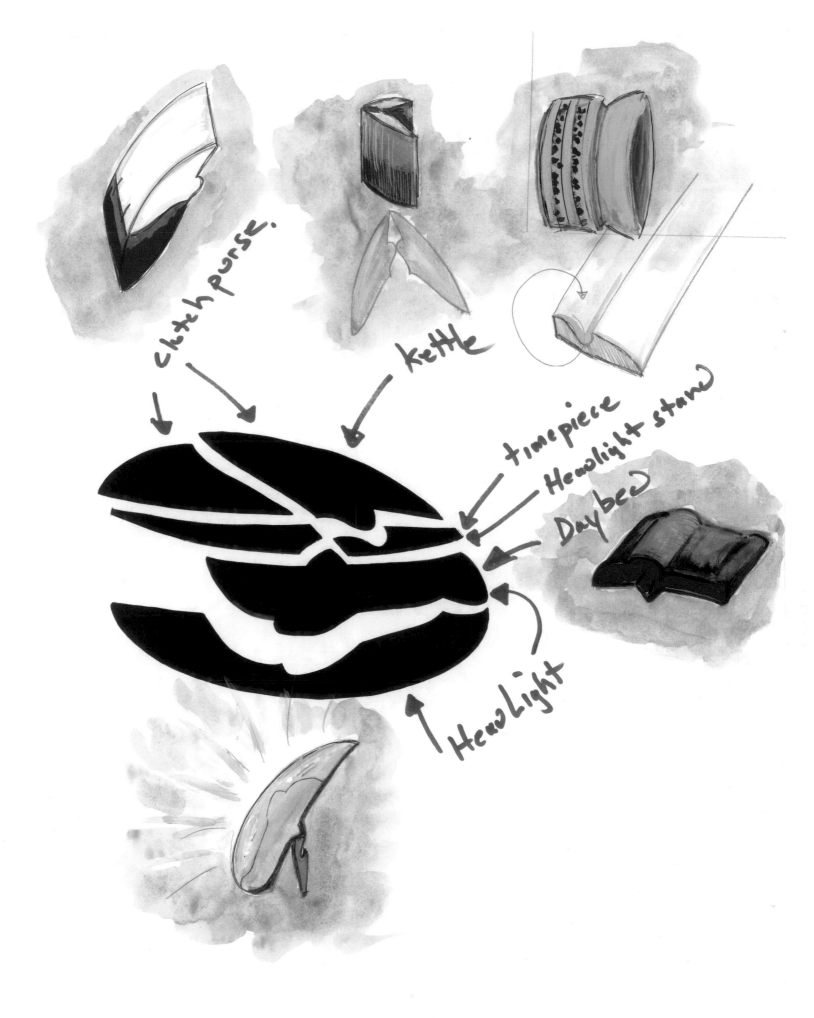

5

6

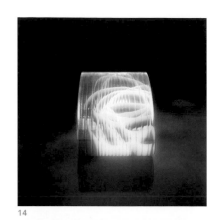

14

7

10

11

15

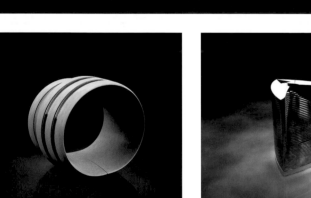

8

12

16

9

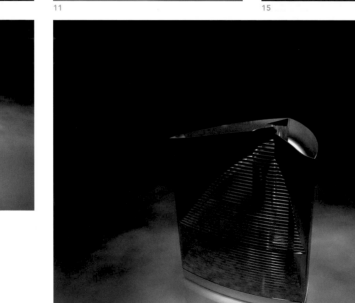

13

17

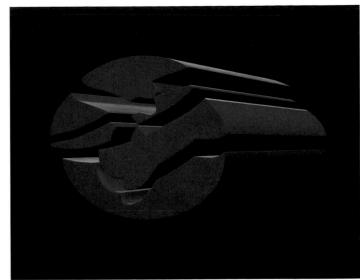

22

18

6–9

Cuff – A Supersystem
Concept for 5 Products
Timepiece concept
2005
3-D computer model

10–13

Little Kettle –
A Supersystem Concept
for 5 Products
2005
3-D computer model

19

14–17

Head Light –
A Supersystem Concept
for 5 Products
2005
3-D computer model

18–21

Rest – A Supersystem
Concept for 5 Products
2005
3-D computer model

*How do a bed, timepiece,
light, kettle and purse
relate? A Supersystem
Concept for 5 Products
is a design proposal that
explores general human
affections such as desire,
comfort, disappointment
and love.*

20

22–25

Supersystem
Exhibition piece
2003
Polyurethane foam,
flock coat

*The Supersystem was
developed for 'body-play'.
The 'play' was conducted
over two days at the
Gertrude Contemporary
Art Spaces in Melbourne.
Using the parts to construct
small architectonic forms,
the 'play' inspired a
vocabulary for imagined
dwellings and furniture.*

23

24

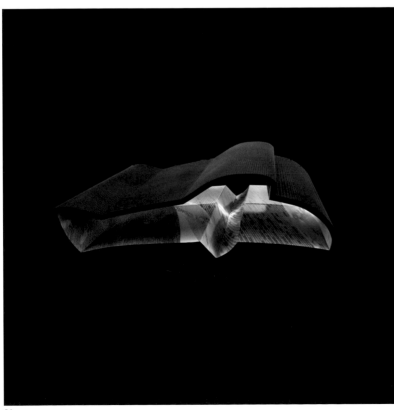

21

25

1

Sang-jin Lee, Seoul

Sang-jin Lee makes his aesthetic 'products' by combining inexpensive and commonplace ready-made components, making the most of their inherent characteristics. Using manufactured products as the main elements, he uses his craft skills – incorporating a kind of laser cutting – to construct his designs, an effective way of overcoming the limitations of working in a small studio. And without the constraints of mass-production, Lee is able to take full control of his work and express his creativity.

When Lee designed Zipper Light (2000), he already had a working knowledge of the material. As well as focusing on the design's functional aspects, he was able to anticipate that he could use extensions so that each unit becomes an element for controlling the strength of light. When the light is turned off it appears no more than a combination of inexpensive objects, but when switched on Zipper is seen as a silhouette, showing textures and patterns and giving a new value to the entire object.

Lee enjoys using cable ties to join plastic objects together. Even with his furniture pieces he uses the same method of assemblage: he ties together bent wood in the same way as he does his plastic boxes, using neither special hinges nor dovetail joints. Lee creates an entirely new object by using the basic method of binding with strings, and thus his products embody both uniformity and ingenuity. The incorporation of ready-made objects is not limited solely to fine art works – Marcel Duchamp's Fountain being the seminal example – as it has also been applied widely to design products (Ron Arad's Rover Chair, for example).

Lee has found a new space in art by tying manufactured products together and this in itself has shown a unique approach that results in aesthetic completeness.
Sang-kyu Kim

1
Bookmark
Light
2004
diwill
Acrylic, bulb

A handy reading light that also doubles up as a bookmark when you are ready to go to sleep.

2–3
EL-butterfly
Lights
2004
diwill
EL

Decorative fairy lights with butterfly detailing.

4–5
Straw Light
2003
diwill
Plastic straws

Unique and fun lighting feature made from plastic straws.

6–8
Lady Light
2002
Artemide
Polyurethane

Portable lighting that can be fastened to many surfaces with the suction pads incorporated into the design.

2

3

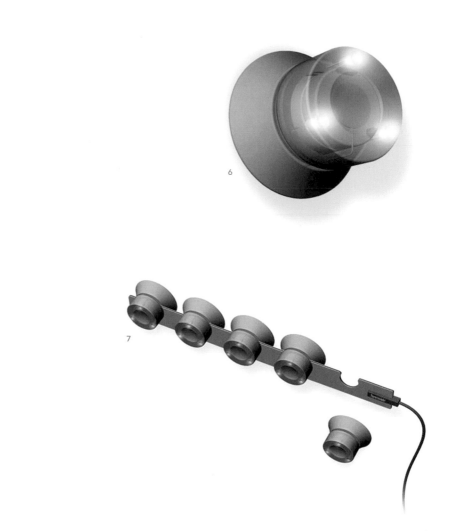

6

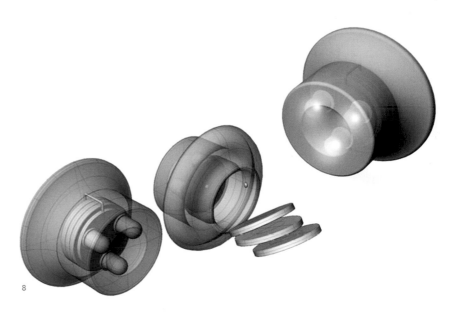

7

4

8

5

Sang-jin Lee

9

Collage
Chair
2006
BENTEK
Plywood

*The chair is cleverly created
from pre-existing seat/back
parts of chairs. When these
parts are connected, new
and various curved lines are
formed, creating a more
interesting formal story.*

10–13

Zipper Light
2000
diwill
Zips, acrylic

*By opening and closing
the zips, different
structures and lighting
effects can be achieved.*

9

10

11

12

13

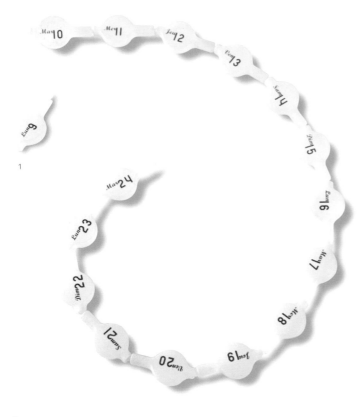

1

Medicine by the Centimetre –
Therapeutic Objects
Collection
2001
Prototype
Polypropylene

*An oestrogen treatment in
the form of a string of
bead-like doses. The chemist
is able to measure out the
length of the string according
to the dose of the medication,
and also print instructions
on each pod.*

2

House 213.6
Cat accommodation shelter
2004
Prototype
Low density rotomoulded
polyethylene

*An accommodation unit
for stray cats that is
both a shelter and a rather
interesting play area.*

3

dB/Elements Collection
White noise diffuser
2006
Prototype
ABS, mini speakers,
electric engine, charger

*If the volume of noise is
too great within a domestic
setting, the dB rolls and emits
white noise, which enables
the brain to adjust to, and no
longer be disturbed by, the
outside noise.*

Mathieu Lehanneur, Paris

Mathieu Lehanneur works outside the traditional boundaries of product design. He explores unusual paths with a very reflective and analytic approach that nourishes his thought-provoking interdisciplinary projects.

Therapeutic Objects Collection (2001), Lehanneur's graduation project, was inspired by an article on the gap between prescribed medicine and patients' behaviours. His solution was a breakthrough. Taking into account the psychological aspects of medication – including placebo effect, faith, fear and attraction – Lehanneur proposed a series of new shapes, packaging and dosage systems adapted to various medicine types. For example, Antibiotics in Layers is a single pill made of several layers that are taken one by one, from the darkest to the clearest – signalling final recovery. Medicine by the Centimetre (2001) is an oestrogen treatment that takes the form of a string of bead-like single doses; the medicine's length is judiciously proportional to the treatment's duration.

In line with Therapeutic Objects, the Elements Collection (2006) was developed for the Carte Blanche award, which Lehanneur won at the French Valorisation de l'Innovation dans l'Ameublement (VIA) in 2006. Lehanneur invented five domestic elements for use in the home, aimed at regulating light, oxygen, temperature, sound level and our immune system. Equipped with very sensitive sensors, 'K.', 'O.', 'C°', 'dB.' and 'Q.' evaluate your needs and compensate accordingly. For instance, in the case of lack of oxygen, 'O.' turns on a light that activates an oxygen-producing algae. 'dB.' wanders around autonomously detecting excessive noise and emitting 'white noise' that obliterates the pollution.

Lehanneur has also proved to be an inventive furniture designer and a clever exhibition architect. His Useful Mouldings (2002) are distortions of decorative mouldings that become shelves. As for scenographies, those he designed for both the Musée des Arts Décoratifs and the Cartier Foundation in Paris speak for themselves. Both cleverly conceived, they were distinctive enough to be remarkable while at the same time humble enough to serve the exhibited art pieces. Currently working on a series of restaurants in Paris, Lehanneur's world seems full of potential dream projects ripe for exploration.
Didier Krzentowski

2

6

3

1

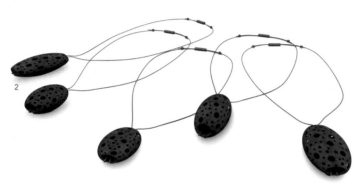

2

3

Arik Levy, Paris

From graphic and industrial design to stage design and even exhibition design, Arik Levy covers almost all fields that a designer could wish to work in. He widened his territory even more by establishing Ldesign in partnership with Pippo Lionni in 1997. Having international companies and museums as its clients, the Paris-based Ldesign deals with a wide range of projects including corporate identity design, product design and merchandising.

The Minishake lamp (2006), which uses stereolithographic 3-D printing, represents Levy's unique combination of technology and craftsmanship. Levy often applies the voids and volumes inspired by star constellations to his designs. The mystical feeling aroused by the irregular voids in the Minishake reaches its peak when the light is on.

Black Honey fruit bowl (2005) is made of epoxy resin. 'The final form is a contrast between hard and soft,' Levy states. Composed of a slightly irregular honeycomb pattern, this bowl seems to have been designed to embrace its contents rather than act as a neutral container.

The Umbilical lamp (2003) is a good example of Levy's design identity and excellence. Although it is based on the simple structure of an electric cable, this lamp creates an abundance of images, and the knotted and woven electric cable delivers a very different feeling from any other electrical appliance. By exploiting the physical characteristics of wire, Levy transforms an industrial product into a handcrafted art piece.

Even though Levy is eager to explore the best use of new materials and methods, his designs are not dictated by high technology. Rather, his designs engage with both technology and craftsmanship. As New York's Museum of Modern Art's architecture and design curator Paola Antonelli notes, 'good contemporary design is therefore an interesting composition of high and low [technology] … and some low-tech materials that respond to ecological needs.'

Sang-kyu Kim

4

5

1–2

**Shaman
Necklace
2006
Materialise.MGX
Translucent epoxy**

A piece of jewellery that makes a statement; a modern take on the tear drop necklace.

3

**Minishake
Suspension lamp
2006
Materialise.MGX
Epoxy-rapid prototyping, selective laser sintering**

Initially produced in a limited edition using stereolithography, the Minishake is now produced by Materialise.MGX.

4–6

**Black Honey
Fruit bowl
2005
Materialise.MGX
Epoxy resin**

The honeycomb pattern of this undulating fruit bowl design is slightly irregular giving the impression of a living organism.

Arik Levy

7

8

7–8

Rock
Coffee table
2005
Mouvements Modernes
Stainless steel or brass

Coffee table with reflective surfaces that makes for an impressive centrepiece in any room.

9–12

Mistic
Vase and candleholder
2005
Gaia & Gino
Borosilicate glass tubes

Intertwined glass tubing that allows for interesting table arrangements combining both flowers and candles.

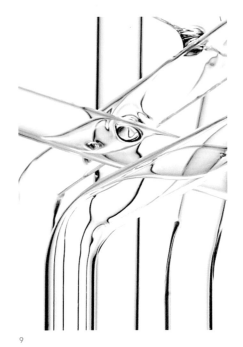

9

10

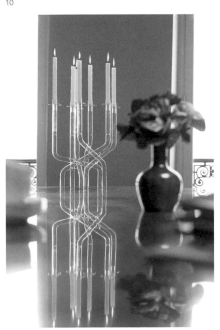

11

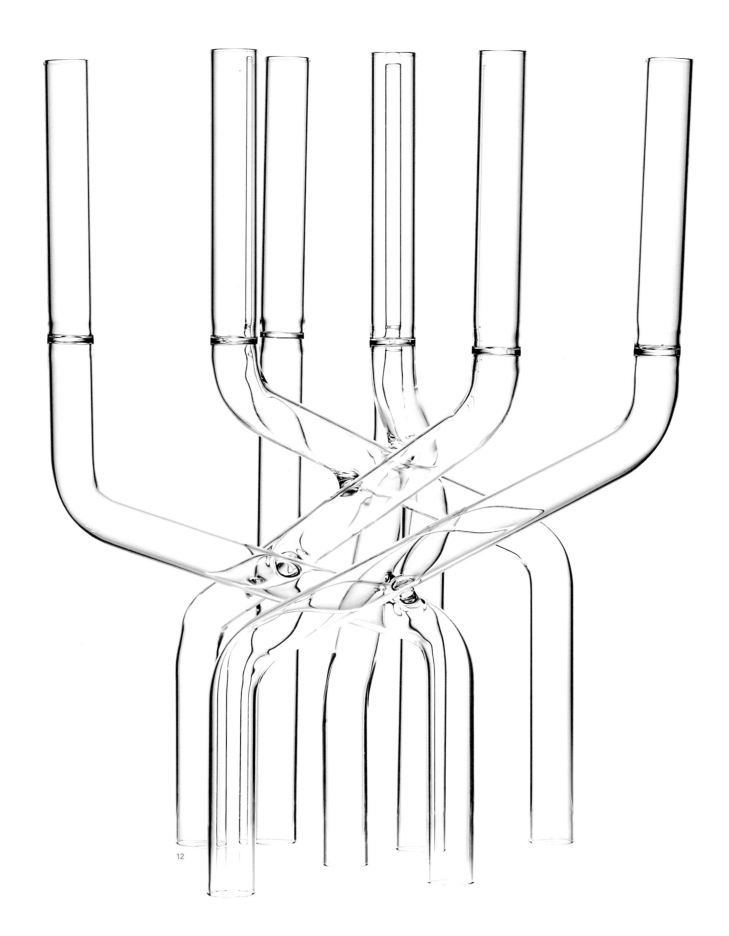

12

1

1–6

Bicycle Brake Sketches
2006
Julia Lohmann
Pencil, Digital Montage

*Concept sketches for
a light box installation
for Velorution, a London
cycle retailer.*

7–8

PET-Pick Bottle Carrier
2001
Zeltec
Polypropylene

*A re-usable bottle carrier
that can fit up to three
PET bottles of 0.5 to 2 litre
capacity at a time.*

Julia Lohmann, London

Julia Lohmann's close relationship with animals fuels her design creations, whose refined aesthetics hide a provocative questioning of our responsibility towards the natural resources we use.

Born in Germany, Lohmann has been based in England since 1998 where she first studied graphic design. Initially, she explored our relationship with waste by redesigning a Kitchen Bin (2001) and found an environmentally friendly way of carrying water bottles, PET-Pick Bottle Carrier (2001), which was put into production by the German firm Zeltec. During the summer of 2002 a stay at a horse farm in Iceland brought a new twist to her fascination for animals. She became aware of 'how disconnected humans are from animals and how complex our relationship is to the creatures that sustain us'.

During her product design studies at London's Royal College of Art, Lohmann introduced her childhood interest in veterinary medicine into her artistic practice by learning the craft of taxidermy. Her series of lamps, Flock and Ruminant Bloom (2004), turned sheep stomachs into delicate and billowing translucent globes, revealing the beauty of organic shapes that went beyond conventional prejudices about what is judged beautiful or disgusting. In this way Lohmann's designs turned the meat industry's waste into sophisticated design material. Her inspired reinterpretation of the ever-so-familiar leather couch was a sculpture of a life-sized cow torso, a single cowhide lovingly upholstered and bearing all the traces of the animal's life. It became her famous series of Cow Benches, which attracted the attention of institutions such as the French National Art Collection and the Droog Design group. With this sleek memento mori, Lohmann achieved her goal of bridging the gap that has grown between the living animal we raise and the end products we manufacture from them.

This is a striking debut that announces a bold designer.
Didier Krzentowski

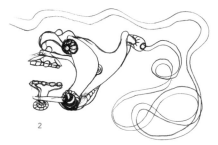

2

3

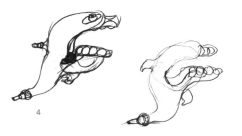

4

5

6

7

8

Julia Lohmann

9

10

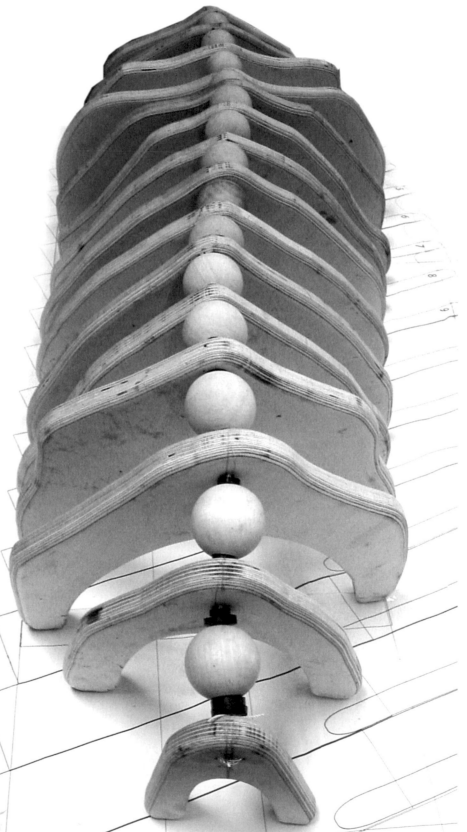

11

12

13

14

15

16

9–14

Cow Bench 'Antonia'
2006
Julia Lohmann/Alma Home
Leather, foam, wood

*A series of handsculpted
pieces each with a different
name and shape. The
benches serve as memento
mori for the cows that
died to make the leather
they are made from.
Each bench is upholstered
with the hide of one cow,
and the leather is placed in
the same position as it was
on the living animal.*

15–17

Ruminant Bloom
2004
Julia Lohmann
Preserved cow stomach,
glass, light fitting

*A round lantern made
of honeycomb-textured
cow stomach.*

17

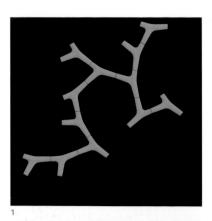

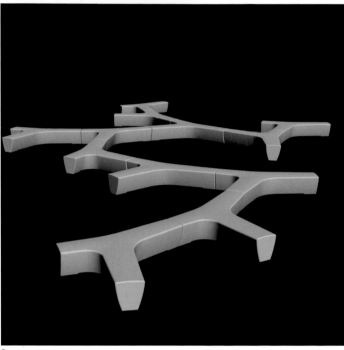

Alexander Lotersztain,
Brisbane and Barcelona

Alexander Lotersztain is a design nomad – his work has allowed him to travel and, in turn, travel inspires his design. Born in Argentina to Polish parents, Lotersztain began studying design in Buenos Aires before moving to Australia as a confident 18 year old looking for new experiences. After completing his studies in Brisbane he took a job in Japan with GK Design & Planning. However, after only two months he met Japanese design guru Teruo Kurosaki and became part of his team, designing for Sputnik and IDEE. For the manufacture of Soft Sofa (designed for IDEE in 2001, and now represented in the collection of the Centre Georges Pompidou, Paris), Lotersztain spent time with skilled workers in the Philippines factory experimenting with wicker.

Since returning to Brisbane after a year or so in Japan, Lotersztain spends at least a month each year in Europe, principally in Barcelona and Milan. His current design projects include a sofa for Italian company Covo, a range of mirrors for ArtQuitect in Spain and tables for Planex in Australia. He has no permanent studio address but feels comfortable and connected working from cafés or airports with his sketchbook, laptop and mobile phone.

Along with his passion for seeing the world, Lotersztain is driven by a desire to contribute to positive change in the world through design. His most ambitious and fulfilling project to date is his ongoing collaboration with the In Africa Community Foundation, in which he works directly with village communities in rural Africa to develop design products that can be made by local artisans, using local materials, for international export markets. This long-term project promotes both economic development and communal self-esteem of the villages and is expected to continue for at least 20 years.
Brian Parkes

1–2

**Twig Concrete Bench
2006
Procast Australia for
Alexander Lotersztain
Studio
Pre-cast reinforced concrete**

Concrete bench designed for the parks and open areas of the Southbank Institute of Technology in Brisbane, Australia.

3–6

**Solitaire Workstation
Prototype
2005
Alexander Lotersztain
Studio
Reinforced plywood**

Stylish hybrid table/chair workstation.

3

4

5

6

7

8

9

10

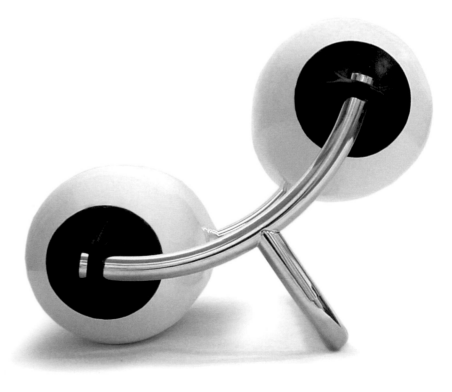

11

7–9

Soft Sofa
2001
IDEE Co., Ltd.
Wicker, cane

Rocking sofa with internal urethane foam padding on back rest area.

10–11

Crusoe Sofa
2002
Derlot.com
Stainless steel, inflatable rubber boat fenders

A stylish limited edition sofa with a nautical twist, Crusoe is waterproof and appropriate for outdoor use.

12–13

DEK range
2004
Street and Garden Furniture Co.
Stainless steel, various wood

User-friendly street furniture that both blends in with and enhances urban environments.

12

13

1

2

Xavier Lust, Brussels

Let us pause to consider for a moment a minimalist, simple, clear, efficient and non-problematic design. Does such a thing exist? And if so, is it less interesting than an exuberant, emotional, symbolic, ambiguous design? When discussing the future of design we must not lose sight of the fact that efficiency and functionality are still our priorities. Naturally, we need objects that fulfil to perfection the purposes for which they were designed. These objects also have to coexist with other items and appliances, which in turn fulfil their own distinct and individual roles, contributing to the highly concerted and contrived job of making our lives run more smoothly. These days, as we organize our lives in an increasingly complex manner, few things are as important as reliable, well-performing objects that communicate succinctly.

Xavier Lust creates such products, his designs conveying immense reliability and measuring up to their visual promises. For Lust, function does not obliterate everything else but is simply the driving force of his work, paired with his acknowledged desire to make objects that are, as he says, 'self-evident'.

Born in Belgium in 1969, Lust has been active in industrial design since 1990, working for clients such as Elixir, Driade, MDF Italia and Aquamass. La Grande Table (2001–2) or Crédence (2002–3) – in terms of form, some of his bolder pieces – fit interestingly into his creative process, which according to him 'is to try to steer clear of the designer's usual stumbling blocks by evolving production systems that result in objects with innovative volumes'.

Lust's minimalism should not be rashly dismissed as cold or stripped of any emotional charge, however. Although he factors in price, comfort and durability, he is just as quick to stress that 'a piece of furniture may give a pleasure or spark an emotion that is linked to the idea of beauty'. Belgian design critic and historian Max Borka has described Lust's work as 'poetic minimal design'. And there is always plenty of room in our world for that approach. **Guta Moura Guedes**

1–2
Paso Doble
Umbrella stand
2003–4
Driade Kosmo
Polypropylene

This umbrella stand is created by using a rotation-moulded plastic technique.

3–4
PicNik
Picnic table and bench
2002
Extremis
Aluminium
In collaboration with
Dirk Wijnants

Made in one single sheet of aluminium, this unmoulded, economical and stand-alone seating achieves its shape through the natural deformation of the material.

3

4

5

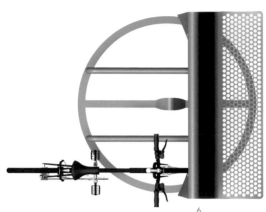

6

5–6
Banc-vélo
Bike stand/bench
2004–6
Elixir
Lacquered steel

*Made of folded steel
sheets, this bike stand is
combined with a public
bench to maximize the use
of space and materials.*

7
Abri-vélo
Bike stand
2004–6
Elixir
Lacquered aluminium

*This bike stand is protected
from the elements, and
staggers the bikes' storage
height to increase ease of
access and storage potential.*

8
Turner
Candlestick
2003–4
Driade Kosmo
Mirror-polished cast
aluminium

*This single, elegant
candlestick is designed
so that it can be used
on its own or in a variety
of combinations to create
stylish centrepieces from
the separate elements.*

7

–1–2

Candlelight Dinner for One
2004
Prototype
Steel

*A candlelight holder with
a table attached.*

3–6

SoftBox
2005
Prototype
Polyester, plastic triangles

*Collapsible bag/soft box
for storage with an original
origami-like folding device.*

7

Caravaggio
Lamp
2005
Lightyears A/S
Steel, high-gloss lacquer

*Pendant lamp that comes
in both black and white.*

1

2

Cecilie Manz, Copenhagen

The Danish designer Cecilie Manz has a taste for radical
simplicity. She has produced glassware called Minima (2006),
a sternly linear lamp named Mondrian (2005) and a chair
formed using two cleverly manipulated rungs of a wooden
ladder, Ladder Hochacht (1999). Her Accessory for Nothing
(2001), a rectangular wooden cage designed, in her words,
with 'as little skin and soul as possible' alone qualifies
her as a minimalist. However, Manz also has an instinct for
creating cosiness that leads her to design, say, a table
for accommodating lonely diners. Asked whether the name
of her Copenhagen-based studio, Manz Lab, reflects a clinical
approach to design, she says it's derived from 'a dream of an
experimental laboratory with many different "soups" boiling
at the stove'.

Manz grew up within a family of artists, so the associations
between hearth and craft come naturally to her. She and
her siblings (a brother is an architect; a sister practises graphic
design) earned pocket money as children assisting their
parents in the studio at home. Today the family members
bounce ideas off one another and sometimes work together.
Several years ago, Manz collaborated with her mother,
the ceramicist Bodil Manz, on Trapped Cylinders (2000),
a series of sculptural meditations on materials and geometry.

Manz bears many of the hallmarks of a Scandinavian
designer, including respect for boldly expressed simple
materials (as demonstrated, for instance, in her Veneer Bags
(2003) made of thin sheets of wood taped together with
colourful strips of canvas) and a pragmatic approach (her
Micado side table (2003) is easily assembled from a circle
of wood balanced on three criss-crossed legs that poke
through the top). She resists geographical classifications,
however: 'It is always interesting and important to know
where from and why,' she says, 'but I think we should focus
more on the quality of the design – then afterwards we can
measure and box it if we really need to.' **Julie Lasky**

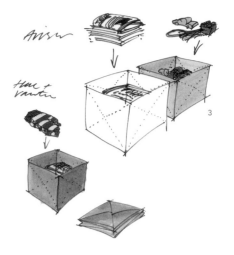

3

4

5

6

8

9

10

8–10

Minima
Glassware
2006
Holmegaard
Glass

*Series of stunningly
simple blown glass
for Holmegaard.*

11–12

Micado
Side table
2003
Fredericia Furniture
Solid ash, cherry or
oak wood

*Collapsible table made
only from 3 sticks and
a table top.*

11

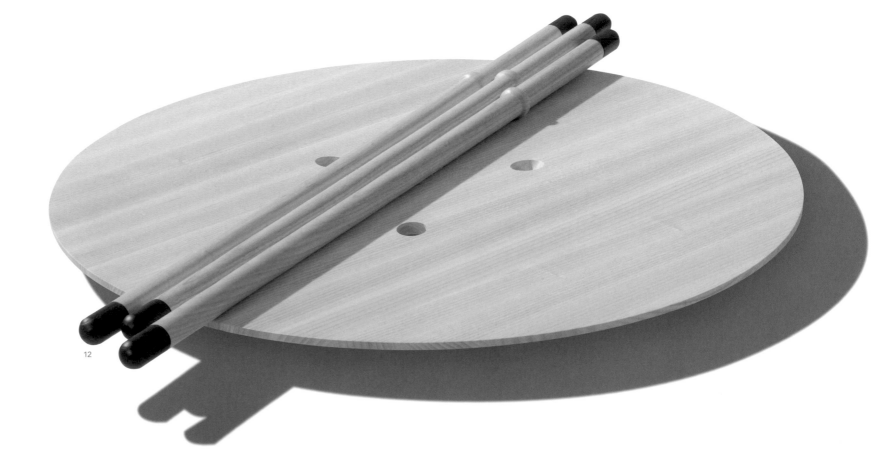

12

1

Tatsuya Matsui, Tokyo

The designer Tatsuya Matsui says that he wants 'to make robots for the happiness of mankind', and the robots he is talking about are already here among us. The Louis Vuitton mannequin Palette (2002) decorating the brand's shop windows, and Posy (2001), a robot imitating a flower girl at a wedding and featured at Robodex 2002, are both Matsui's creations.

After studying design at university, Matsui joined the office of the world-famous architect Tange Kenzo. It was a great opportunity to use Tange's teachings as a basis for understanding modernism. Using these concepts, he searched for the intersection between the city and information before moving to France to study computer technology. He returned to Japan to research the interface between computers and the real world, finally discovering robots and joining up with the scientist Hiroaki Kitano at Japan Science and Technology Corporation, where he participated in the development of SIG and PINO before establishing his own company of engineers and architects in 2001. Although he has been working on both company and government robotics projects since then, he has also explored other design areas. Among his shop and event-space designs, his most remarkable body of work is Starflyer (2006), which comprises over a hundred different designs for an airline, including aeroplane bodies, logos, uniforms, check-in counters and TV commercials.

Although this work seems unrelated to robotics at first glance, in fact Matsui is working all the time to bring together the world of robotics and the contemporary and commercial design world. In the same way that Apple Computers introduced us to the life-changing computer, Matsui's aim is to integrate the innovations of the robotics industry with our everyday lives. The objective of his company Flower Robotics is to build robots that have a special meaning for people. **Chieko Yoshiie**

1

Upper Torso Palette
2005
Flower Robotics, Inc.
FRP

A mannequin-type robot that can transform the space in a shop window into a new body of media.

2

Palette
2002
Flower Robotics, Inc.
FRP, aluminium

A mannequin-type robot that reacts to the movements of those around it.

3

P-noir
2001
Flower Robotics, Inc.
FRP, aluminium, silicon

*This is a robot that
receives and analyses
data concerning a
person's bodily movements,
and then mimics
that person's movements.*

4

Posy
2001
Flower Robotics, Inc./SGI
Japan, Ltd.
FRP

*A humanoid robot based
on the image of a flower
girl at nuptials.*

5

Platina
2003
Flower Robotics, Inc.
FRP, aluminium

*An environmental-
interfacing robot
whose purpose is to
create a concept for
combining the three
different environments
of people, machines
and their networks.*

Tatsuya Matsui

3

4

5

Matthias Aron Megyeri, London

The talented curator Paola Antonelli put all her energies into the design exhibition 'Safe: Design Takes On Risk' held at the Museum of Modern Art in New York in 2005, and aroused a lot of attention when she chose Matthias Aron Megyeri's work, Mr. Smish & Madame Buttly – a cute but also suitably vicious security razor wire – for the front cover of the exhibition catalogue. At only 32, artist Megyeri exhibited five of his Sweet Dreams Security™ works (of which Mr. Smish & Madame Buttly is a part) in the show.

Megyeri, of German-Hungarian descent, is based in London. He studied visual communication art in Karlsruhe, Germany, and has been working as art director for the prestigious Swiss architectural magazine *werk, bauen + wohnen*. He studied product design at the Royal College of Art and Sweet Dreams Security™ is the result of his 2003 graduation project, which investigated how the rigorous needs of security systems can be integrated artistically into daily life.

As well as his work being sold in Paul Smith and the well-known Parisian Colette boutique, among others, it has featured in many prestigious design magazines. The public awareness that these features raise are part of Megyeri's objective – to make people see that paranoia and an exaggerated desire for cuteness is taking over our lives. His animal designs on railings, bear-faced padlocks and heart-shaped chain links give an appealing and ironic face to security issues. But not only is the exterior design appealing, the high-level functionality also continues the long traditions of good, old English-quality lock-making. 'Form follows function' is at the heart of Megyeri's work, but he extends this dogma by offering functioning products that also improve our mental well-being. The huge impact of his first major project has raised our expectations for whatever he does next. **Chieko Yoshiie**

1

Sweet Dreams
Security™ Billy B.
Old English padlock
2003
HY Squire & Sons Ltd
Brass

Sweet Dreams Security™ introduces a fresh and friendly face to the old English padlock by offering Billy B. to grace your back garden or private space. This fun product boasts the quality workmanship expected from an old favourite.

2

Sweet Dreams
Security™ – Heart to Heart
Chain
2004
Megyeri & Partners Ltd
Iron

Heart to Heart enhances a classic iron product with a sprinkling of sweetness.

3

Sweet Dreams
Security™ – R. Bunnit,
Peter Pin and Didoo
Victorian railings
2003
Whitton Castings Ltd
SG iron

This contemporary take on traditional Victorian street furniture, with its smiley faces, proud beaks and floppy ears, provides functional safety as well as mental well-being.

4

4

Sweet Dreams
Security™ – Mrs Welcome
Nottingham lace curtain
2005
Lace Dimensions Ltd
Polyester

*Woven to look like a
metal security shutter,
this traditional lace curtain
with a twist is a placebo
product designed to keep
burglars at bay, without
making you feel imprisoned
in your own home.*

5

Sweet Dreams
Security™ – CityCatTV
CCTV attachment
2003
Megyeri & Partners Ltd
Steel, aluminium

*These quirky CCTV
camera attachments add
a welcome touch of humour
to our increasingly
security-conscious lives.*

6

Sweet Dreams
Security™ – Mr. Smish &
Madame Buttly
Razor wire
2003
Megyeri & Partners Ltd
Steel

*Standard razor wire has
been transformed from
a purely threatening object
into one that is made out
of cute butterflies and
fishes; yet it is every bit
as effective.*

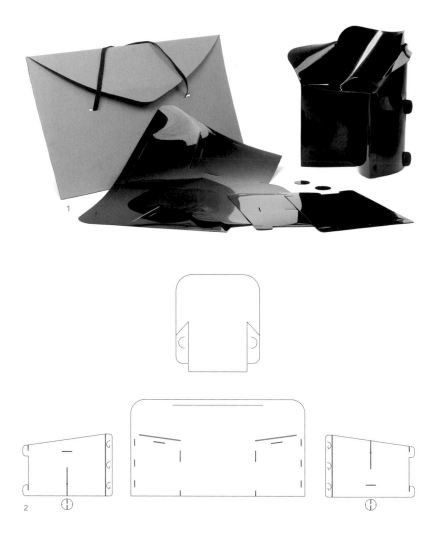

Marcelo Moletta,
Rio de Janeiro

When Marcelo Moletta starts a project, his aim is to avoid repeating what already exists; he returns to form, function, simplicity and intuition for his goals and tools. He combines this approach with new technologies, materials and concepts. His objects are either fully portable or can be disassembled for ease of transport, combining functional aspects such as a simple folding mechanism with designs that incorporate comfort.

For Moletta, choosing the most appropriate material is crucial as this decision will determine the final outcome. The choice of material also determines where and how the object can be used, what weather conditions it can withstand, how often it can be handled and whether it will be light enough to transport. Another factor influencing this choice, as well as the manufacturing process, is the price and availability of the material. Simple and efficient manufacturing processes that make use of as little material and structure as possible demand a precisely planned design, a process that can be speeded up using intuition and computers. An example of this approach can be seen within the clever design of his Envelope Chair (2000), which is sturdy yet portable.

According to Moletta, 'Any type of artistic expression is a cultural manifestation, and my projects are always linked to the surrounding world, such as the city where I live, the places I have been to, nature, my friends, my family, the girlfriends and teachers I have had, people who influence and inspire me daily, the music I listen to, paintings, sculptures, architecture, movies, books ... The designer's role is to make people's life easier, more comfortable and prettier.' **Maria Helena Estrada**

1–9

Envelope Chair
2000
Marcelo Moletta and Itapeva
Polypropylene

This chair can be disassembled to become a series of polypropylene puzzle pieces that can fit into an envelope. It's an easy way of storing a large number of chairs.

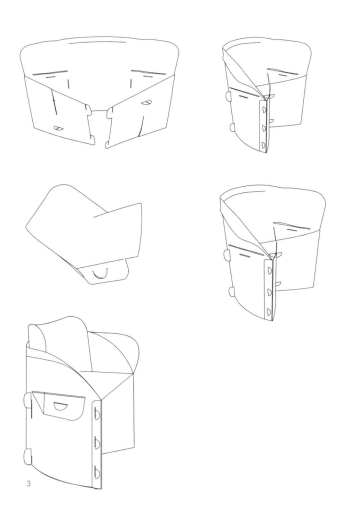

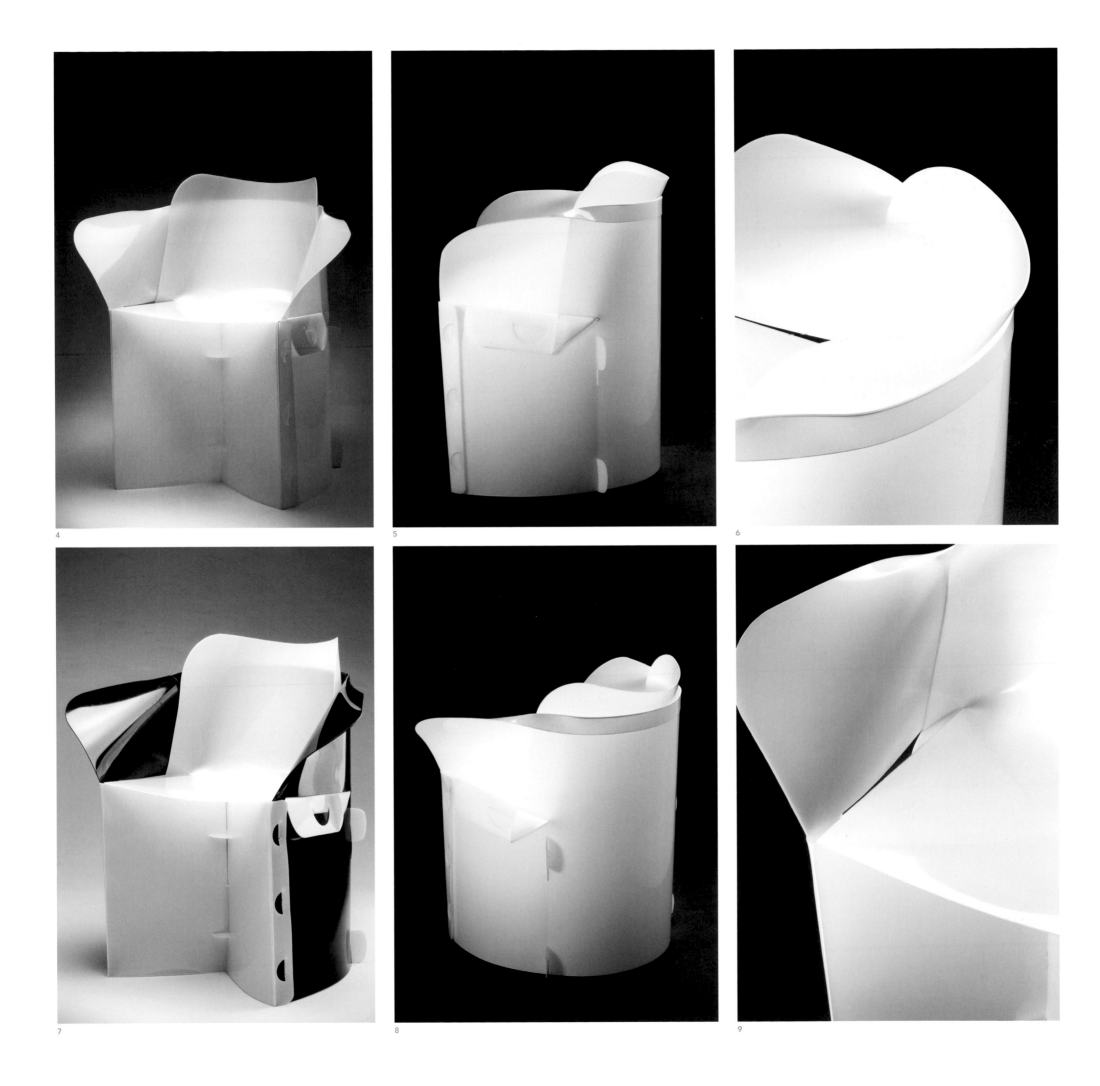

4

5

6

7

8

9

10–16

Solar Battery Charger
2003
Prototype
Polystyrene plastic

Solar-powered battery charger for electronic equipment such as mobiles, digital cameras and MP3 players. Charging speed: 135mA/h at 7 volts.

17–21

Carioca Chair
1997
Prototype
Aluminium, canvas

Beach chair that can be disassembled to fit neatly into its own bag.

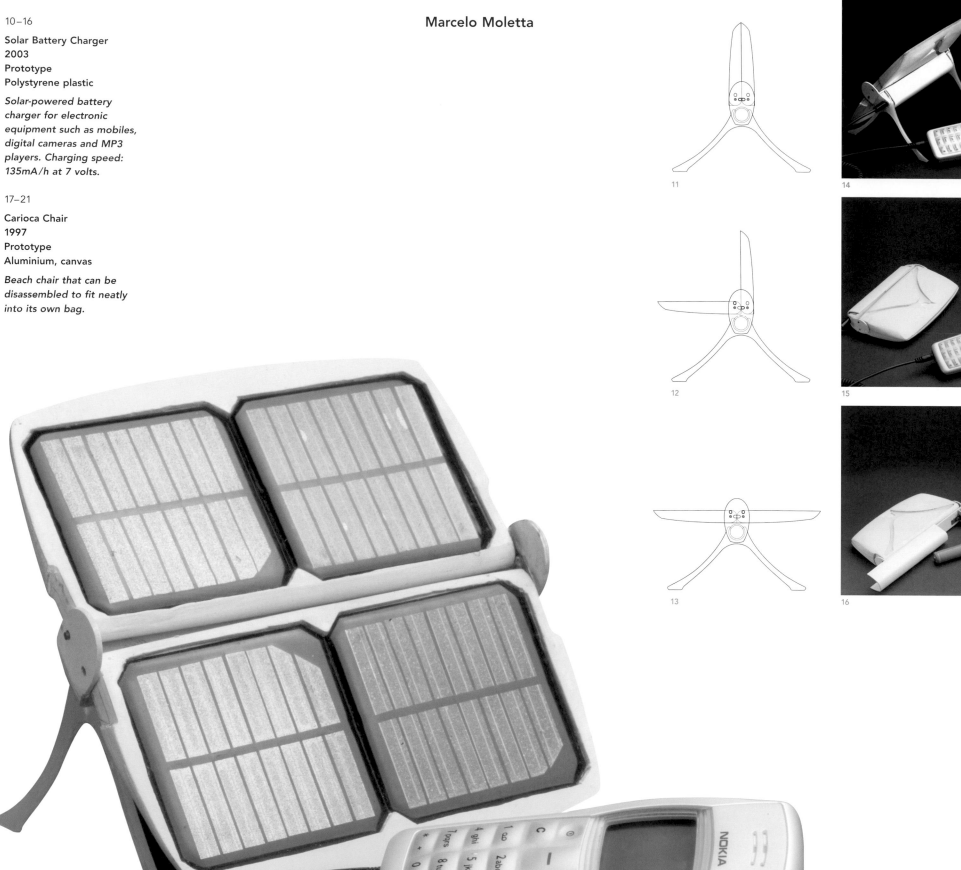

10

11

12

13

14

15

16

17

18

19

20

21

1

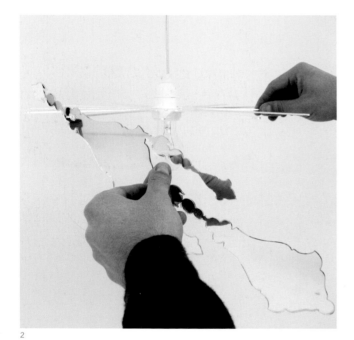

2

3

Nada Se Leva, São Paulo

Formed in 2005, the design studio Nada Se Leva (literally, 'nothing is taken away') emerged as a result of the common approach shared by designers André Bastos and Guilherme Leite Ribeiro. Its first collection was launched in autumn 2006 in São Paulo's Firma Casa, a store that seeks to integrate Brazilian and international design.

Whereas Bastos worked for ten years in fashion and then opted for product design, Ribeiro graduated from Fordham University at CLC, New York, in arts and communication and then worked in publicity and design for ten years before becoming art director at Schell/Mullaney, New York. Returning to Brazil in 2003, with the know-how acquired from his New York design experience, Ribeiro developed his first furniture collection.

This twosome's union emerged from a desire to revisit the past and translate a world of holiday references, cinema and art into objects and furniture. Using modern technology, Bastos and Ribeiro's designs bring what is playful and nostalgic forwards to the present. Their unification of technology and design is demonstrated in the combination of laser cutting, laminate and glass printing with a diverse array of materials such as wood, glass, acrylic and melamine.

Inspired by baroque style, Bastos and Ribeiro designed the Antique side table (2005) (and would later be inspired to create the Ligero Small Side Table (2006)). The Bastos' playful element is manifested in the print, held between glass sheets, which is generated from a digital image of a real-life baroque table. It is an unexpected contemporary reinterpretation of classic furniture. With their Ishi line (2005), the designers made use of digital printing on melamine sheets. Inspired by English designer Paul Smith's colourful stripes, the series of lamps, benches and tables evokes the simplicity of Japanese design.

Taking as their inspiration luxurious baroque designs or fashion pieces, the duo translates historic and contemporary elements into sumptuous chandeliers and mirrors, only this time produced with cutting-edge technology. These designs, achieved via multiple influences, are currently highly desirable and display the connection between past and present.
Maria Helena Estrada

1–4
Ligero Ceiling Lamp –
Crystal Motif
2006
Acriresinas
Acrylic, aluminium

The chandelier is made of crystal-shaped acrylic pieces that can be unmounted to fit in a flat envelope.

4

6

7

8

9

10

5

5

Ligero Small Side Table
2006
Acriresinas
Acrylic

Based on a baroque design, this side table is made from one continuous piece of 1cm acrylic.

6–10

Wax Stool
2006
Fiorini
Wood, denim

3 denim plates, normally used in floor waxing machines, are slotted onto the stool's wooden base to form the seat cushion of this appealing utility object.

11–12

Bull's Eye Table
2006
Tergoprint
Formica, chromed steel, MDF

The multicoloured bull's-eye Formica table top is derived from a digital print, and sits atop simple steel-chromed legs.

11

12

Incognito – Rococo & Empire,
Hidden Wealth Project
Dinnerware
2003–4
Porzellan-manufaktur
Nymphenburg
Porcelain

*A range of tableware whose
distinctive detailing marks
it as a luxury item. However,
the twist is that only the
owner knows as this detail is
on the base of the pieces.*

Khashayar Naimanan,
London and Munich

Khashayar Naimanan is a designer who has turned famous designers into products. His Designstars project (2001–), which includes a collection of 'action figures' based on world-famous designers such as Ron Arad, demonstrates his enthusiasm for star designers. While the project as a whole provides design information, Naimanan's action figures are highly collectable.

Born in the UK to Persian and Japanese parents, Naimanan's designs seek to explore the different sides and possibilities of familiar objects. For example, in the Hidden Wealth Project (2003– 4), he plays dramatically with an object's perceived use or value. One design outcome is a common-or-garden nail presented in a yellow paper box – only the nail is made of gold. Another subverts the idea of expensively decorated dinner plates: the patterns on Naimanan's Incognito dinnerware (2003– 4) are all on the underside. Nobody would know that the dinnerware is expensive because all that is on show is the white plate. Only the owners know the true value, achieving satisfaction when they turn it over.

Although these designs might be interpreted as Naimanan accommodating capitalism, projects such as Robin of Fulham (2004) and Sleeving Jacket (2003) playfully show the opposite. Robin of Fulham enables you to purchase goods more cheaply by substituting the original tags with homemade barcodes that register lower prices. Comic Thief (1988, 2004) and Sleeving Jacket are devices that aid the theft of comic books and wine, respectively. These may seem to go against the project's title – 'Design against Crime' – but Naimanan's intention was to raise awareness of the existence of crime.

Naimanan's designs present interesting approaches based on users' experiences that, as a result, become specific features of each product. **Sang-kyu Kim**

1

2

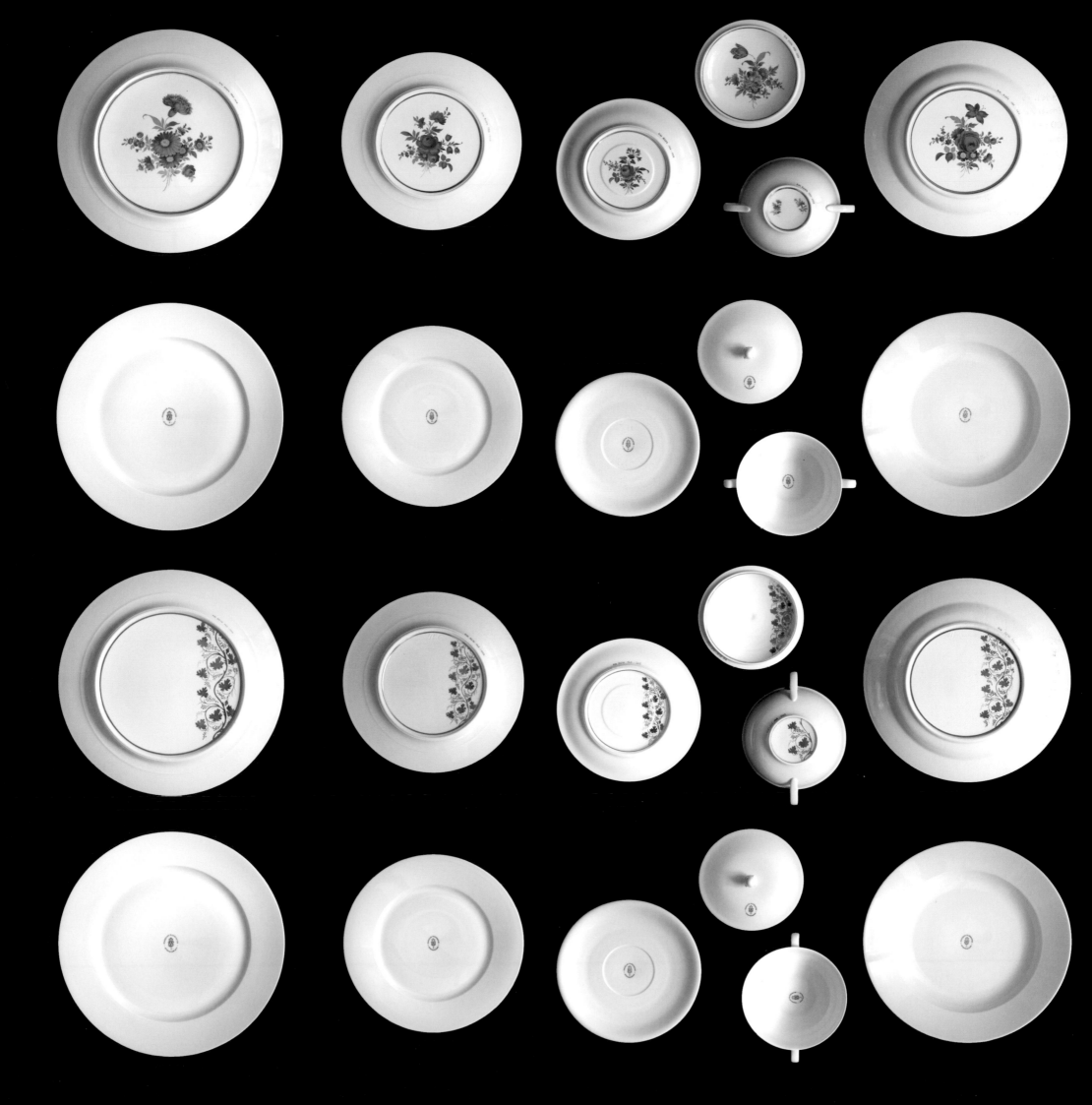

3–4

Robin Board V04 – Robin of Fulham Project
2004
Self-production
Polystyrene

A plastic component attached to the skateboard conceals a comic series of one-off models.

5

Ça Ne Vaut Pas un Clou – Hidden Wealth Project
Round nails
2003
Self-production
Gold, cardboard

Re-evaluating traditional ways of hiding wealth by melting precious jewellery into objects with little perceived value. Here the nails are made of gold.

6–7

Designstars: Ron Arad with Little Albert
Limited edition action figure
2001
Khashayar Naimanan and Chris Fitzgerald
Resin, card, polystyrene

The first collection of Naimanan's set of famous designers realized as action figures.

3

4

100 mm 4 "ROUND NAILS

5

"THINGS PEOPLE DON'T REALLY NEED (BUT CAN'T AFFORD)"

RON ARAD

DATE OF BIRTH: 1951
PLACE OF BIRTH: ISRAEL
STATUS: DESIGNER/ ARTIST/ ARCHITECT
AFFILIATION: RON ARAD ASSOCIATES
CILENTS: VITRA, MOROSO, DRIADE, KARTELL, FIAM, ALESSI,

Design & Concept: Khashayar Naimanan
Modelled and painted: Chris Fitzgerald

CONTINUE YOUR COLLECTION WITH THESE DESIGNERS

ACHILLE CASTIGLIONI   ISSEY MIYAKE   KONSTANTIN GRCIC

Item 0001
RON ARAD

CE

6

COLLECTION 1
DESIGNERS

DESIGN PRODUCTS

7

1–3

Hanabi
Lamp
2006
Prototype
Shape-memory alloy

*The heat of the bulb makes
this shape-memory alloy
lamp 'bloom' whenever the
light is turned on.*

4

Sasae
Coat stand
2006
Gebrüder Thonet
Wood

*Unable to work by itself,
the Sasae only becomes
functional when it is
propped up against a wall.
This 'weakness' allows it to
melt into its environment as
if it had always been there.*

5–7

Polar
Nesting table
2006
Swedese
Glass, steel

*When the tables are
stacked, the polarizing film
inside the table tops makes
flower patterns appear
through the table's clear
glass surfaces.*

nendo, Tokyo

The nendo design office, headed by Oki Sato, takes its name from the Japanese for 'plasticine', suggesting the 'free, flexible and diverse stance' apparent in Sato's architecture, interior design, event organization, graphics, corporate identity and other creative projects. Since its establishment in 2002, nendo has received widespread acclaim, winning a Salone Satellite Design Report Award and has become one of Japan's most vibrant designer groups.

Sato established his studio soon after leaving Waseda University Postgraduate School. The present ten-member group shares the workload of nendo's diverse range of products, with Sato coming up with the initial concepts.

Although nendo's activities might seem glamorous, its designs are quite conservative. Sato says, 'There are many good designers for luxury goods, so I try to design differences into daily items.' He invests a lot of thought in his appealing designs, such as Hanabi (2006), meaning 'fireworks' in Japanese, has a lampshade that opens gently with the heat of the light, and Sasae (2006), a series of coat stands that are useless without user interaction, only becoming functional when propped up against a wall. His designs are based on the notions of 'kire', issuing from right-brain emotional feelings centred on fun and beauty, and 'koku', from left-brain rational ideas encompassing cost, performance and meaning. Although interior and architectural design are slightly more glamorous than product design, Sato says that his design policy remains basically the same. The nendo designers aim to emphasize everyday differences, making the resultant designs very dynamic.

Having established an office in Milan in 2005, nendo is hoping for global growth. Although its designs at first glance seem reserved, it will be interesting to see how their 'kire' and 'koku' concepts are received. **Chieko Yoshiie**

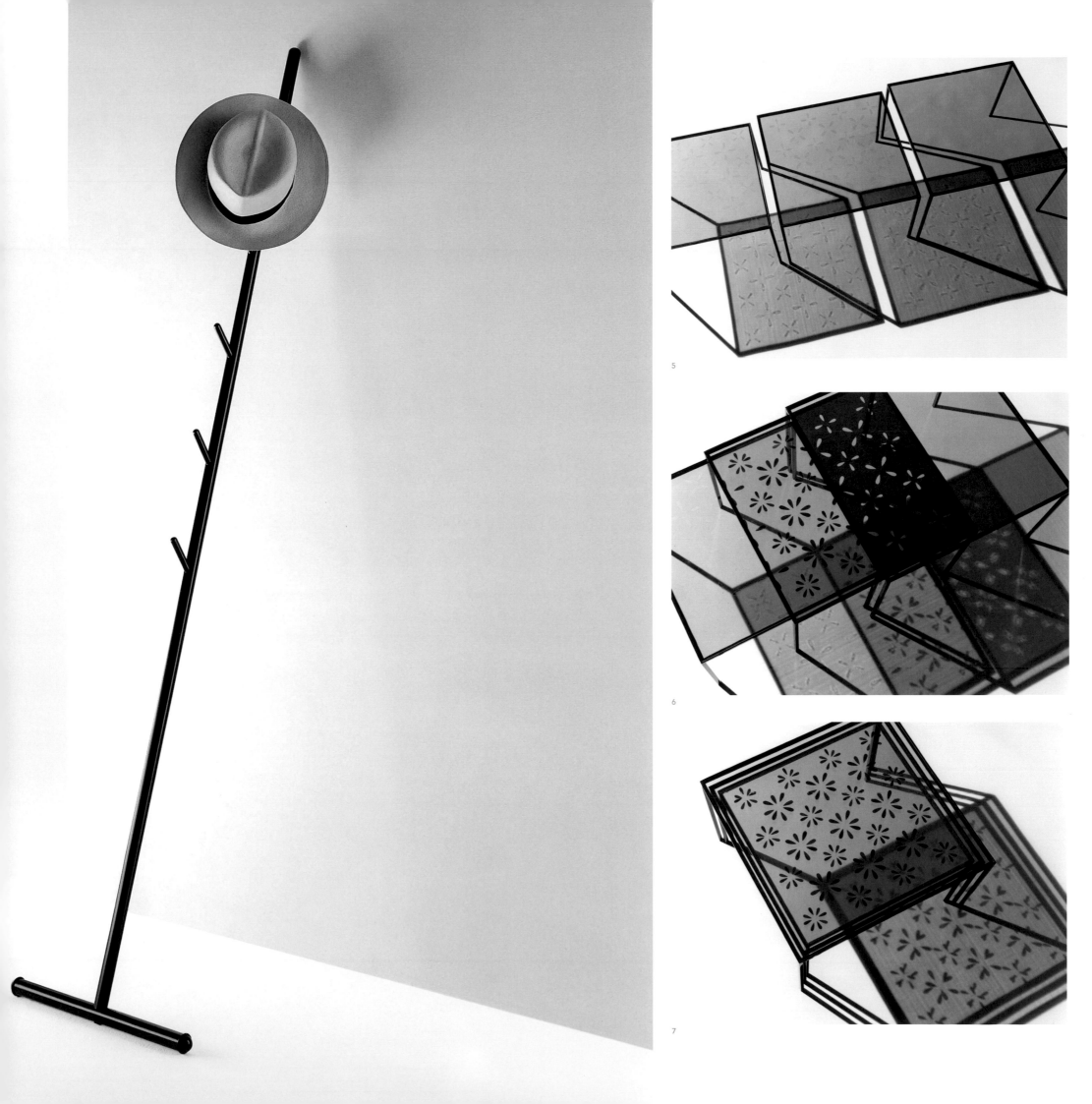

8

8
Chab-table
Side table
2005
De Padova
Beechwood, steel

The Japanese Chabudai is a multifunctional low table that is equally useful in the dining room, living room or bedroom. The Chab-table draws inspiration from the flexibility of this traditional and versatile design.

9
Yuki
Screen
2006
Cappellini
Plastic

'Yuki' means snowflake and the screen is made of simple snowflake-shaped plastic pieces. It is formed by piling the separate pieces up as if it were snow, and as a result the size of the screen can be altered easily.

10
Fruit-Tree
Fruit holder
2006
Prototype
Steel

The Fruit Tree delicately and decoratively holds the fruit in its branch-like prongs.

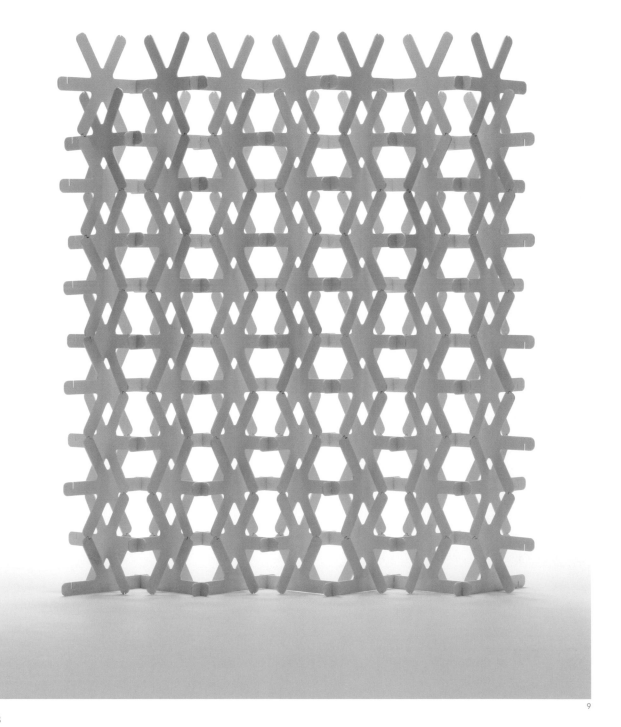

9

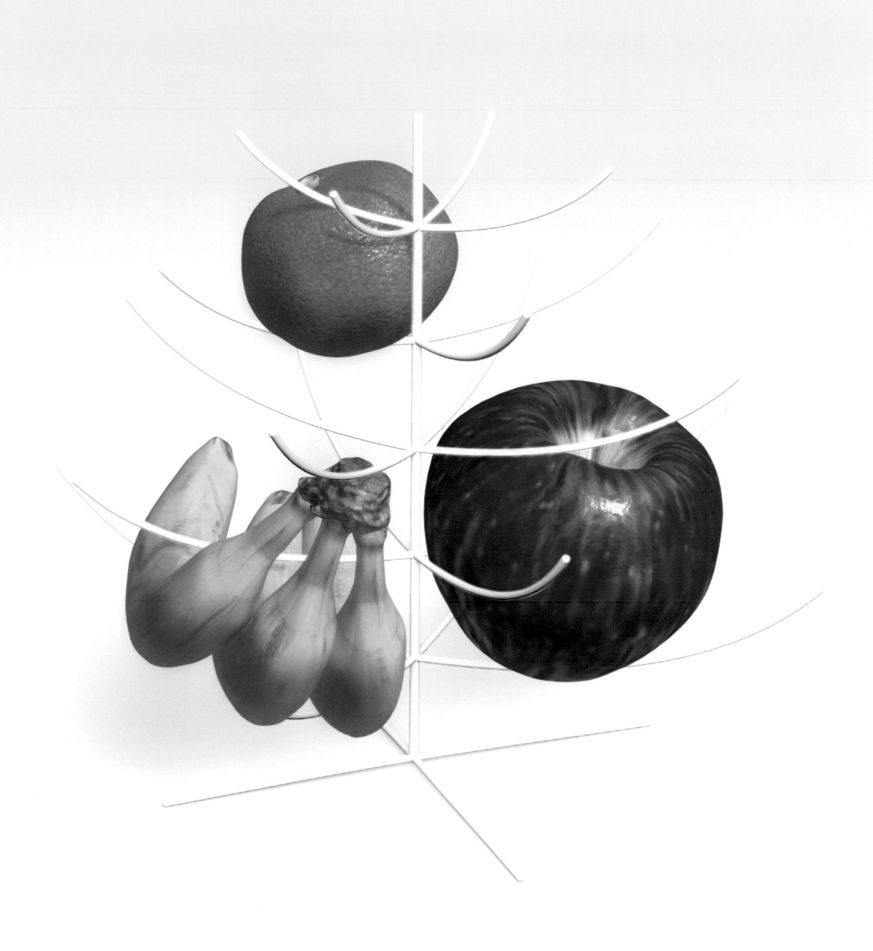

Nido Campolongo, São Paulo

Nido Campolongo, who originally worked as a typographer with his father in São Paulo, is a graphic and product designer as well as a plastic artist. He uses paper, cardboard and their derivatives to construct his objects, which evolve from graphic pieces into unusual pieces of furniture. His designs include small objects, such as cushions, bracelets and bedcovers all made from a 'fabric' composed of tiny rolled paper tubes, and larger projects, such as his Translucent walls (2001) and domestic flooring called Cardboard Flooring (2006), completely made out of treated cardboard.

Campolongo recycles the industrial waste and cellulose by-products left over from the manufacture of paper and cardboard. The original function of these materials is barely visible in his work. Only with close observation is it possible to notice, for instance, the transformation of discarded paper tubes (jettisoned by the paper industry) into modular shelves made up of cardboard rings.

The designer works from the tiniest to the largest scales: from small objects such as bracelets to total installations of exhibitions and public facilities. An important project is Casa do Leitor (Reader's House), São Paulo (2001); with four different environments covering a total area of 40 square metres, the house is constructed out of paper and cardboard, and includes furnishing, flooring, soft furnishings and lighting.

Campolongo is deeply involved in socio-environmental issues, promoting links between sustainable practices and the creation of products. He also works in several Brazilian cities devising projects for street communities and has developed, in collaboration with its inhabitants, a housing complex in São Paulo made from cardboard and MDF residues. In 2006 Estúdio Nido Campolongo created a model for an outstanding alternative architecture project, using 6,000 cardboard cones to build a housing shelter called OCA (Brazilian Native Indian shelter). **Maria Helena Estrada**

1–2

Educational Work of
Craftsmen Project
1993–2006
Galeria do Papel Gráfica
Editoria Ltda.
Cardboard, waste paper

*Nido Campolongo is one of
the first Brazilian designers
to work with societal and
ecological values.*

3

Avestruz Lamp
2001
Galeria do Papel Gráfica
Editoria Ltda.
Cardboard cones, wax
paper, resin paper

*The form of this lamp is
made possible with the
manipulation of recyled
waste materials.*

4

X Stool
2006
Galeria do Papel Gráfica
Editoria Ltda.
Cardboard

*These elegant stools
are made from
recycled cardboard and can
be placed on Campolongo's
Cardboard Flooring (2006).*

4

5

Nido Campolongo

Casa do Leitor
(Reader's House)
2001
Galeria do Papel Gráfica
Editoria Ltda.
Cardboard, paper, paper
with resin, other
complementary materials

*The interior of this
house was designed to
explore the use of unique
materials such as paper
and cardboard.*

6

Cardboard Rings Shelf
2006
Galeria do Papel Gráfica
Editoria Ltda.
Cardboard tubes, wax

*By re-using cardboard rings
recycled from the paper
industry these shelves offer
a creative and sustainable
shelving system.*

7–8

OCA
Model for a paper house
2002
Galeria do Papel Gráfica
Editoria Ltda.
Wax paper, cardboard
cones, cardboard, paper
with resin and other
materials

*A model for a paper house
inspired by indigenous
Brazilian habitation.*

9

Cone Table
2002
Galeria do Papel Gráfica
Editoria Ltda.
Chipboard cones, glass
and paper resin tubes

*This recycled table is made
completely out of resin-
coated tubes and finished
with a glass table top.*

6

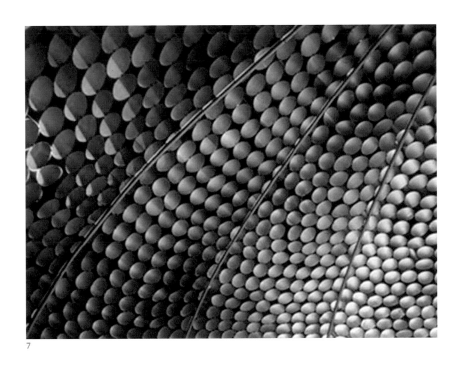

7

8

9

1

2

3

Nódesign, São Paulo

Nódesign was created in 2001 by a group of college friends who began exchanging ideas and discussing the possibility of starting a studio based around the connections between their working philosophies. They started by designing pieces to be sold in product design stores but, due to the low-scale production of their designs, the costs were too high and sales were disappointing. However, in 2002 their products began to appear in exhibitions, publications and competitions, raising the studio's profile and earmarking them for success.

Today the studio works through several different segments, collaborating with market leaders as well as small companies. It believes that the creative value of its design is due to its diversity of fields. For instance, a project developed in the area of perfumery brings up ideas that can be applied to other designs. When working on a new creation, Nódesign starts from the object's purpose rather than pre-existing versions: when designing a glass it thinks about the act of drinking, not the common 'references' associated with the object. Companies often approach Nódesign to develop consumer products, confident that the studio will see as crucial the interaction between the object and its user. Nódesign offers an integrated approach to its clients over the whole life of the process – from the creation and development of the product to its release, including the graphic development of brands, manuals, packaging and sales outlets.

The team believes that, more important than meeting the requirements of a specific function with perfection and rationality, an object must interact with its surrounding environment and through utmost flexibility should seek to be self-regulating within its own eco-system. Multifunctional, interactive projects are new ways of satisfying human needs and these are the motivating factors behind their creations.
Maria Helena Estrada

1–11
Blob
Mobile phone
2006
Easy Track
ABS

The alternative mobile phone designed especially for children, parents and the elderly. Numbers can be programmed in and dialled at the touch of a single button. Simple and easy to use, parents are able to locate their children's whereabouts through the internet or an Information Service Centre.

4

5

6

Nódesign

12

13

14

15

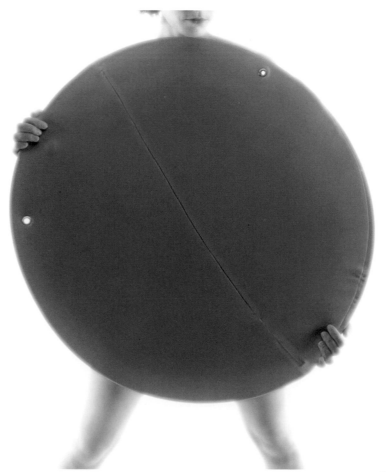

16

12–16

Black Hole
2003
Nódesign
Lycra and steel

A neat solution to storage, this flat-packed product transforms into a handy laundry bag.

17–23

Tangram Chair
2004
Nódesign
OSB plate

Using a material that can be 100% recycled, this CNC-manufactured chair is cheap and efficient to produce, and offers an attractive aesthetic for an eco design.

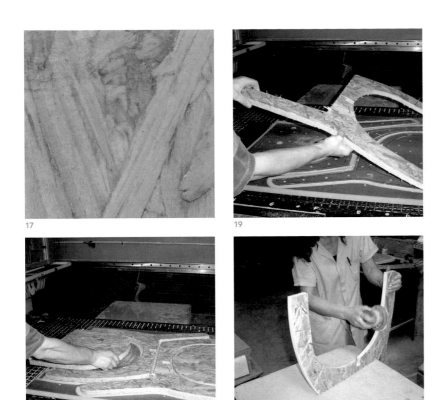

17

19

18

20

21

22

Johannes Norlander, Stockholm

Swedish-born Johannes Norlander believes that 'utility furniture is beautiful' and it's easy to agree with him when you see his designs. For him, simplicity and balance are all important. His design language is always one of restraint and he has a sense of proportion and structural transparency that links him to the best traditions of Nordic design. He likes clear and natural lines and these factors are always evident in his work. It's no surprise, therefore, that his designs have been produced by international companies such as E&Y, Asplund and Collex.

The ORI Chair (2006) – which his company, Johannes Norlander Arkitektur AB, produces – is a case in point. The sturdy wheeled base and unfussy flat-plane seat look unassuming. Look more closely, however, and you can see the intense care with which it has been ergonomically engineered and the thought that has gone into the creation of such an apparently 'simple' idea. And that's what Norlander does best. He takes a single idea, often based on something that exists already, and makes it more effective, both aesthetically and practically: the beauty of the products essentially comes from the success of their function.

In this way Norlander is from the same school of thought as Konstantin Grcic and Jasper Morrison. He keeps design really basic and uncluttered. His Cano office system for E&Y (2005) is a perfect example. The use of industrial typologies – sheet metal, prominent bolts and crossed metal detail – may look simple, but Norlander is a thorough thinker and knows a great deal about production and materials. He makes it look easy, when to realize these designs successfully is actually amazingly difficult. His design always appears to be effortless. **Tom Dixon**

3

4

5

1

2

1–2

Cano
Shelving unit
2005
E&Y
Powder-coated steel

*Height-adjustable
shelving unit with cross-
over supports.*

3–7

ORI Chair
2006
Johannes Norlander
Arkitektur AB
Powder-coated steel

*Elegant swivel chairs
that come in both grey
and white – perfect for
a modern office.*

6

7

8

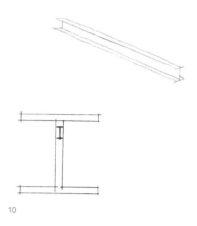

10

9

11

12

8–11

Beam Table
2006
Johannes Norlander
Arkitektur AB
Ash, glass, steel

*Using a mixture of wood,
glass and metal this
table possesses a unique
aesthetic. The table is
fully collapsible for ease
of storage.*

12–13

Kyparn Stacking Chair
2003
Nola
Steel

*Sleek and simple these
chairs stack away to take
up minimal space when
not in use.*

13

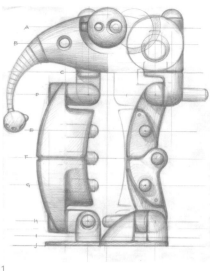

1

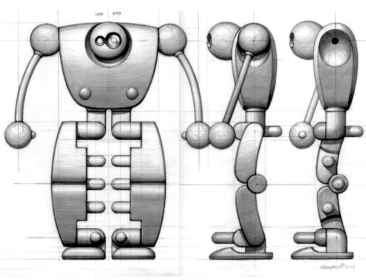

2

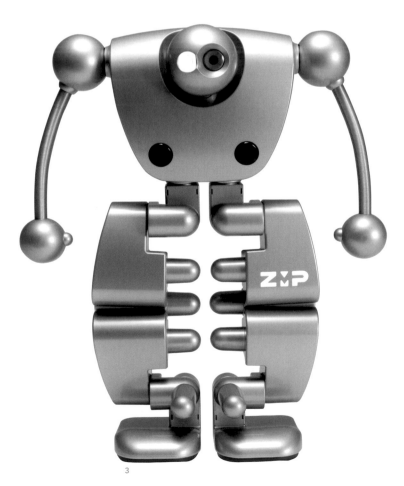

3

Ken Okuyama, Turin

Kiyoyuki Okuyama, known in the west as Ken Okuyama, is currently based in Turin, where he joined the well-known design house Carrozzeria-Pininfarina as creative director in 2004, until he became independent in 2006. Although he was one of a team of designers responsible for the world-famous Ferrari and Maserati sports cars, he has expanded his design projects to include his own brand of eyeware and furniture, and has widened his interests to include robotics and city planning.

Okuyama's life has been a series of challenges during which he has achieved many design 'firsts'. After studying at the Los Angeles Art Center College of Design (ACCD), he worked at both General Motors and Porsche before approaching Pininfarina, where he had long wanted to work. He became the first non-Italian chief designer at Ferrari. He subsequently returned to his Alma Mater, ACCD, as the first foreign dean. In 2004 he returned to Pininfarina, where he became the first outsider to hold the post of creative director, with responsibility for designing Nuvo (2003), the world's first bipedal urban humanoid robot. Okuyama says, 'In the end, cars are all about proportion, but in reality massive capital is required to manufacture them.' However, he continues to be fascinated by automotive design and is currently working on the next generation Ferrari.

His strengths as an industrial designer exceed those of a simple designer and, making use of these skills, he has recently been focusing his powers on revitalizing local industries through design in his ancestral region of Yamagata Prefecture, Japan. He has just launched the Ken Okuyama label and we eagerly await the next range of design 'firsts' from this multi-faceted global designer. **Chieko Yoshiie**

1–3
Nuvo
Humanoid robot
2003
ZMP/Seiko Instruments

A 38cm (15-inch) functioning 'home robot' that can walk, play music and take photos. It can also be used as a security device by relaying photos of your home on command via the internet.

4–5
Maserati Birdcage 75th
Concept car
2006
Maserati

Based on the road racing chassis of the Maserati MC 12, this car was designed to commemorate the 75th anniversary of Okuyama's design company, Pininfarina.

6–9
Isu
Folding chair
2005
Tada Mokko
Laminated wood

A folding chair that comes as a single or double seat. When upright it can act as a clothes stand or movable partition.

4

5

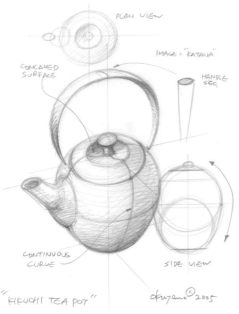

10

10–11

Mayu
Teapot
2005
Wazuqu
Cast iron

A teapot that combines a classic Katana form with modern functionality.

12–13

Spiral
Coat hanger
2005
Tada Mokko
Laminated wood

This decorative spiral coat stand provides hooks at differing heights.

14–15

Albero
Coat hanger
2005
Tada Mokko
Laminated wood

A sculptural coat hanger that takes inspiration from the bare branches of a tree.

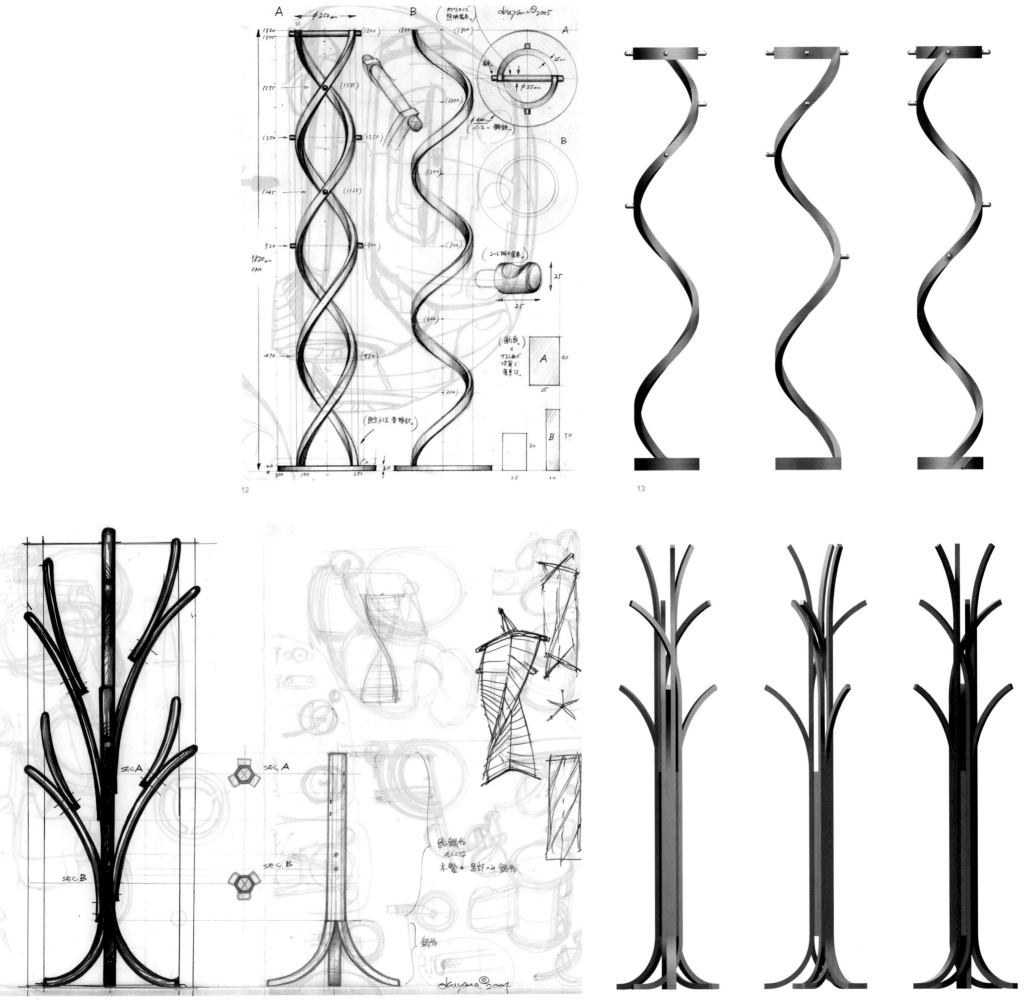

12

13

14

15

OVO, São Paulo

OVO ('egg', in Portuguese) was founded in 2002 by Luciana Martins and Gerson de Oliveira, both graduates from the film faculty at ECA/USP, who started working as a design team in 1991.

Although OVO has achieved a significant number of awards for its equipment and product designs in particular – for instance, first prizes for its Huevos Revueltos hanger (2005, Museu da Casa Brasileira Design Award) and its Mientras Tanto table (2000, Brasil Faz Design competition) – its activity stretches far beyond this area. The duo has also been responsible for several installations and site-specific products shown in exhibitions at leading Brazilian cultural venues, including Rio de Janeiro's Museu de Arte Moderna and São Paulo's Vermelho gallery. Endowed with acute powers of observation and intense creative vitality, OVO edits most of its own pieces and its studio doubles as a showroom – clear signs of its pragmatic, proactive and resourceful spirit.

Although Brazilian design has left its mark on the international scene through the particular and personalized language of the iconic Campana brothers, OVO's work follows a thoroughly distinct approach, in terms of both creation and production. Its design bears the deeper imprint of a formal language that is at times somewhat more European and more urban. The impact of São Paulo – Brazil's largest city, one of the world's biggest urban conglomerations, a 'concrete jungle' heaving with inter-woven networks and dynamic circuits – can be seen in a certain abstract, geometric, complex quality. Another prominent feature of OVO's work is its expressive strength and the subtle influence of associated art and artistic practices, further enriching it with additional layers of symbolic meaning.

The consistency and coherence of its production and its growing recognition make OVO one of the most exciting and noteworthy creative teams of the present time.
Guta Moura Guedes

1
Terceira Chair
2004
OVO
Illustration

2
Terceira Chair
2004
OVO
Wood, straw

The chair was inspired by Brazilian design from the 1950s, especially the work of Joaquim Tenreiro and Sergio Rodrigues.

3–4
Box in the Box
Set of boxes
2005
OVO
Plexiglas

In these boxes, everything is visible. The transparency of the Plexiglas allows you to see everything that is inside them.

5 & 9
Cadé (armchair) & Feriado (bookcase) – in situ shot
Feriado
Bookcase
2002
OVO
Wood

The Feriado bookcase is related to the calender and the idea of one special day that is different from other days. The word 'feriado' roughly translates as 'holiday', and the different section acts as a kind of formal holiday from the rest of the bookcase's pleasing simplicity.

6–8
Cadé
Armchair
1995
OVO
Elastic fabric, metal

The Cadé contradicts the design rule 'form follows function'; here, instead, form hides function.

3

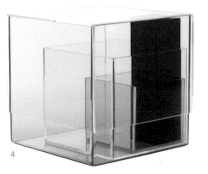

4

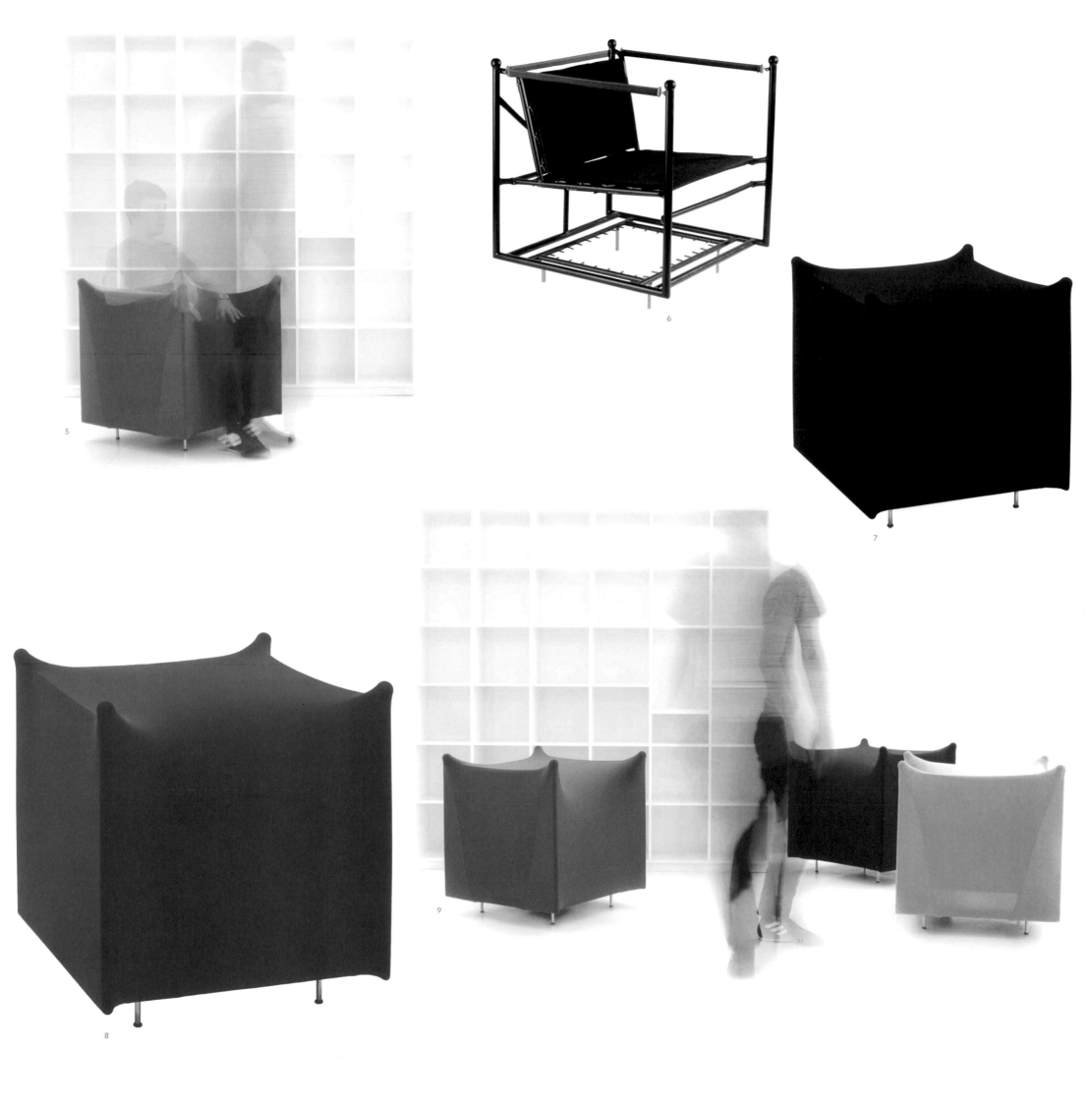

5

6

7

8

9

10

12

13

11

14

15

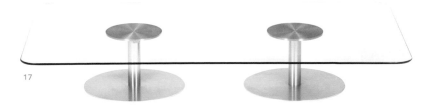

16

18

10–11

Tobogã (chair and chair with armrests) and 2.88 (dining table)

Tobogã
Chair
2002
OVO
Stainless steel, upholstery

Tobogã
Chair with armrests
2006
OVO
Stainless steel, upholstery

With the joints carefully hidden and the designers using the thinnest materials possible, OVO worked to create a chair that looked as if it had always effortlessly existed.

2.88
Dining table
2006
OVO
Lacquered wood

The table's thin base was designed to give the visual impression that it might not be able to support the top.

12–15

Box in the Box
Set of boxes
2005
OVO
Plexiglas

16

Entre
Console
2006
OVO
Lacquered wood, Plexiglas

The back of the console has a transparent acrylic surface to give the impression that there is no structure to the furniture.

17

Disco
Coffee table
2006
OVO
Aluminium, glass

This table can be square, round or rectangular and can be painted in many different colours. The base, however, is always the same.

18

In Vitro
Lighting fixture
2005
OVO
Glass, flexible metal lamp

The light is emitted from a light-bulb tape that can be twisted into different forms.

Flávia Pagotti Silva, São Paulo

On graduating from São Paulo University in 1997 with a degree in architecture, Flávia Pagotti Silva decided to go first to the United States and then to the United Kingdom to continue her studies. It was during this period that she started designing furniture and went on to study at London's Royal College of Art for a further two years. After leaving the RCA, Silva worked for the design department of Foster and Partners, also based in London, where she was responsible for the development of furniture and interior projects. Finally, in 2001, she returned to Brazil and began designing her own furniture and objects for domestic use.

With Silva's designs, shape is never the main objective but rather the result of some other factor or combination of elements. The development of each product includes research into its uses, materials, historical references and the psychological needs of the user. The final product is an expression of these combined elements. Her research is based on the observation of the way people use furniture and objects – how they interact with them and what they expect from each piece. Embodying simple needs, unexpected movements and unusual shapes, Silva's projects explore different emotions and attempt to encourage people to see objects from different viewpoints and to understand the underlying meanings of daily activities.

Another aspect of Silva's design is the combination of handmade and industrial processes (both traditional and contemporary) with natural and artificial materials. This is evident in projects such as Bate-Papo Stool (2002), based on country benches and made from the contemporary materials metal, plywood and rubber, and Rendeira (2003), a chair made from handcrafted lacework 'sandwiched' between layers of acrylic.

Silva's designs retain elements of handicraft while, at the same time, they are adapted for industrial production by using materials, techniques and processes that allow for mass-production. **Maria Helena Estrada**

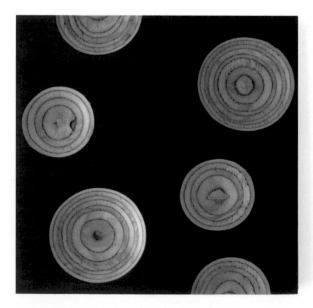

1

2

3

1–3

Impressöes Trays
2006
Self-production
Naval plywood, Formica,
wood veneer

*Part of the Impressöes
series, a special tool
was used to carve the
depressions on the
trays. They function as
decoration and a practical
place to put glasses.*

4–5

Impressöes Table
2006
Self-production
Naval plywood, Formica,
stainless-steel tube

*The dark glue used in the
Impressöes series gives
the illusion that the carved
round patterns are coming
out of the surface.*

4

5

6

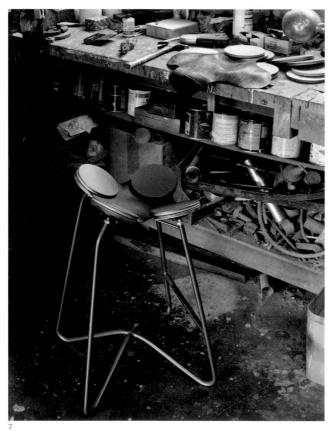

7

8

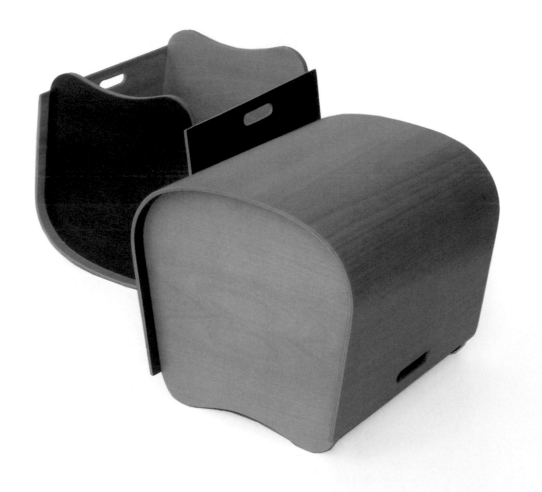

9

10

11

6–8

Bate-Papo Stool
2002
Self-production
Plywood, metal tube,
re-used rubber from
air tyres

*This playful stool is
based on wooden stools
traditionally used in the
Brazilian countryside.*

9

Bau Stool
2006
Self-production
Curved laminated wood,
Formica or wood veneer

*The stool can also be
used as a rocking box for
magazines and toys when
it is turned upside down.*

10–12

Laminado Magazine Rack
2005
Monte Azul Wood Workshop
Curved laminated wood

*The magazine rack is
part of the Laminado series,
a variety of products
designed to be formed from
a pre-existent mould.*

13

Laminado Stool
2004
Monte Azul Wood Workshop
Curved laminated wood

*This combined stool
and magazine rack was
the first product in the
Laminado series.*

14

Porta-Tudo Por Enquanto é Só
Shelf
2006
Prototype
PVC, metal

*The series, which translates
as 'That's all for now',
experiments with the
flexibility of PVC to create
innovative storage.*

13

12

14

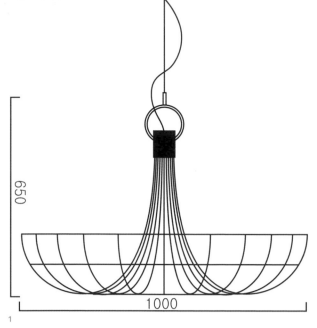

650

1000

Large (21 bulbs)

Zinoo Park, Seoul

Zinoo Park's work is very much engaged with popular culture. For example, he has often worked on ideas surrounding the Coca-Cola brand. His Coca-Cola classic (1999–2004) takes an old-fashioned coke bottle and transforms it into an art/craft item. When a familiar object is placed in a different context and made in an unfamiliar material – in this case, silver – it provokes interest and prompts reflection.

Other Park designs are informed by other areas of mass media. The Erotic Dishes (2003) are a reinterpretation of the relationship between sex and food, made explicit in the Japanese movie *The Realm of the Senses* (1976, directed by Nagisa Oshima). They function independently as dishes, but can also work as a sensual covering to hide parts of the body, delightfully associating the body and food with 'freshness', in seeming contradiction to the pessimism of the film.

Another work, 5 Minute Candles (2004), develops and plays with the functions of matches and candles. Here, miniature cake candles are held in portable cardboard match holders like those found in bars and restaurants. But this product has a playful social function that goes beyond just lighting up. Park has turned an everyday object into a special-occasion product: the pocket-sized book of candles can be brought out for a forgotten birthday or anniversary at the very last minute!

The concept of altering the original function of materials can also be found in the Spaghetti Chandelier (2005), where Park transforms industrially used red-coloured electric wire into an ornate domestic light source. The 'spaghetti' or wire can be shaped as you like.

Park always keeps up with pop culture in his work. He knows which elements will excite the public and uses them to highlight and consider issues within society. **Sang-kyu Kim**

1–3

Spaghetti Chandelier
2005
Park Plus
Powder-coated steel,
electric wires

*A random decoration
pendant light, the
chandelier exemplifies
Park's rich visual humour.*

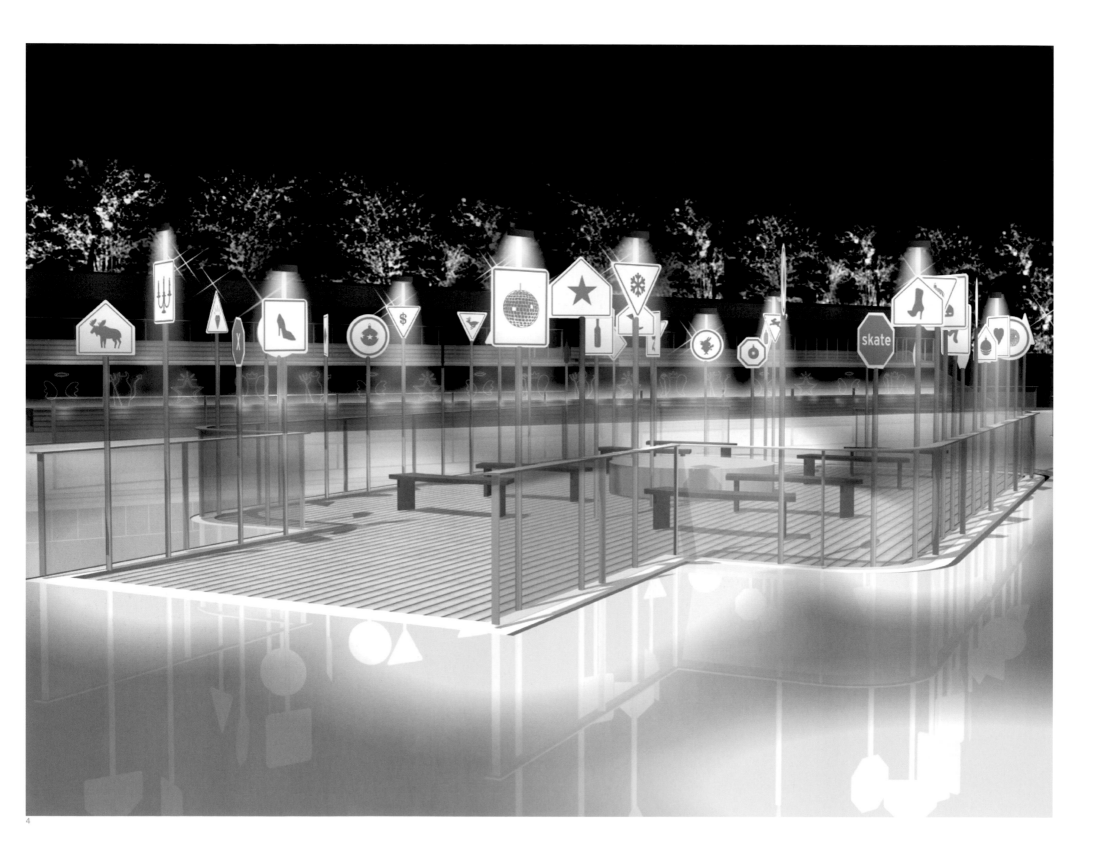

4

4

Winter Wonderland
© Walkerhill ice rink
Ice-skating park
2005

*An ice-skating park, designed
to cover the swimming pool
at the Walkerhill hotel during
the winter season.*

5–6 & 11–12

5 Minute Candles
2004
Park Plus
Design drawings

7–10

5 Minute Candles
2004
Park Plus
Paper, aluminium, candles

*These ten small pocket
candles are designed to
celebrate happy moments,
and can be brought
out at the last minute if
you have forgotten
someone's birthday!*

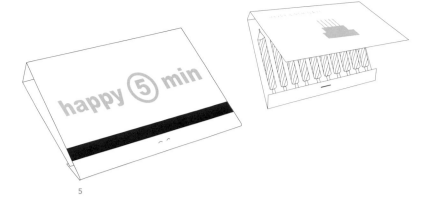

5

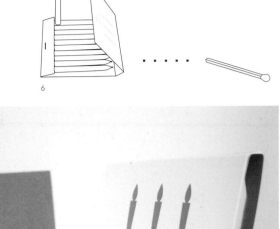

6

7

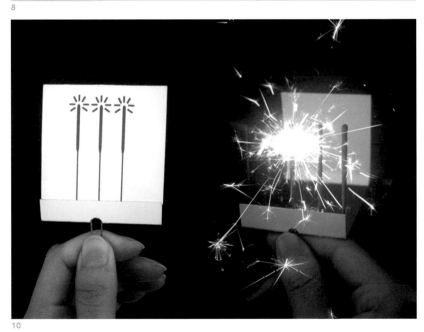

8

9

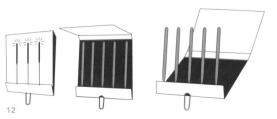

10

11

12

Russell Pinch, London

1

Russell Pinch's most iconic and photographed design to date is the TWIG bench (2005). It is a bench made from thin coppiced hazel branches held together with unseen metal pins. The cut-off circular ends of the piled wood forming the cube is both natural, irregular and visually pleasing, and makes it look as if a never-ending strip of stacked hazel twigs has merely been cut into manageable sections. As a piece of furniture it has a density and a presence to it without being at all showy.

Pinch stands out from the design crowd for the craft skills and knowledge he brings to his work, skills that used to be a mainstay of British furniture-making. His work uses natural materials, and subtly melds together an English traditional style with a more modern aesthetic. His unforced sense of proportion and understanding of process has gained him many admirers. In fact his talent was spotted as soon as he graduated from Ravensbourne College of Design in London. In his early twenties he became Terence Conran's design assistant and was soon a senior product designer for the Conran Group. Following this he co-founded a brand design agency before finally setting up PINCH in 2004 to concentrate on furniture, products and interiors.

PINCH now designs products for companies such as Neal's Yard, Linley and Heal's. And the link with Conran has not been severed either: Pinch's WAVE sideboard (2004) is part of the Content collection at the Conran shop. This piece is a wave-fronted oak cabinet with three drawers in the centre and cupboards with adjustable shelves on each side.

Pinch is using his developing and individual voice to let us know that there is still a strong place for crafts-based natural products in the design world. **Tom Dixon**

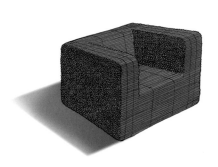

2

3

1–3

TWIG
Bench and cube
2005
PINCH
Coppiced hazel

This piece is handmade from what is normally a waste product. With cleverly hidden nails, the TWIG bench offers a unique aesthetic.

4

Maiden
Table and stools
2005
Benchmark Furniture
Solid green oak

The stools and the base of the table are turned from a solid piece of green oak, which will age over time, adding further character to this set of furniture.

Russell Pinch

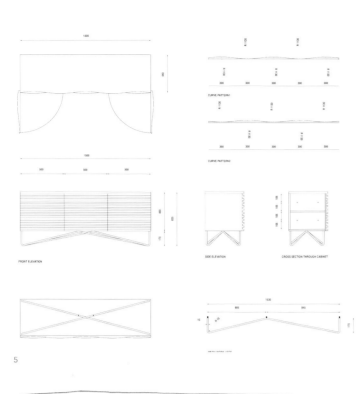

5

8

6

7

5–8

WAVE
Sideboard
2004
Content by Conran
Spray-painted
beech/oak/walnut

*Chest of drawers with
wave detailing on the front
to disguise the doors.
Comes with either ski or
straight legs.*

9–11

Avery
Chair
2006
SCP
Oak

*Simple yet sophisticated
black dining chairs made
with great attention
to proportional detail.*

9

10

11

1

2

3

4

POLKA, Vienna

Q: *Tell me what you do.*
A: *We do product design, we call it product pleasure, and we work in different areas, from furniture and consumer products to interiors.*

During the course of an interview, Marie Rahm and Monica Singer came up with this answer, which provides, in a way, a key to their work. At the risk of opening up a minefield of political correctness and gender issues, this could be regarded as an ultimately feminine attitude, but it has to be acknowledged that the work of Rahm and Singer is imbued with a clear sense of femininity.

Rahm was born in Munich and Singer in Salzburg, both in 1975, and they studied at the University of Applied Arts in Vienna. Rahm also studied at the Royal College of Art in London, but they are now currently based in Vienna. In 2004 they set up their own design company, naming it POLKA: 'We like the idea of being a little bit irritating. And the polka is a very fast and energetic dance.' Their intention to act as agents provocateurs runs through every one of their pieces, many of which promote interaction on the part of the consumer. The Cut light (2002), includes a pair of scissors, inviting the owner to cut the lamp's shade into whatever size and form he or she desires. Likewise, Breadbasket (2006) allows the user to freely shape his or her own basket, whose main, structuring element is the napkin usually placed over a standard breadbasket.

Another piece suggesting exchange and pleasure is POLKAchair (2005), a rocking chair that responds to the user's body movement, with its three regular and one shorter feet. In a more subtle and ambiguous but also slightly disturbing approach, POLKA developed Tattoofurniture (2004). Tapping into customization and decoration, and making the surface an equivalent for skin, the result was, as the designers commented, a kind of 'irritating familiarity'.

POLKA wants to uncover the beauty and the extraordinary hidden beneath the banal and the ordinary. As POLKA casts its own special light on the everyday, our surroundings seem to take on a liveliness that is as amusing as it is unsettling. **Guta Moura Guedes**

1–4
Tattoofurniture
2004
POLKA
Natural leather

This range of furniture uses the tattooing process as a means of decoration, exploring the idea of combining old and new techniques.

5–7
Josephine
Water jug
2006
Lobmeyr
Musselinglass (mouthblown)

This elegant water jug comes complete with a glass that fits perfectly into its neck to make for an interesting sculpture on any desk or bedside table.

8–10
Breadbasket
2006
POLKA
Fabric, wire mesh

A napkin that can be transformed into an attractive breadbasket.

5

6

7

8

9

10

11

15

16

12

13

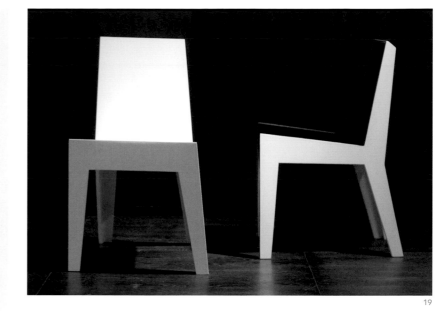

19

11–16

Short Set
2005
Herend
Handpainted porcelain

A 4-piece set of mug, bowl and two plates that is decorated with chinoiserie-inspired detailing.

17–18

POLKAchair
2005
Disguincio
Varnished plywood

19

Quattropolka
Chair
2005
Disguincio
Varnished plywood

The POLKAchair is a 'dancing chair' that allows you to shift around, as one leg is shorter than the others. Quattropolka is POLKAchair's sibling chair that leans towards it, but keeps all four feet firmly on the ground.

1

ransmeier & floyd, Eindhoven

ransmeier & floyd was founded in 2003 by fellow Americans Leon Ransmeier and Gwendolyn Floyd, who originally met in the hardware store at Rhode Island School of Design, where Ransmeier was a student. When Floyd decided to study design at the Eindhoven Design Academy, they both moved to Europe. 'The rich design culture and central location of The Netherlands have made it an ideal working place for us. Although design awareness is growing among American consumers, there is still a lack of high-quality design manufacturers there, relative to Europe. For that reason alone it is best for us to be based here.'

A closer look at the work of ransmeier & floyd, whose projects have been taken on by Danese, Droog, Authentics and Artecnica, among others, immediately yields three mainstays of their creative process: observation, clarity and pragmatism. Their designs are based on penetrating and detailed observations of various phenomena and behaviours in contemporary life. Likewise, one can sense a strong, implicit desire to simplify procedures and clarify functions. The D.I.Y.M. lamp (Do It Yourself Modern, 2005), for instance, is a striking and beautifully simple lighting solution. It plays with the idea of the sizes and shapes of standard light bulbs and lengths of electrical wire found in each and every house-hold, allowing users to come up with their own do-it-yourself product, effortlessly and effectively.

In projects such as the Gradient dishrack (2005) and the Worn In doormat (2005), an ingenious, enquiring look at our everyday life is the basis of alternative, streamlined solutions capable of bringing truly extraordinary functionalities to objects. In Snow side tables (2006), for example, this intense scrutiny is taken to a more poetic level, though never to the detriment of the object's functional capacity.

ransmeier & floyd knows exactly where it is heading: 'We want to draw people in towards an engagement with our products through their conscious consideration of them, whether their ruminations are based on visual, material or cultural elements. We strive to do this without heavy-handed pictorial symbolism or quickly passing witticisms. To create designs that describe themselves inherently without resorting to overt semantics.' **Guta Moura Guedes**

1
Crop
Bookshelf
2005
Prototype
Birch, birch plywood, laminate

The horizontal flaps attached to this bookshelf visually homogenize the different book sizes and can be lifted to retrieve the desired volume.

2
Worn In
Doormat
2005
Prototype
Polypropylene, plastic bristles

The profile of this doormat is cut to resemble a brush that was once uniform in length but has been worn in, creating a pleasing visual effect. The shorter bristles are for everyday use, while the longer ones are useful for removing extra dirt when necessary.

3

3

Gradient
Dishrack
2005
Prototype
Polypropylene

*The dense concentration
of plastic rods that is useful
for holding cutlery and
thin glassware, gradually
and evenly fades to
a spacing large enough
to accommodate bowls,
plates and pans.*

4

Snow
Side table
2006
Limited edition
Aluminium

*Thick, highly reflective
road paint is applied to
these small side tables,
evoking a blanket
of freshly fallen snow.*

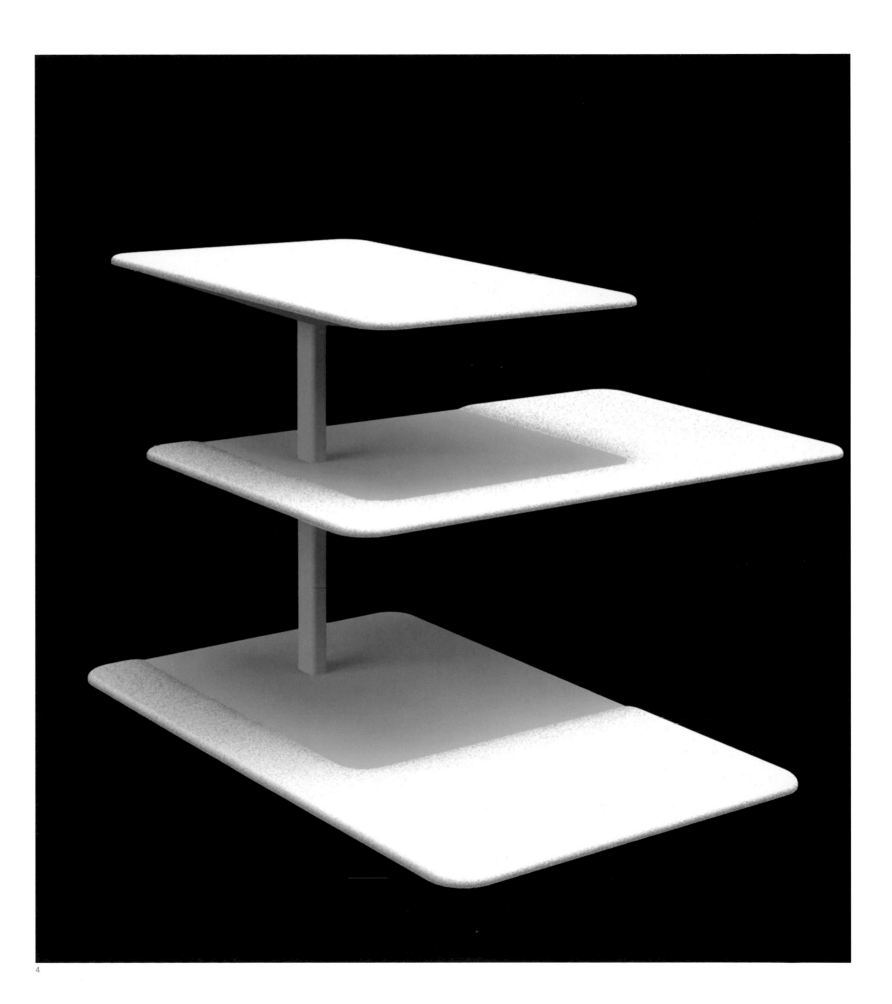

1

2

3

4

Adrien Rovero, Lausanne

Adrien Rovero's brain is teeming with ideas. Observing everyday life with humour and distance, he captures iconic situations and infuses them with poetry. Rovero is one of the most gifted designers to come out of the École cantonale d'art de Lausanne (ECAL), brilliantly embodying the playful and clever approach that is one of the school's trademarks. His projects have a distinctive visual aesthetic which makes them immediately recognizable.

After studying as an interior designer, Rovero received graduate and post-graduate degrees from ECAL, regularly winning design competitions and exhibiting collective and solo projects. A relentless worker, he now undertakes numerous projects under his own name.

Rovero's personal projects reflect the very particular way in which he perceives people and objects. Drawing most of his references from current habits and behaviours, he then works and plays on incidental gestures, analysing and distorting them. With *Dés-ordre*, 2004 (Un-tidiness), the theme for his graduation project, he designed a series of unusual objects that refer to the small domestic incidents that happen in our daily lives. For instance, the end of his sofa simulates the fall of a chair while the surface of a side table is imprinted with the shape of objects left lying about. Slippers come in threes, just in case you lose one.

Rovero's work with carpet for his *Dés-ordre* project inspired further research on woollen flooring, resulting in Size? (2006), a series of modular carpet patterns and assemblage. The Chain modular carpet (2006) can be extended endlessly in a surrealis-tic way; likewise, Belt (2006) offers multiple combinations that can be adapted to the floor size. Other examples of Rovero's witty distortions are the Brush Couple (2003) and VD 003 (2006) – a bike rack shaped like a car. Understandable at first sight, both 'design puns' are cleverly developed to become usable, practical objects.

Rovero epitomizes the model of a single designer who is able to create industrial projects that are inventive and iconic, and yet can be successfully manufactured. **Didier Krzentowski**

1–3
Belt/Chain
Model/Carpet
2006
Tisca France
100% tufted wool

*Design-stage models
of the carpets*

4
Chain
Carpet
2006
Tisca France
100% tufted wool

5
Belt
Carpet
2006
Tisca France
100% tufted wool

*This range of stylish
carpets is not just a fashion
accessory; it uses the idea
of linkage either in chain or
belt form to create carpets
that can adapt neatly to
the space in which they are
being used.*

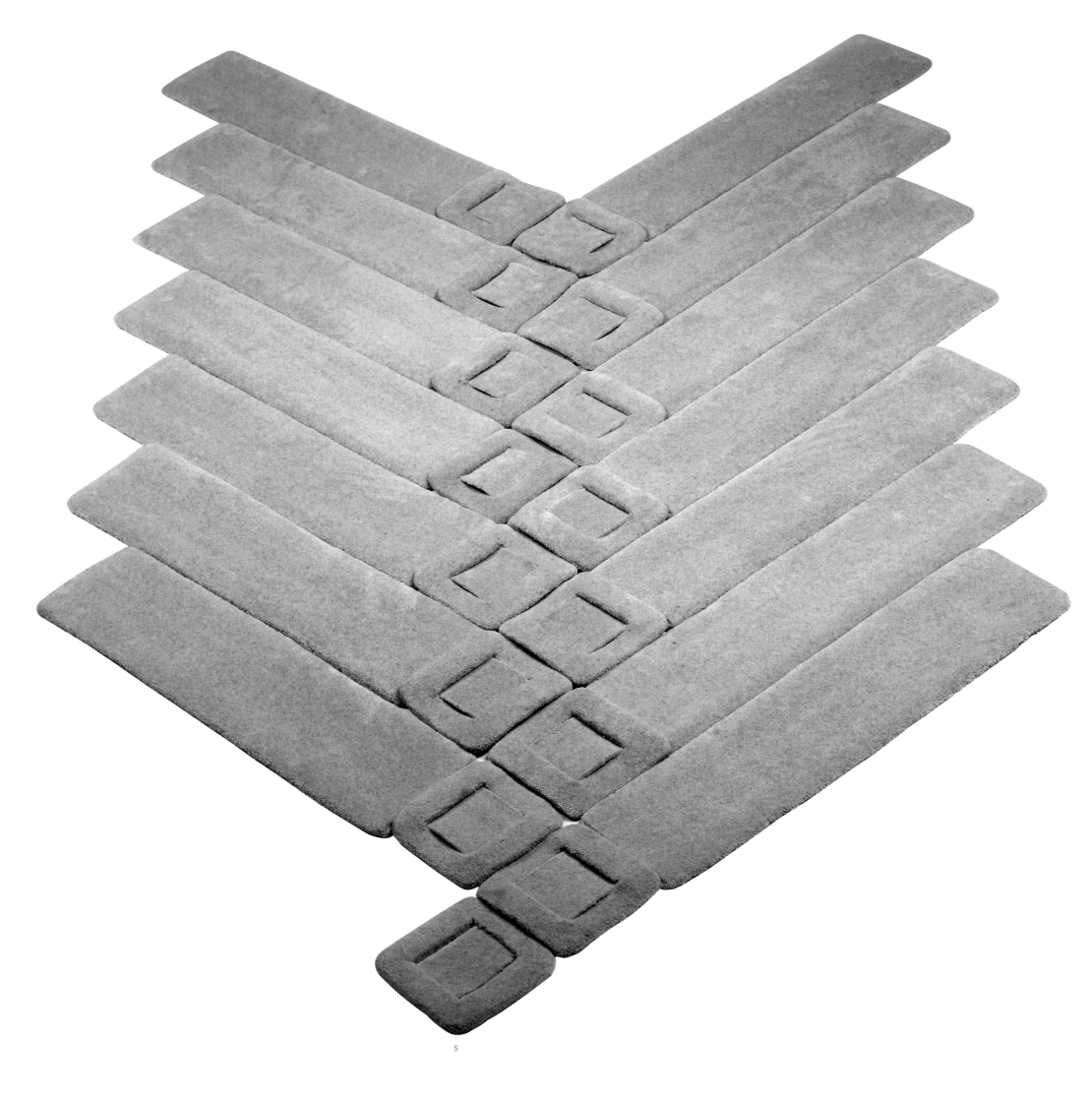

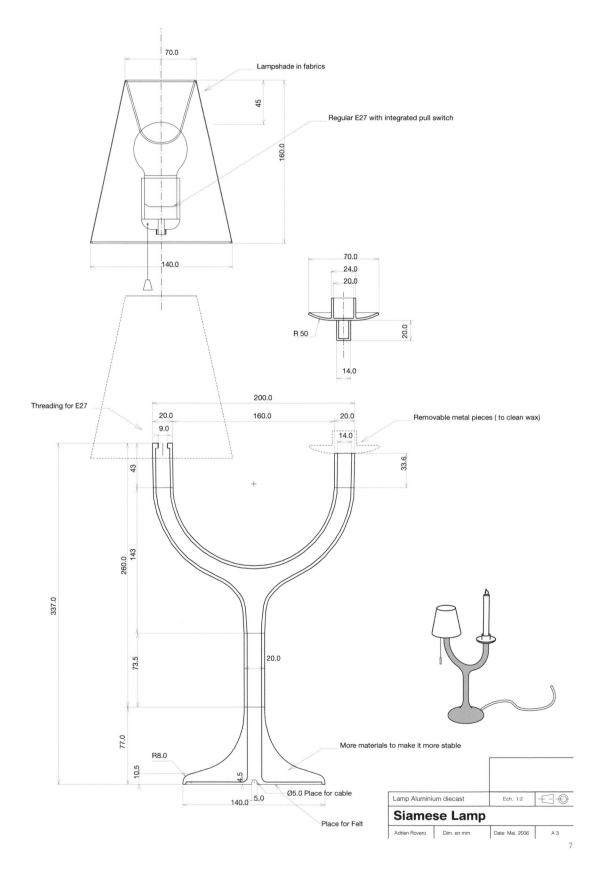

Lampshade in fabrics

Regular E27 with integrated pull switch

70.0

45

160.0

140.0

70.0
24.0
20.0

R 50

20.0

14.0

Threading for E27

200.0

20.0   160.0   20.0

9.0

14.0

Removable metal pieces (to clean wax)

43

33.6

260.0   143

337.0

20.0

73.5

77.0

R8.0

10.5

4.5

140.0   5.0

Ø5.0 Place for cable

More materials to make it more stable

Place for Felt

Lamp Aluminium diecast		Ech.: 1:2	
Siamese Lamp			
Adrien Rovero	Dim. en mm	Date: Mai. 2006	A 3

6–7

Candlelight
2003
ECAL
Ceramics, fabrics, metal

A product that does exactly what it says, this is both a candle holder and a lamp, depending on what mood you are in.

8–10

Basketstool
Stool
2005
Prototype
Chrome-plated steel

Part of the series Supermarket, this handy stool is made from an upside-down shopping basket.

11–12

Brush Couple
2003
Die imaginäre manufaktur
Wood, metal

This is a brush and shovel combo that fits neatly together using magnets to ensure you never lose the other half.

8

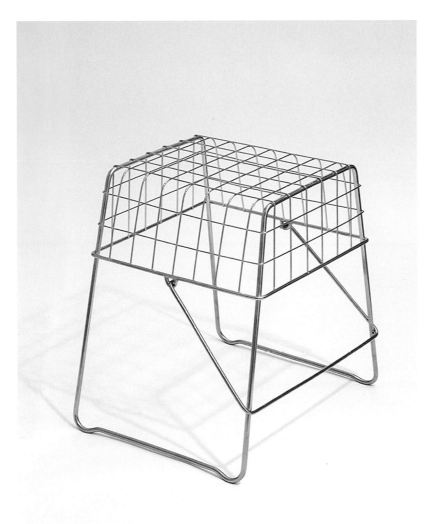

9

10

11

12

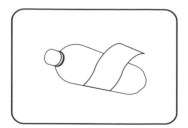

Alejandro Sarmiento, Buenos Aires

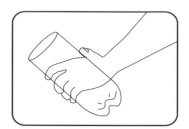

Argentinian designer Alejandro Sarmiento graduated from La Plata National University, Argentina, with a degree in industrial design and has worked independently ever since. His projects encompass different design fields, from product research to development and implementation. He is the official chairman of Production Studios I and II, and academic assessor for design at Palermo University, Argentina. He also oversees Torcuato Di Tella University's postgraduate Aluminium Technology and Uses programme.

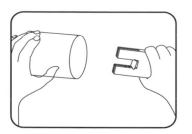

The products developed by Sarmiento can be divided into two groups: those made from standard components that are readily available; and those made up of disposable materials such as PET and inner tubes. Sarmiento investigates the nature of the product or material in order to discover its essence and preserve its nature in the new piece. 'I do not change the material's essence. I grant it another possibility, through design,' he explains. This is the case with the Araña lamps (2004) created from pieces of a sugar bowl, and Miniufo (2005) made from parts of a lorry hazard light. Out of context and with a new function, these components acquire a different identity until, within the final design, it's almost impossible to recognize their original use.

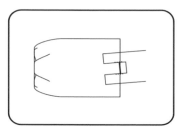

The same applies to Sarmiento's use of disposable materials, especially in the case of PET. Sarmiento developed a simple tool that allows plastic bottles to be transformed into ribbons of several widths that have an excellent finish. The final products are most varied and include chairs, bags and placemats.

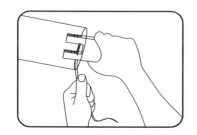

In 2005, again recycling plastic bottles, Sarmiento created Querencia (2004), an installation where standard plastic balls are linked to the necks of the bottles, forming a semicircle. The aim of this project was to recover lost personal space at home by creating an intimate 'temple' or 'nest' area. This project, very much on the border between art and design, was presented in Rio de Janeiro's 'Panorama Internacional do Design' exhibition, curated by Paola Antonelli and Nicola Goretti. **Maria Helena Estrada**

1–3

Supernova
Lamp
2003
IndustriAL Standard
Recycled PET bottles, bulb

Uniform strips of old plastic bottles are used to create innovative lighting designs.

4–7

Querencia
Dodecahedron modular system
2004
IndustriAL Standard
Recycled PET bottles, inflatable balls

Unique sculptures created from recycled materials.

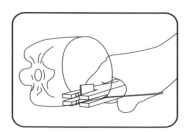

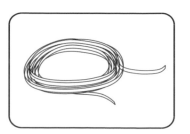

4

5

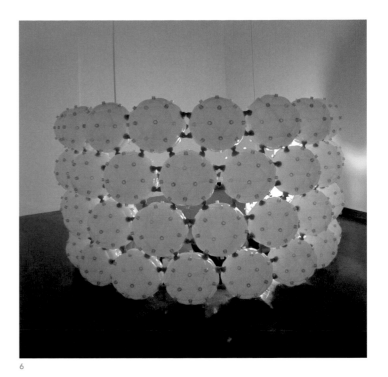

6

7

8

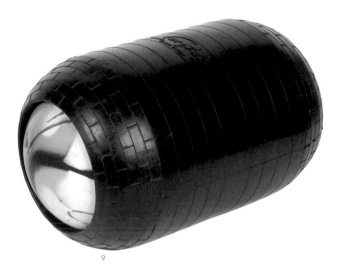

9

10

8–11

Cilindro Puf
2003
IndustriAL Standard
Recycled rubber,
inflatable balls

*A comfortable seating
arrangement can be made
out of this product, or it
can be used on its own as
a casual rest.*

12–16

Ruberta Iron
Home furniture
2003
IndustriAL Standard
Recycled rubber, steel

*An unusual family of iron
stools, chairs and tables.*

11

12

14

15

13

16

1

Tables LaChapelle
Side table
2006
David design
Polywood, lacquered steel

*The table was inspired by
the litter bins in the Jardin
du Luxembourg in Paris,
where the designer spent
time in her childhood.*

2

Suspension Plissée
Lamp
2003
Cappellini
Metalized double-sided
fabric, incandescent bulbs

*This pleated hanging lamp
was originally created for
a personal exhibition at the
Museé des Arts Décoratifs
in 2003.*

3–5

Grande Lampe Plissée
Lamp
2002
Cappellini
Fabric, wire structure

*Standing at over 2 metres
(6.5 foot) tall, this sculptural
pleated lamp emits a soft
light through its diaphanous
hanging fabric.*

Inga Sempé, Paris

Inga Sempé's world is elegantly puzzling. A certain timeless classicism emanates from her stunning objects: simplicity, wit and refined humour with a hint of provocation seem to be a few of the mysterious elements behind her design approach.

Born in Paris where she still lives and works, Sempé started out as an assistant to Marc Newson and then to André Putman before setting up on her own in 2000. Her first main projects included an articulated lamp, and an analogue and digital clock that she developed with the French Valorisation de l'innovation dans l'ameublement (VIA). After a year spent at the French Academy in Rome, where she began working for Cappellini, she returned to Paris and was later awarded the Grand Prix de la Création in 2003. The same year, she had her first solo exhibition at the Musée des Arts Décoratifs, giving her the opportunity to display her works as a coherent whole.

Sempé's world is inhabited by a series of intriguing projects: her ghostly pleated lamps – Grande Lampe Plissée (2002) and Suspension Plissée (2003) – dressed with folded fabric, seem to live a stylish life of their own; enigmatic shelves, covered by industrial brushes that simultaneously open and close, seem to absorb your hands as you place objects inside the space; her Upholstered Metal Chair (2003), ingeniously padded with foam and upholstered in leather, provides unexpected comfort; Long Pot (2004), a flower pot with four legs, looks as if it could stand up and walk away; Table Lunatique (2006) rises up and down according to your wishes; and iconic Baccarat wine glasses are turned upside down to become mischievous candleholders.

For Sempé, ideas and drawings always precede materials and ingenious systems. Yet materials that traditionally have industrial uses find new, sophisticated and seemingly appropriate lives in her designs. Her unexpected and witty technical solutions, elegantly understated, leave a lasting impression. **Didier Krzentowski**

3

4

5

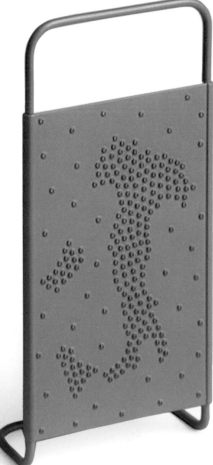

6

Égouttoir et Rape à Fromage
Colander and cheese grater
2005
Prototype

*These items were made or
the exhibition 'Souvenir
d'Italie' in 2005, organized
by Casa da Abitare and Alessi.*

7–9

Table Lunatique
Telescopic oval side table
2006
Ligne Roset
Lacquered MDF, steel

*The height of the table can
be adjusted by means of
a jack that is operated by an
articulated ring set into the
centre of the table top.*

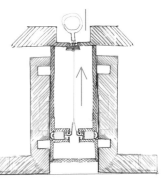

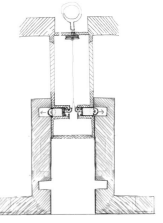

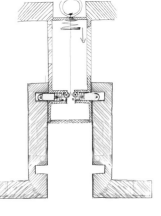

9

7

8

Wieki Somers, Rotterdam

Wieki Somers' objects have an enchanting beauty and dreamlike quality. The sophistication of Somers' aesthetic is a demonstration of her impressive craft skills and of her taste for experimenting with materials, both contemporary and ancient.

After graduating from Eindhoven's Design Academy in 2000, Somers investigated the properties of two kinds of polystyrene foam to produce Muffin (2001), a humorous stool that is rigid at one end and soft at the other. She also started collaborating with Droog Design, for whom she designed the Trophees (2004) based on glasses by Arques International, and in 2005 was rewarded with a solo exhibition at the National Glass Museum in Leerdam.

On a conceptual level, Somers' use of materials influences our perception of her designs as she shifts the connections between patterns and materials, thereby confusing our reference points. For example, her Mattress Stone Bottle (2002), with its flowery print imprinted on the ceramic, encourages us to believe it is a soft, embroidered mattress.

Stories also play a major part in Somers' realizations, inspired by common associations we all have with objects as well as her own very vivid imagination. Once such notion is that 'floating on water or bathing in water deal[s] with so [many] similar feelings'. As a result Somers created two boat-shaped bathtubs, Frozen Bath (2001) and Bathboat (2005), that allow one to drift away on dry land. And in her vase Blossoms (2005), seeds are planted in the ceramic flower buds, which then germinate and become vegetal extensions of the ceramic, as though in a fairy-tale where vases would grow their own plants.

Somers' High Tea Pot (2003) brilliantly epitomizes her cleverly extravagant spirit. Its porcelain skull, so finely cast that it becomes transparent, and its fur cover express the paradox of the upper-class rite of killing animals as a civilized hobby. It is simultaneously exquisite and barbaric. For all her fantasy-based ideas, Somers none the less keeps a critical watch on our society and its mores. **Didier Krzentowski**

1–3

Bathboat
2005
't Vliehout boatbuilders
Wood

*By turning a small boat
inside out, it becomes
a bath and as a result the
experiences of sailing
and bathing are conflated.*

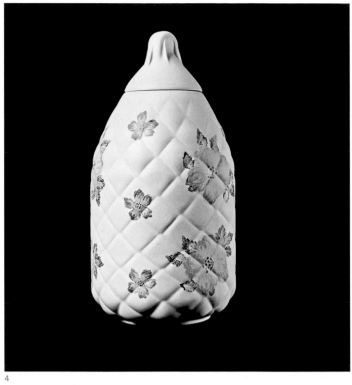

4

5

6

7

8

9

10

11

4–5

Mattress Stone Bottle
2002
E.J. Kwakkel
Ceramic

A fishnet tied around a balloon created an unexpected shape that was the inspiration for this bottle. It visually refers to an embroidered mattress or water buoy.

6–11

Blossoms
Vase
2005
Cor Unum
Ceramic glass

This vase allows living flowers to blossom through its ceramic flower-bud openings.

12

Trophees – Serie 2
2004
Arques International
Crystal glass

One of Arques' common drinking glasses hangs in a cylindrical vase with the bottom cut off. The two glasses are then joined together at the top to form a bizarre drinking vessel.

13

Trophees – Serie 1
2004
Arques International
Crystal glass

Two standard drinking glasses have been ground off in such a way that they fit precisely together.

12

13

Alexander Taylor, London

Alexander Taylor established his eponymous, now highly successful, London design studio in 2002 and since then has produced products for several high-profile manufacturers, including Zanotta and E&Y Tokyo. His work is not only practical and functional – although these qualities are essential to him as a designer – but also displays a distinctive lightness of touch, and is infused with a quiet and very British sense of humour.

Tank (2006), his folded aluminium suspension light, is an apt example. The largest version of the light has an imposing base diameter of 0.72 m x 0.5 m (29 in x 19 in) giving it a real presence. This bold approach is offset by the obvious quality of the object, and the way that Taylor has used the interlocking surface of the aluminium, strong colours and the illumination from the bulb to create a surprisingly subtle light source. This clever use of materials shows the craftsmanship that is a signature of Taylor's work. He understands the physical properties of his chosen materials and is essentially good at making things, a quality often absent in today's young designers.

His Fold light (2005) plays with ideas of form and function. The lamp employs an archetypal silhouette, which from a distance makes it look like a 2-D flat shape, but is in fact a working 3-D object complete with a visible, brightly coloured electric wire that conversely highlights its functionality.

His interest lies in creating 'classic' objects that are still decidedly modern, qualities you can also see in his wooden furniture pieces like the Husky bench (2005) for E&Y. Made of natural and black-stained wood (nyatoh), inspiration for the bench came from the traditional toboggan, which Taylor simply adapted. He is versatile, ingenious, smart and productive, and most of all understands the manufacturing possibilities of the materials he uses. But he doesn't shout about it; his instinctive restraint makes his product design both individual and accessible. **Tom Dixon**

1

Antlers
Coat hook
2003
Thorsten van Elten
8mm steel wire

*These antler-shaped
wire hooks are a fun way
to hang up your coats.*

2–3

Butterfly Table
2006
Zanotta/E&Y co-edition
Bent plywood (natural
or wengé stained),
10mm toughened glass

*This table was developed
out of the designer's
research into producing
a modular table base
that could support a heavy
glass top.*

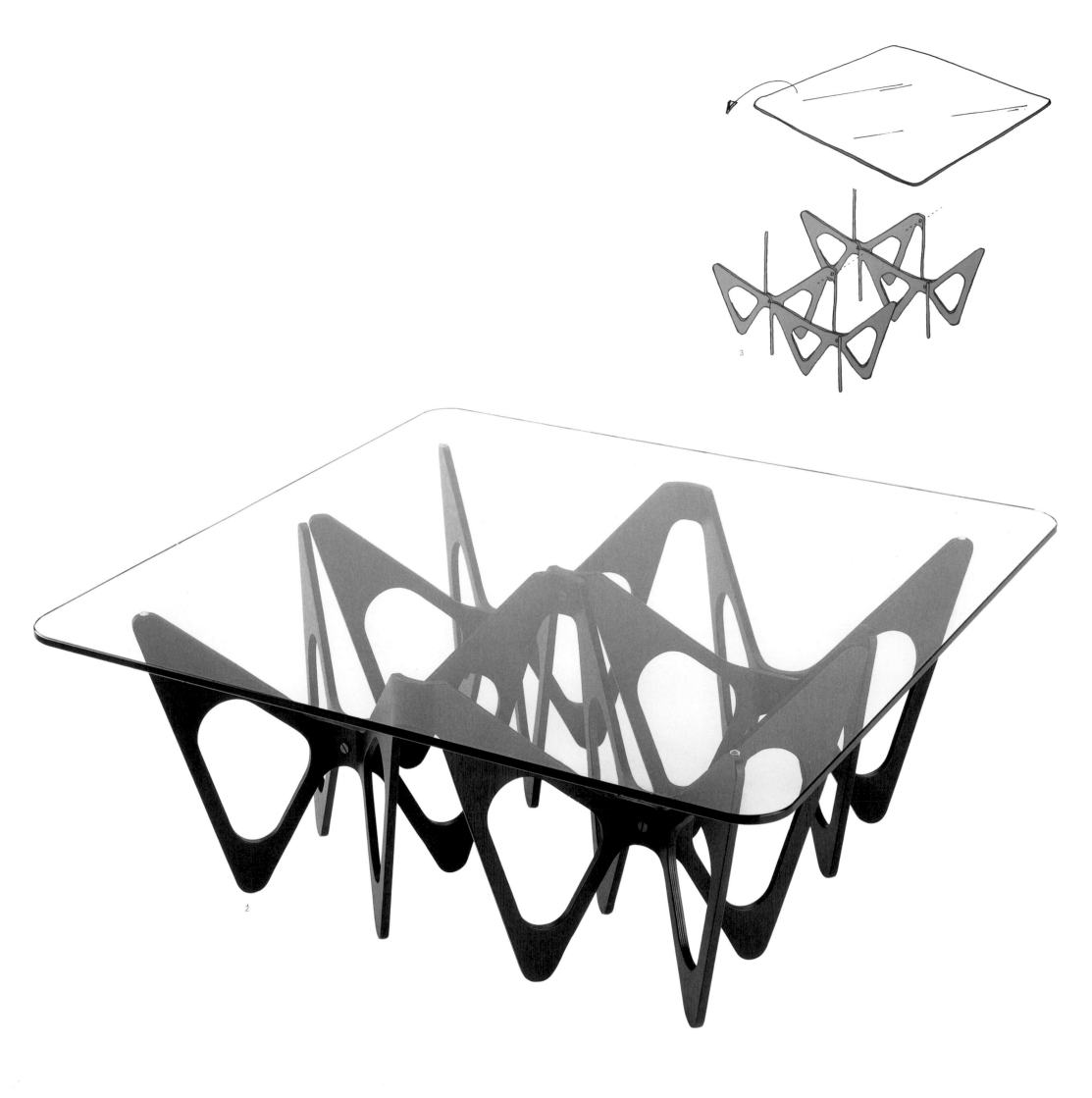

Husky
Bench
2005
E&Y
Natural or black-stained
wood (nyatoh)

*Inspiration for this bench
came from the traditional
toboggan.*

6–9

Fold
Light
2005
Established & Sons
Powder-coated mild steel
(small); aluminium (medium
and tall)

*The smallest model of this
lamp is folded from a single
sheet of steel.*

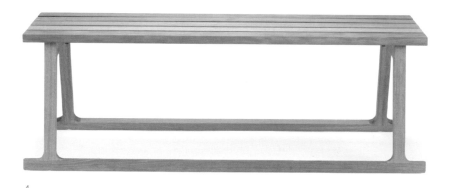

4

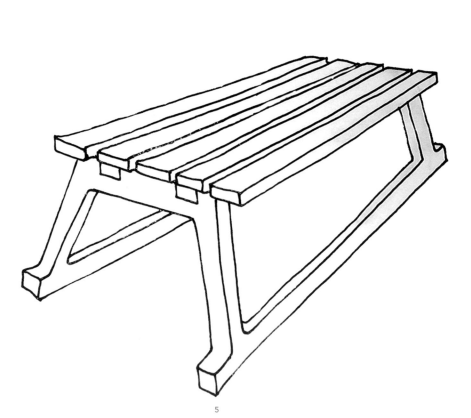

5

6

7

8

9

Carla Tennenbaum, São Paulo

1

2

Carla Tennenbaum's research focus is the aesthetic reassessment of non-recyclable residues and other traditionally under-utilized materials for the development of products with visual impact. Her research causes her to reflect on the possibilities of creating objects that are 'useful' to the environment and, as a result, EVA (ethylene vinyl acetate) immediately became an interesting material for her design practice.

Kinetic Paintings and Kinetic Spirals are made by hand from residues left over from producing EVA. The manufactured product is created through breaking down and re-forming the residue through a mortise system designed especially for the purpose. Colour variations are due to the unpredictable nature of the residue, making each piece unique.

The amalgamation of colours from the EVA particles generates abstract visual patterns, impregnated with rhythms and meanings. Kinetic Paintings suggests an interface between design and plastic arts, by promoting an aesthetic play in which possibilities for composition are infinitely multiplied. They are designed without the use of glue or other external agglutinates. Parts are mortised on the canvas, combining an economic production process with positive environmental impact.

Initiated in 2000, the Festa Pill and sofas (2004) were developed to create a product that would incorporate and optimize massive amounts of EVA wastage. The handmade Kraft armchair and cubes (2003) form a self-assembling, lightweight and comfortable furniture line that easily accommodates the varying dimensions and weights of different people. The pieces can be covered by a corn-based resin or left in their natural paper state, emphasizing the paper's texture, softness and warmness.

Tennenbaum proposes to continue exploring common areas between graphic arts, craftsmanship and object design, deconstructing and re-aggregating industrial residues – especially multicoloured EVA – in her work. She has already received a Latin American Design Foundation award for her 2003 exhibition held in Holland, and in 2005 she took the Design 21: International Design Award's first prize, as well as participating in exhibitions in Canada and Portugal.
Maria Helena Estrada

1
Kraft
Chair
2003
Self-production
Kraft paper, organic glue

Seen here with the Festa Pill (2004), a seat made from, ground white EVA refuse with a PVC cover.

2
Kraft
Chair
2003
Self-production
Kraft paper, organic glue

Pictured with the Kraft Cube (2003) and the Kinetic Tapestry (2006).

3
Kraft
Chair
2003
Self-production
Kraft paper, organic glue

Self-supporting chair made entirely out of paper and glue. It can support the weight of a person up to 115 kg (253 lbs).

4

5

Carla Tennenbaum

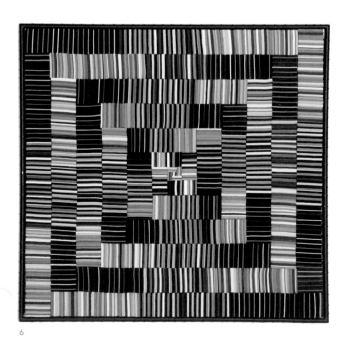

6

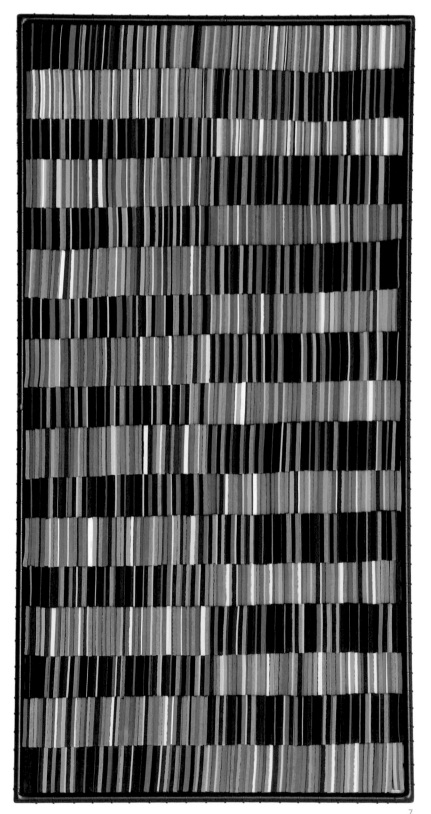

7

Kinetic Tapestry
2006
Rai Mendes
EVA industrial refuse,
metallic screen

*Compositions created
through the cutting and
modular attachment of
EVA refuse over a screen.*

8

Transoptic Screen
2003
Self-production
Paper tubes, organic glue

*An opaque wall when
looked at from the front,
the Transoptic can actually
be looked through from a
transversal (diagonal) angle.*

9

Transoptic Screen – on wheels
2003
Self-production
Paper tubes, metal
structure, wheels

*Formally identical to the
Transoptic, the addition
of wheels and a metal
frame make this version
easy to move and increases
the object's durability.*

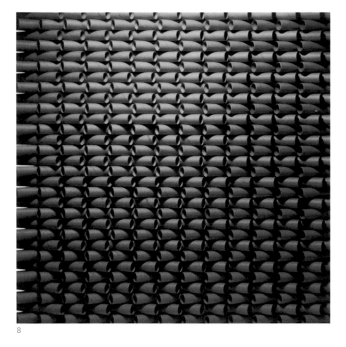

8

9

TONERICO:INC., Tokyo

At the 2005 Salone Satellite in Milan the design world's attention was caught by this up-and-coming design office when it unveiled its numerical light sculpture, MEMENTO (2005). It is formed from reflective (or white) laser-cut steel numbers, which surround a central light source. The resultant intriguing play of light through the overlapping numbers engages with concepts of light, shadow and reflection.

MEMENTO went on to win the prestigious Design Report Award, which was the start of exciting opportunities for its young designers. The main design office members are Hiroshi Yoneya, Ken Kimizuka and Yumi Masuko. The design work is done by Yoneya and Kimizuka, while Masuko is the interior co-ordinator/conceptual thinker whose role is to stand back and take an objective view. For example, the numeric concept for MEMENTO was her idea.

One of the key influences for TONERICO:INC.'s design work and approach comes from the Japanese designer Shigeru Uchida of Studio 80. Yoneya and Kimizuka were both working as designers at Studio 80 when they met Masuko and decided to set up as an independent design office in 2002. At the time, they wanted to acknowledge and put into practice Uchida's teachings about the basic usefulness and functions of design elements. Although their designs still rely on these ideas, more recently they have been experimenting with ideas that cross boundaries, resulting in designs such as MEMENTO. They intend their work 'not simply to be useful and functional but also to influence people emotionally'.

While a good number of TONERICO:INC.'s projects are centred on award-winning shop designs, those offering a modern interpretation of 和 (Japanese aesthetic sense) have received wide acclaim. Although it has been designing furniture to match pre-existing spaces, the recent award for MEMENTO has stimulated a desire to express its true essence in future work. The name 'TONERICO:INC.', taken from the Japanese for 'ash tree', was chosen by the group because ash is their favourite working material. **Chieko Yoshiie**

MEMENTO-LINK
Hanging light
2005
Yamakin Co. Ltd
Laser-cut stainless steel

Exhibited at '100% Design Tokyo' in 2005, this piece shows the designers seeking to visualize the ambiguity of human memory.

3

GUSHA
Side chair
2004
Minerva Co. Ltd
Painted black, white and
burnished stainless steel

4

GUSHA
Stool
2004
Minerva Co. Ltd
Painted black, white and
burnished stainless steel

*Experimenting with
folded paper, the designers
accidentally crushed it. There
was an unexpected beauty
to this that became the
inspiration for the GUSHA.*

5

Hands
Bowl
2003
Ceramic Japan
Ceramic

*The bowl is based on the
idea of hands being the
origin of all tools and utility
objects; by holding out
one's hands, things can
be obtained.*

6

TONERICO: FOR TWO
Side table and cups
2003
Maki Nakahara
Ceramic

*Having tea is a sign of
love and friendship and
this work embodies
those feelings by defining
the act of drinking tea.*

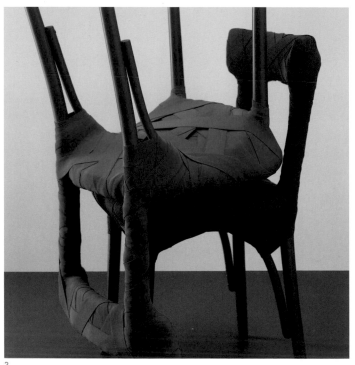

Peter Traag, Antwerp

From his studies in 3-D design at the Hogeschool voor de Kunsten in Arnhem and at the Royal College of Art in London, Peter Traag has acquired undeniable formal skills, a fascination with visually powerful objects and a conceptual approach to furnishings. Following up the research interests of designers such as Gaetano Pesce and Roger Tallon, Traag fosters a proactive involvement in technical and industrial processes in order to be better placed to change and improve them.

Traag also enjoys introducing a certain formal randomness to his designs. For example, when a reinforced fluorocarbon polyester thread fabric is filled with polyurethane foam, as is the case with his Sponge Chair (2004), the foam oozes through the fabric in an unpredictable way. The chair emerges from the mould creased and compressed, creating a unique and distinctive object with each production. His recent series of chairs and tables likewise accommodate accidents and modifications during production and use, which he sees as an integral part of the design.

While seeking to define forms that can be made under industrial production conditions, Traag also seeks to replace the generic constraints of industrial rules and limitations with specific qualities that result from mastering (or even subverting) them. The 130% armchair (2003) and the Rubber Chair (2003) demonstrate this perfectly and show the diversity of Traag's aesthetic typologies and expressions.

Traag's Mummy beach chair (2005), for Italian manufacturer Edra, shows Traag both utilizing and playing with traditional manufactured forms. A standard beechwood chair is wrapped in 60 metres (196 feet) of brightly coloured fabric, and polyurethane foam padding is used in the seat and backrest. The resulting and surprising softness of the chair references his self-expressed interest in upholstery as a crucial part of his work: 'I think of upholstery as a design skin.'

Customization, the creation of one-off pieces, industrial processes, historical references and contemporary subcultures are just some of the sources from which this talented young designer draws impetus and inspiration. **Pierre Keller**

1–4

Mummy
Beach chair
2005
Edra
Padding, fabric

A plain wooden chair wrapped in elastic ribbon to give it softness.

5–13

Sponge Chair
2004
Edra
Sponge, fabric, polyurethane, steel

A chair made by making use of a specially constructed mould that enables the (de)formation of the seat during polyurethane injection.

14

15

14

Rubber Chair
2003
Peter Traag
Rubber, miscellaneous
materials

*This chair is made out
of discarded materials and
objects, and finished off
with a thick-dipped coating
of rubber.*

15

Dip
Table
2006
Pallucco
Aluminium, steel
In collaboration with
Sebastien Noel

*This table explores the
possibilities of multiple
anodizing to create
stunning colour overlays.*

16–20

LTD Collection
Chair
2006
British Council
Fabric, polyurethane, steel

*Chairs derived from
sucking multiple layers of
fabric through a 2-D circle
or hourglass template.*

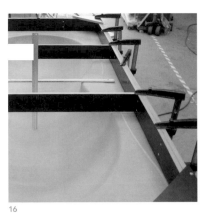

16

17

18

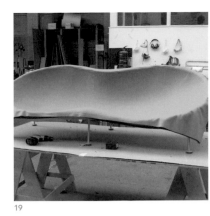

19

20

1–2

Waterproof Vase
2002
Qubus
Porcelain

*A vase in the shape of
a Wellington boot provides
a unique and fun way of
keeping plants.*

Maxim Velcovsky, Prague

Maxim Velcovsky mirrors the viewpoint of a whole generation across Central Europe who witnessed the fall of the Iron Curtain while dreaming of the freedom epitomized by the icons of American consumer society. 'It was as if we wanted to experience the possibility of choice through a Western-style supermarket,' he says. Velcovsky, who happens to possess the kind of looks perfect for an edgy media campaign, applies his intelligence and sharp critical sense to comment on or even modify our relationship with the amazingly visual and glittery world that surrounds us.

A ceramics graduate from the Academy of Arts, Architecture and Design Prague, Velcovsky is also an artist and a communications designer. In fact, communication is a key feature of his work, as is indicated by his position as editor for *Blok*, the Czech Republic's first fashion-arts-design magazine.

Out of the tensions between a personal framework closely attuned to the information society and a strong attraction for consumer culture, and a background of Czech art and industrial tradition, Velcovsky delivers complex work where ambiguity and intellectual games are quintessential. With adept precision, he puts forward both correlations and questions, giving objects contradictory functions or identities. He revisits other people's work, using it as raw material, explaining, 'I don't copy these things. I work with them.' In fact, its like a conceptualized version of sampling, so popular in contemporary music, video and art, and a critical take on 'Nothing is lost, nothing is created, everything is transformed.'

Co-founder of his own company and store, Qubus, in Prague, Velcovsky is making his mark on today's design through his discerning personality and unadulterated eye. 'I live in a fast time when the same form is given different labels in the name of evolution. I'm just reacting,' says Velcovsky. We hope he doesn't stop. **Guta Moura Guedes**

3

4

3–4

Air Vase
2005
Qubus
Ceramics, glass

A decorative vase that sparkles in the light.

5–8

Moneybox
2005
Qubus
Porcelain

A series of ornamental animals, perfect for any mantelpiece, that also serve as a useful place to stash those extra pennies.

9–10

Norwegian Wood Candleholder
2005
Qubus
Porcelain

A faux driftwood candleholder makes a unique centrepiece for any table.

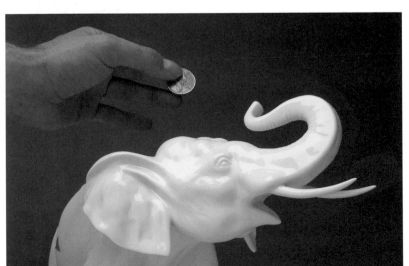

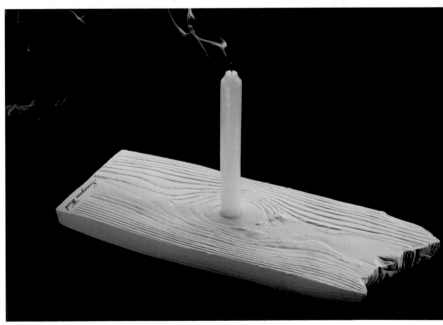

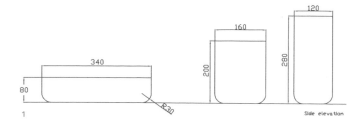

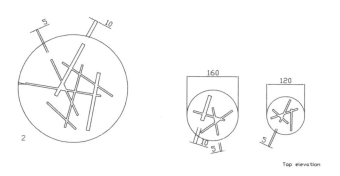

voonwong&bensonsaw, London

Voon Wong and Benson Saw began their successful collaboration in 2001 designing furniture, lighting and products from their London studio. As it is, they are already very prolific, but their work will develop and progress further in the future. Their style could almost be termed minimalist but is far more expressive and sensual than that would suggest. There is a sense of balance about it, and an interest in natural forms, although the end result is distinctly modern.

Considering their recent start they have achieved a great deal, perhaps most notably being nominated for the Compasso d'Oro for their Loop lamp (2001), manufactured by Fontana Arte. This light piece, which is made from stainless steel or painted nickel, has a mesmeric quality and subtly plays with the possibilities of a single light source. The upper and lower halves of the lamp are connected by a spindle on an eccentric axis and as the spherical halves rotate around the spindle so the light is distributed in an ever-changing manner.

Another stand-out piece is the Ribbon Coatrack (2004), where a single piece of painted mild steel is twisted and turned to create four rising columns where coats can be hung. A ring holds the whole piece together and allows umbrellas to be stored there just like a traditional hallstand. It is strikingly architectural in form and at the same time very practical.

Perhaps this conjunction can be explained by the duo's mix of design disciplines: architecture, product design and engineering. The two have pooled their experience and knowledge to create elegant manufactured products that have a striking structural presence and, best of all, work beautifully. **Tom Dixon**

1–4
Landscape Vase
2005
Self-production
Silicone rubber

These vases derive their function from the act of carving into generic cylindrical forms. The complex intersecting cuts are only visible when viewed from above.

5–8
Twig
Flower frame
2003
Self-production
Powder-coated stainless steel

Stainless-steel rods are soldered together in a random fashion to form a frame for flowers. Cylinders serving as flower holders are interspersed between the rods.

5

6

7

8

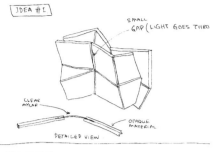

ORIGAMI LAMP

IDEA #1

SMALL GAP (LIGHT GOES THRU)

CLEAR MYLAR

OPAQUE MATERIAL

DETAILED VIEW

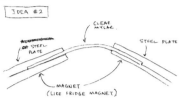

IDEA #2

CLEAR MYLAR

STEEL PLATE

OR STEEL PLATE

MAGNET (LIKE FRIDGE MAGNET)

STEEL PLATE

9 ✳ - THE ENTIRE STRUCTURE IS FLEXIBLE.
- USER CAN "WRAP" AROUND THE LIGHT SOURCE
AT HE/SHE WISH
- FRONT PIECE AND BACK PIECE ADHERE TO
EACH OTHER BECAUSE OF THE MAGNET PLATE

Voon,

Per Vivian's latest rendering.

10

Latest proposal.
(just a crude trial).

Let me know what you think!
Too complex?

Benson.

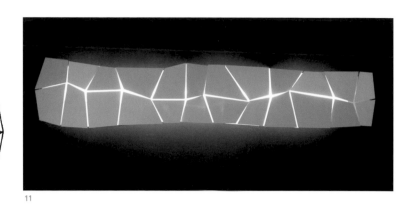

11

12

9–12

Origami Light
2003
Self-production
Painted mild steel

Twenty-four laser-cut steel plates are held together creating an intimate lighting effect through narrow chinks between the plates.

13–17

Ribbon Coatrack
2004
Self-production
Painted mild steel

The coatrack is a continuous band of steel that bends at different heights to form four arms. These arms end in tapered points that serve as the coat hooks.

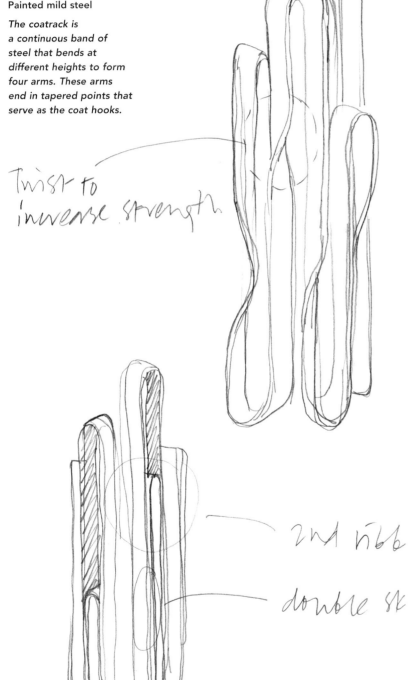

Twist to inverse strength

2nd ribb

double sk

13

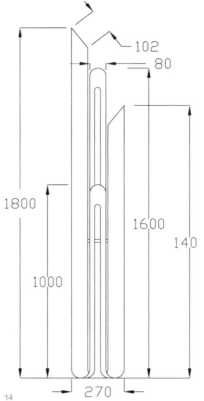

1800
1000
1600
140
102
80
270

14

15

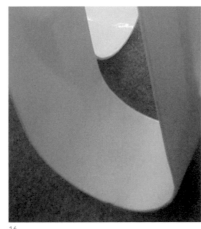

16

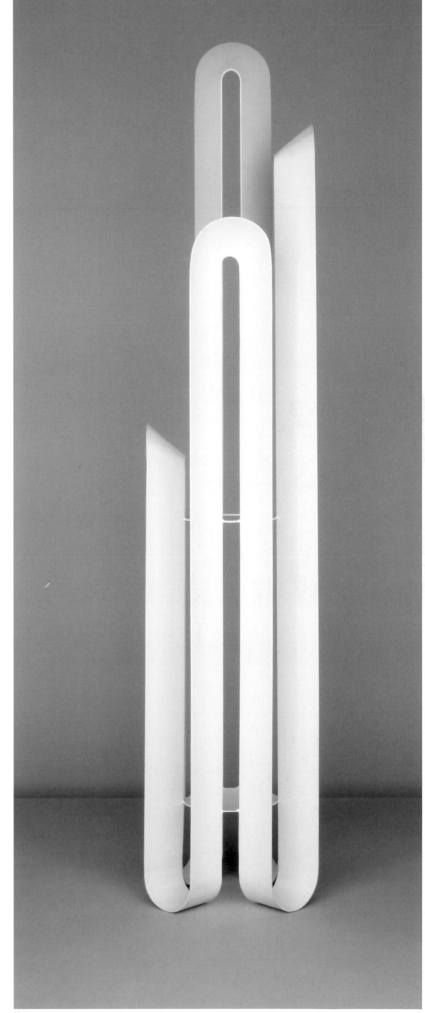

17

Dominic Wilcox, London

1

While studying art in the late 1990s, Dominic Wilcox deserted the flatland of paper and canvas to build a bed whose mattress was carved into the silhouette of a sleeper. He went on to study product design at London's Royal College of Art and continues to turn out objects that are largely about the body. In many cases, this is Wilcox's own: he made a lamp by inserting a tiny LED and battery into his nose – Nose Light (2004), cast his hand in rubber for gloves – The Glove (2002), and transferred his enlarged thumbprints to shoe soles to produce curious treadmarks – Foot Print (2002). Wilcox has also focused on the bodies of others. With fellow RCA alumnus Steve Mosley, he designed a product collection – including Bowie tables, Queen mirrors and Blondie shoes – that revolves around photos of rock stars.

The Wilcox design that has caused the greatest sensation, however, is War Bowl (2002–6), a pile of plastic toy soldiers melted and fused together into a lacy basin. An admiring press judged War Bowl variously as a homage to childhood, an audacious exploration of found objects and an anti-war message. It may well be all those things – Wilcox is deliberately vague on the point. The product evolved, he has noted, from an 'aim to make a material that had a voice of its own, not just a passive, lifeless thing'. He is fascinated not just by the structure of bodies but also by their movements – or, equally, their paralysis. Early works include a birdcage with the size and contours of a canary and a full-size swing arrested in mid-air. More recently, he embedded insect legs in resin to produce a faithful interpretation of the phrase 'the bee's knees' – The Bee's Knees (1999).

Lately, Wilcox has returned to his 2-D roots with a series of Honesty Stamps (2002–6), rubber letters fused on to classic wooden blocks that take away the tedium of producing repetitive messages. Sentiments that in his view require this type of mechanical treatment include 'I sincerely apologize for all the trouble I've caused' and 'I know in the past I've found it difficult to say these words but I LOVE YOU'. **Julie Lasky**

1

Ivy Shelf
2002
Dominic Wilcox
Vacuum-formed plastic

A clear plastic shelf that ivy can grow inside and so create an attractive wall unit.

2–5

War Bowl
2002–6
Thorsten van Elten
Plastic soldiers

A bowl created by melting toy soldiers together, this version represents the battle of Waterloo depicting both French and British infantry.

2

3

4

5

1

2

3

WOKmedia, London

Michael Cross and Julie Mathias, partners in the London-based WOKmedia, launched their international reputations with a shock – or at least, the threat of one. Invited to exhibit with other young British product designers at the 2005 Milan International Furniture Fair, they submerged electrified wires in water to produce a glowing glass lamp called Flood (2004). Gut-like tangles of circuitry nestled in the vessels, which were arrayed on shelves like a mad scientist's experiment. And yet, Flood, which began as Cross and Mathias's joint degree-show project at London's Royal College of Art, was harmless. The designers figured out how to mix electricity and water according to a principle they say is ridiculously simple (although they decline to explain how it works). They merely toyed with the public's fear of electrocution, casting a shadow on a domestic object that brought it into thrilling relief.

A terrible beauty was born – and born again in other WOK products. The studio's electric fan, Blow (2004), turns on with a puff of breath and turns off only when a finger is stuck, harmlessly, into a moving blade. WOK's Lunuganga bookshelf (2005), which is shaped like a tree branch, was inspired by the sight of vegetation breaking through village houses in Sri Lanka. An art project commissioned by Contrasts Gallery in Shanghai is a vast assemblage of porcelain crushed like eggshells.

It is apt that the studio's name, which refers to its diverse range of projects, also recalls a frantic preparation of chopped and sizzled foodstuffs; the designers evoke forces that rip through nature and make short work of custom. It is also fitting that WOK locates such anarchy in early youth. In an *I.D.* magazine profile, Cross disclosed the studio's fascination with 'killing things, breaking things, kicking – basically the dark, violent side of childhood'. What could be more authentic?
Julie Lasky

1–3

Flood
Light Installation
2004
Wires, light bulbs, glass
containers, water

*A light that explores
and triggers our
nervousness. Though
entirely safe, this product
gives the appearance
of danger by mixing
electricity and water.*

4–6

Hidden
Table and benches
2005
Project sponsored by
Dupont Corian
Corian, cotton string

*This set of furniture
explores the childish notion
of hide-and-seek, yet at the
same time hints at other
things that can go on when
no one can see.*

7

Sprinkle
Rug
2004
WOKmedia
Alpaka wool, rubber

*A rug in hundreds of
parts that can be sprinkled
on the floor to form
a disjointed carpet. Each
piece has a silicone bast
to prevent it from slipping
when walked on.*

8–9

Lunuganga
Bookshelf
2005
WOKmedia
Resin

*This shelf was inspired
by the jungles of Sri Lanka.
The culture shock
experienced by the group
on a trip there is translated
into this branch-like shelf
that seems to invade
the home, juxtaposing
the domestic realm
and the natural world.*

7

8

9

4–6
Cyclops
Robot
2002
Leading Edge Design
Corp./
Nichinan Corp.
Photosensitive resin,
aluminium, acrylic resin,
air muscles, steel

*A robot that can display
human-like qualities.
Possessing attributes such
as a spine and an 'eye', this
highly innovative piece of
machinery is able to detect
and follow the movement
of people around it.*

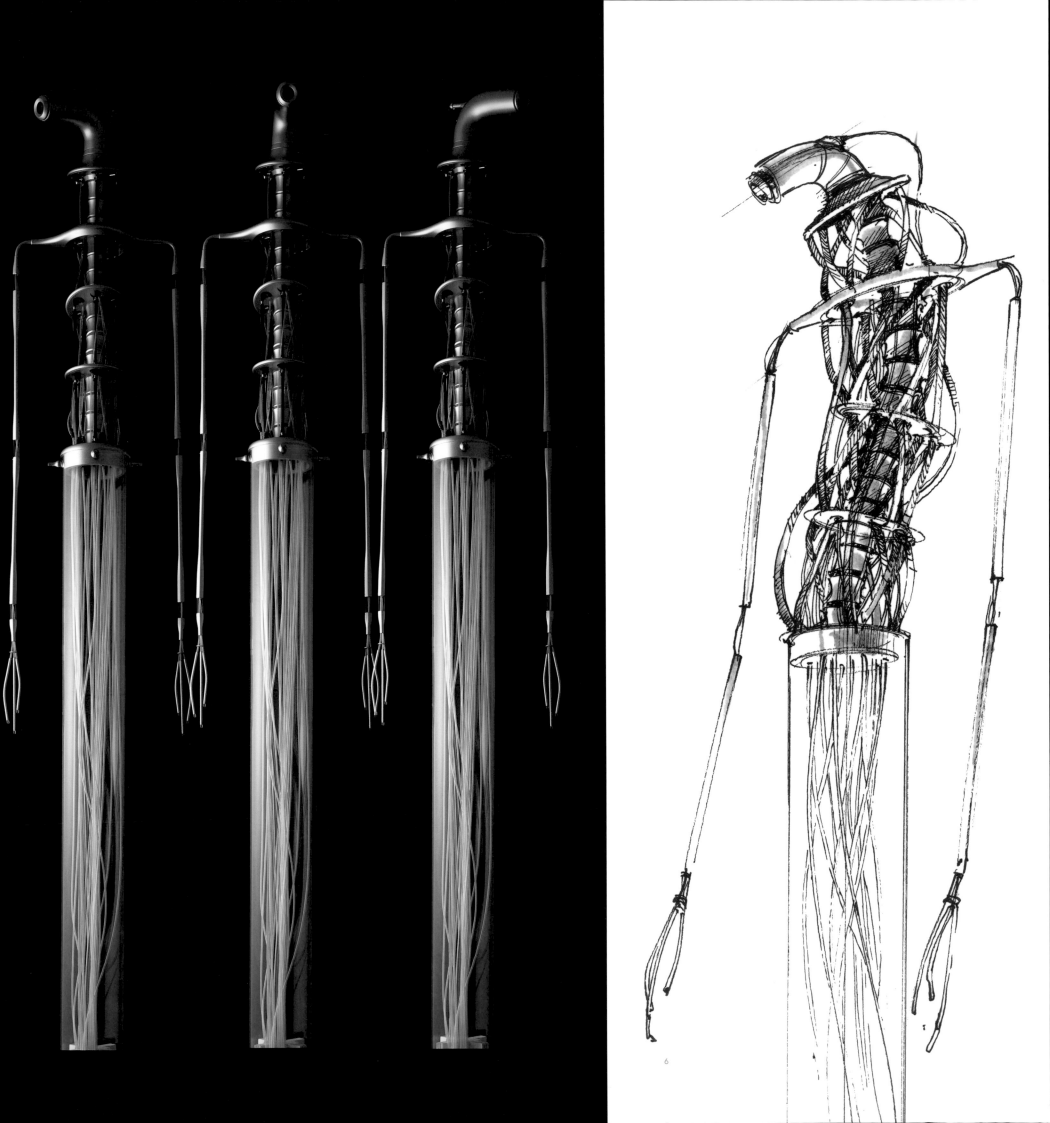

Noriko Yasuda, Tokyo

Noriko Yasuda is a designer with feelings – both a sensitivity towards objects per se and an interest in exploring them tactility. She feels a true empathy for these 'companions' of our daily lives.

Yasuda worked in New York and studied product design at the Parsons School of Design. Here she became acquainted with Droog Design, who immediately included some of her products in their catalogues and exhibitions. She graduated from Parsons in 2001, before returning to her native Japan.

Yasuda would like her designs to awaken our senses to a more considerate way of looking at ordinary objects in our surroundings. For instance, her Bulbshell (2001) is a series of ceramic covers in a variety of evolving shapes that contain and are illuminated by a single bulb. With Wan (2000) she plays on the great range of forms resulting from the combination of two elementary bowls and achieves a subtle poetry through restricted means. Two bowls of different sizes are attached back to back and the resulting object can stand on either side, enabling play with its proportions. Concerned with the fate of objects in a consumer society where they are available in great numbers only to be used and discarded, Yasuda came up with the idea for Plastibear (2003). Made from salvaged plastic shopping bags, Plastibear becomes as lovable as a cuddly teddy bear when reshaped by her imagination.

Emphasizing the special moments that can be shared with objects, Yasuda's Teacuptea (2000) vanishes, as the moment does, after use. A cup, saucer and spoon all made out of tea and solidified with sugar, the Teacuptea is both the contents and the container itself. Thus, the cup melts away as it is filled with hot water and consumed. The sensory aspect of Yasuda's work is also central to her Footrest (2002). Ceramic bowls shaped like footprints and filled with pebbles, sand or grass, enable our feet to feel different surfaces. Yasuda offers us new and intimate experiences as a way of discovering unexplored facets of ourselves. **Didier Krzentowski**

1–2

Teacuptea
2000
Self-production
Sugar, tea

A teacup that functions as a drink of tea: simply pour hot water into the sugar cup and drink it as it dissolves.

3–5

Footrest Pebbles/Grass
2002
Droog Design (Footrest Pebbles); Self-production (Footrest Grass)
Ceramic, pebbles or grass

To remind tired feet what it's like to walk barefoot on the beach or on grass.

3

4

5

7

8

6

404

9

10

11

6–9

Bulbshell
Light
2001
Self-production
Porcelain, light bulb

A sequence of porcelain pieces to cover light bulbs, which create an intimate and warm ambience.

10

Cremated Teddy Bear
2004
Self-production
Teddy bear, porcelain, gold glaze

Once-loved teddies that have been abandoned are transformed. Dipped in porcelain and fired, the process acts as a sort of funeral for the discarded bear.

11

Plastibear
2003
Self-production
Plastic bags, rubber bands

Instead of disposing of these plastic bags they have been lovingly shaped into teddy bears by Yasuda.

zuii, Melbourne

Industrial designer Marcel Sigel and graphic designer Alana Di Giacomo originally met while studying at Curtin University in Perth, Western Australia. They established their cross-disciplinary design partnership in Melbourne in 2003. The name 'zuii' is derived from a rare manuscript (Karakuri Zuii) detailing mechanical puppets and clocks from Japan's edo period. zuii's debut showing at the 2004 Milan International Furniture Fair's Salone Satellite earned it widespread acclaim in the international design media.

Among the products it presented were the Woodland table lamp (2004) and the Carbon Chair (2004). Reminiscent of a cluster of trees in a forest, the Woodland lamp draws on archetypal lamp design and construction but is infused with wit and irony. It is both practical and familiar yet evokes some fairy-tale narrative or inherent personality. The Carbon Chair, with its 54 faceted planes echoing those of a cut diamond (i.e. pure carbon), contains a similar playfulness. The inward angle of the bulky, squared-off legs makes the chair rest on its outermost corners, creating a sense of poised, tip-toed lightness, while actually improving its stability. Each new product represents a conversation between the two designers and is the result of a lengthy process of refinement and disarmingly frank critique.

zuii is most interested in commercial design and has impressed many companies with its obvious design ability and the distinctive character that it brings to its work. zuii, along with other contemporary designers such as Erik Magnussen, was invited by Royal Selangor, manufacturers of fine pewter wares since 1885, to develop a range of new products. zuii's fresh and idiosyncratic designs, consisting of bowls, vases, photo-frames, tray and set of salt and pepper shakers, were launched worldwide by Royal Selangor in 2007. **Brian Parkes**

1

Woodland
Table lamp
2004
zuii
Aluminium

This light was inspired by a cluster of trees in a forest. Whilst the soft glow enhances the ambience and mood, the aesthetic appeal and iconic design make people look twice at what is an overlooked, familiar object.

2–3

Skinny
Table/floor light
2005
zuii
Aluminium

Skinny's source of light is the shade itself. Its curious stoop makes the object appear to be swallowing itself, exploring notions of exterior and interior form.

4

Carbon Chair
2004
zuii
Fibre composites

A chair with a sense of grandeur and magnificence; its physical appearance is based upon a cut diamond (pure carbon).

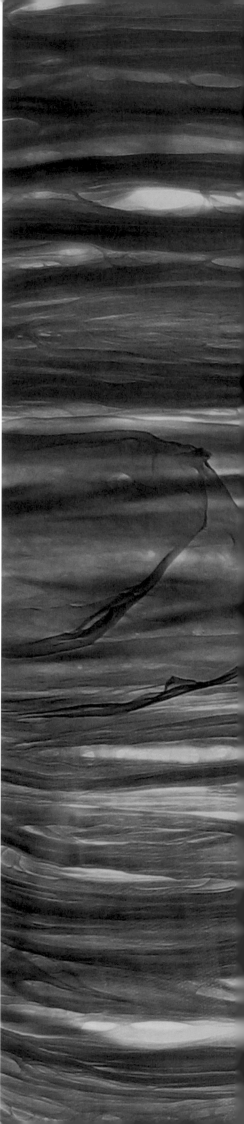

7

5
Henry's Collar
Fruit bowl
2005
zuii
Moulded polyester fibre

The bowl was inspired by the fashionable ruffs worn in the late sixteenth century. Henry's Collar allows a variety of fruits to sit in its decorative contours.

6–7

Mummy Vessel
Vase/light
2004
zuii
Recycled HDPE plastic bags

The Mummy is made from spirally wound plastic of varying thickness and transparency. When backlit (or lit from within with a small LED bulb) the vase is transformed into a subtle light piece, its light being enhanced by the random pattern of its construction.

6

10

Curators' Choices

Armchair 41 'Paimio'
(1931–2)
Alvar Aalto
Artek
Natural lacquered birch,
lacquered bent birch plywood

Armchair 41 'Paimio', 1931–2, by Alvar Aalto

In his work, the architect Alvar Aalto combined inspiration from his trips to Europe, the Bauhaus and the work of other practitioners of modernism, with the natural materials, processes and warmth of spirit of his native Finland. This approach is epitomized by his internationally acknowledged masterpiece, the Paimio Sanatorium (1928–33) in western Finland. With specially designed furniture, lighting, interiors and utilities, the building was an acclaimed example of international modernism: a structural portrayal of the form through function ideal, and yet still a very practical, comfortable and psychologically beneficial environment for its patients. Possibly the most famous furniture piece Aalto designed for the sanatorium was the Armchair 41 'Paimio'. The chair's use of the relatively cheap material, plywood, and its inspired form are often seen as the starting point for Aalto's interest in designing and producing his own furniture, which in turn led him to establish Artek in 1935.

As a part owner of Artek and its creative director I am currently engaged in restructuring the company, which has experienced very little product development or innovation over the past 50 years. As a result of my new role, I have had the most fantastic insight into the processes and thinking involved in the creation of this chair; we have access to the Aalto Foundation archives, and also own some 10,000 technical drawings from the 1930s onwards showing the development of Aalto's ideas.

Since the early 1930s Aalto had experimented with innovative methods for bending and splicing wood, looking for techniques that would allow him to realize his ideas for organic furniture forms. In collaboration with local factory owner Otto Kohonen, Aalto spent several years experimenting with different glues, wood thicknesses and mould shapes to achieve these radical forms … they must have looked like they had landed from some distant planet in 1927. As well as breaking technological ground and creating a pleasing form in the Paimio, the chair's structural ingenuity – its hard seat and gently sloping forms contrasted with its rectilinear base and back – were designed to assist the posture and breathing of patients afflicted with TB, making them feel more comfortable.

The chair encapsulates the primary thinking of the modernist movement with its machine aesthetic, form following function, and its material and structural innovation – the graceful curves of the Paimio belie Aalto's obsessive desire to push shape and structure to the extremes of the technologies and materials of the day. But at the same time, Aalto completely rethinks these key ideas to fit with his local materials palette and humanist ideology.

Aalto's most important contribution to product design was making modernism accessible and acceptable within the furnishings of the home. They spread to the United States, and to a lesser extent to pre-war England. Testament to the object's durable, practical design and structural beauty is that some of these objects – which are still made in the same factory, using some of the same tools, using the same wood from the same forests as in 1927 – are still in everyday use 80 years later. **Tom Dixon**

Poltrona Mole (Sheriff) armchair
(1961)
Sergio Rodrigues
LinBrasil
Wood, leather

Poltrona Mole (Sheriff) armchair, 1961, by Sergio Rodrigues

Architect and designer Sergio Rodrigues was born in Rio de Janeiro in 1927. He graduated in architecture from the Federal University of Brazil, Rio de Janeiro, in 1952 and in the same year began creating handmade contemporary furniture in Paraná. Returning to Rio de Janeiro in 1955, Rodrigues started making authentic Brazilian furniture based on his own research, which became well known for its design interpretation and materials. In this period he also founded Oca, a company that produced and marketed this new furniture. It became a gallery and meeting point for architects and intellectuals in search of a unique language in the most varied forms of cultural expression, and a way of disseminating contemporary Brazilian craftsmanship and design. 'Oca is the indigenous house – pure and structured. In it, everything articulates and integrates with formal precision, in function of life,' commented architect Lucio Costa, mastermind of Brasília's urban plan.

During the 1950s there was a growing need to create a type of furniture that would suit the spaces of the new Brazilian architecture. It needed to abandon the traditions of the colonial style and be as original and simple as the architecture, of which Oscar Niemeyer was the main exponent. It was during this period that Costa saw one of Rodrigues' chairs made in jacarandá wood (rosewood) with a straw seat. He enthusiastically hailed it as 'the first modern chair with traditional spirit'. Still produced today, it was named the Lucio chair after it was championed by Lucio Costo.

Rodrigues' wish to express a national identity was his inspiration for the first prototypes of the Poltrona Mole, a robust piece of furniture made of solid wood that broke all the current standards for chairs. Poltrona Mole was designed to provide comfort and encourage relaxation – very different from the modernist straight-posture form and closer to traditional Brazilian style. Finished in 1961, the armchair was shown at Italy's Cantù International Furniture Competition, along with 438 other projects from 17 countries, and took first prize. An Italian company, ESA of Bergamo, also started to produce and export it immediately under the name Sheriff.

In the 1960s Rodrigues opened another store, Meia Pataca. Unlike Oca, which had been forced to close after 13 years, the new store specialized in furniture for mass-production at more accessible prices. By the 1970s, Rodrigues had started an immensely productive phase of work. His furniture is no longer made of jacarandá wood, now considered to be an endangered species, but of eucalyptus, cinnamon, cedar, Brazilian walnut and ivory wood. Wood and leather are the basic materials of his pieces, which now comprise over 50 styles from different periods and are all produced in a semi-industrial way under his direct supervision.

Rodrigues could be claimed as the creator of Brazilian modern furniture, already recognized as such in 1959 by Giò Ponti who published Rodrigues' work in Milan's *Domus*. The recipient of several awards and the feature of important exhibitions, Rodrigues' Poltrona Mole still remains one of the great designs of today. **Maria Helena Estrada**

Ulmer Hocker, 1954, by Max Bill

Simplicity, unity of form and function, elegance and modernity: these were the watchwords of the Ulm School of Design of which Max Bill was director between 1952 and 1956. The famous Ulm stool, developed in 1954 with Hans Gugelot and Paul Hildinger, illustrates – just as much as the school building which Bill designed, or his many written works on the aesthetics of that period – the spirit and heritage of the Bauhaus, the paradigm of modern design.

First made for the school and initially produced on the premises, the stool was manufactured again between 1990 and 2001 by Wohnberdarf (Zurich) and since 2002 by Vitra. It is timeless, in that it is just as up-to-the-minute today as it ever was. It can be used as a seat, an occasional table or a storage unit, and is designed to be economical and multifunctional. The wooden panels are assembled by a comb-jointing system, its static state is ensured by a simple round dowel rod, and its stability by two pieces that prevent the whole surface of the panel from resting on ground that may not be level. Encouraging a straight sitting posture, its mobility (both physical and functional) and its 'transparency' (both aesthetic and industrial) are evidence of the modernist ideal in which Bill's work originated and which he continued. Not so much because form had to follow function, but because it had to synthesize that ideal and carry those values to a degree of functional and symbolic visibility. Thus the upright posture suggested by the stool, the possibility of combining the elements and the rationality of the design go hand in hand with a concept of modern life based on progress, rationality and democratic access.

Max Bill was an artist, graphic designer, product designer and architect. He lived his life without being constrained by the boundaries of the different fields in which he practised. For in the field of the visual arts, concrete art carries the same values as those from which his industrial or graphic works were derived; and similarly he was able to integrate his different talents into architecture when it came to the functional programming of a space. Thus, apparently without difficulty, and while continuing to assert the specificity of the fields in which he worked, he achieved what so many designers and artists are again trying to do today. To my eyes, he represents a model – undoubtedly still underestimated – for the understanding of both graphic and industrial design in the wider cultural field. **Pierre Keller**

Ulmer Hocker (1954)
Stool
Max Bill (in collaboration with
Hans Gugelot and Paul Hildinger)
Zanotta / Vitra
Birchwood

*Zanotta produces this stool
under the name Sgabillo Stool.*

Sella stool (1957)
Achille Castiglioni and
Pier Giacomo Castiglioni
Zanotta
Leather, pink-lacquered steel,
cast iron

Sella stool, 1957, by Achille and Pier Giacomo Castiglioni

Achille Castiglioni and his brother Pier Giacomo Castiglioni's chair Sella (meaning saddle) raised two important questions about the stereotype of a chair. The first question was 'Why does a chair have to be supported by four legs?' which had been explored by Mart Stam and Marcel Breuer in making their cantilever chairs, in which they proved that two legs were enough to make a chair. What Sella showed us was that one leg could be enough. Mezzadro of the same year also had one leg, and so did Eero Saarine's Tulip (1955) and Stefan Wewerka's Einschwinger (1982). However Sella is different from these chairs; Sella is very unstable. The stool finally becomes stable with the addition of the sitter's legs; it's a chair that was made to be leaned on or straddled. These characteristics are reflected in Sella's seat that was made from a bicycle.

The second question that the Sella stool raised emerged from the seat itself: the question of comfort. In their early years the Castiglioni brothers redesigned a lot of objects, and this stool, which has been described as ready-made artwork, was among them. According to this estimation, Sella was a radical case where a designer assembled industrial products to make a piece of artwork. However, more importantly Sella made us wonder if it was actually necessary to design a new seat to make a brand new chair; it proved that using an already familiar part could be innovation enough. Obviously Sella's seat reminds you of the act of sitting and moreover it tells you which direction to face when you are seated. It goes without saying that this semiotic approach can never be a perfect solution to the all everyday objects, but it can be an alternative when it comes to an excessive sense of unity or completion. It is meant as an escape from the obsession of styling.

What Sella did was invent a new posture. Of course it might not really be comfortable to be in a position that is neither sitting nor standing. However, just as Castiglioni said, it was a perfectly proper and comfortable chair when you wanted to talk on the phone, or take a rest for a short while.

It seems that this new posture was not well accepted at first. It was not until 1983 that Sella, which was designed in 1957, was finally put into production. Even though it was in the market, the Sella stool was still too symbolic and conceptual to be a common everyday product. It took another 10 years for Sella's concept to be realized in an everyday object through mass production by a German company. Wilkhahn created Stitz (a combination of 'Steh'– standing, and 'Sitz'– sitting). The round-shaped seat didn't have any sitting direction like that of Sella, and was visually more comfortable. It was well-tempered compared to the Sella.

With its rough and intense symbolism, Sella still remains very attractive, and it tells you different and abundant stories each time you sit on it. Maybe Castiglioni wanted to tell us that a designer can also be a storyteller. **Sang-kyu Kim**

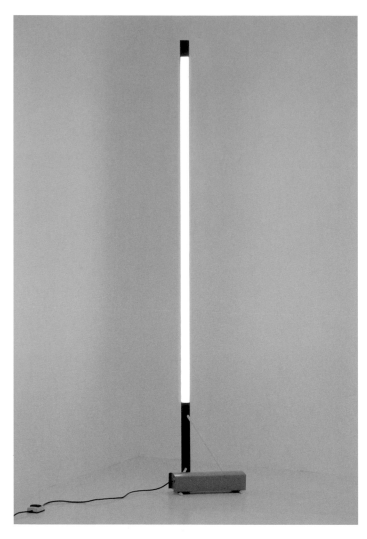

Lampadaire '1063' (1954)
Lamp
Gino Sarfatti
Arteluce
Black-lacquered metal, neon

Lampadaire '1063', 1954, by Gino Sarfatti

Lampadaire '1063', designed by Gino Sarfatti for Arteluce in 1954, is one of the masterpieces of my lighting collection, which I have been gathering for about 15 years. My deepest interest is in lamps that explore the thin membrane between art and design, especially when technological innovation is part of the process. One of the first lamps to use neon in industrial lighting history, there is nothing superfluous about this lamp. Its simple form delicately serves its function. The base achieves a perfect equilibrium: it is a geometrical play with two lines linked to the almost disappearing electrical device used as a weight counterbalance. A few years before Dan Flavin, Sarfatti plays with the hypnotic power of neon, allowing a fabulous light to emanate.

This piece was my choice because Sarfatti is one of the most inventive and surprising – yet sometimes forgotten – designers of the last century. Specializing in lamp design, each and every project he conceived can be considered as a technological breakthrough and a piece of art at the same time. Simplicity, ingenuity and elegance characterize his work.

Born in Venice in 1912, he studied aero-naval engineering at the University of Genoa before choosing the lighting sector. Shortly before 1939, he laid the foundations for the Arteluce Company that soon became a national and international reference point for lighting in the modern architecture movement.

Arteluce functioned as an outstanding and lively research laboratory for Sarfatti. Any luminous idea was immediately tested, improved, filtered and eventually produced for the market. During his 30 years of activity, Sarfatti himself designed and produced over 400 luminaries. He carried out relentless research on innovation in typology, materials, production technologies, light sources, technical lighting effects and design aspects. Investigating the maze of lighting devices, he came up with solutions to increasingly complex problems, solutions that were easy and obvious for the final user.

Always ahead of his contemporaries, the untiring experimenter was among the first designers to use new materials such as Plexiglas in 1951, neon lighting in 1954 and halogen in 1971. His formal proposals finely bear the marks of influential artists and architects: architectonic rationalism on the one hand but also stylistic influences of concretists and spatialists, with amoebic forms inspired by Arp and Miró, as well as clever structures close to Calder's mobiles or Moholy-Nagy's kinetic sculptures.

In the course of their activities, Sarfatti together with Arteluce won numerous prizes and awards including the Compasso d'Oro in 1954 and 1955, and the Honorary Diploma of the Milan Triennale. Arteluce Company was also a lively meeting place and a forum for many of the leading Italian designers in the 1950s and 1960s. Among those who took part in this seething adventure were Albini, Belgioioso, Santi, Viganò, and Zanuso. As the company began to encounter some difficulties in the 1970s, Sarfatti bitterly sold it to Flos in 1974. The legendary designer finished his life selling rare stamps by Lake Como in Italy before his death in 1984.

The emerging designers I have gathered for this selection share a similar ingenuity to Sarfatti's, and they all have great potential. **Didier Krzentowski**

Schwinn Sting-Ray, 1963, by Al Fritz

When middle-class Americans flocked to the suburbs after World War II, they took comfort in tidy new neighbourhoods with identical houses – refuges from the urban grit and chaos they had left behind. Messy vitality still found its way to these communities, however, in the form of a florid, curvaceous bicycle that quickly became an icon.

Introduced in 1963, the Schwinn Sting-Ray was greeted as an odd-looking successor to the sturdy upright bikes that Americans had been pedalling for decades. Yet, like the Aeron chair of three decades later, it won a loyal following despite its unorthodox (for many, even repellent) styling, because it made the idea of nonconformity comfortable.

The Sting-Ray's origins lay in southern California's custom-car culture, where drivers reconstructed old automobiles for drag-racing. Motorcycle riders copied the look. And enthusiasts who were too young or impoverished to drive followed suit with bicycles, which they retro-fitted with small, densely spoked wheels, long front forks and wide handlebars. Because these custom jobs were labours of love and vanity, the bikes, like the hot rods and choppers they imitated, also featured symbols of the rider's identity in the form of vivid paint jobs, chrome accents and decals.

All these features made their way onto the Sting-Ray. The agent was Al Fritz, a Schwinn engineer, who, according to corporate lore, visited the West Coast in 1962 to look at the customized bikes first-hand. Fritz specified 20-inch wheels for manoeuvrability and the oversize curved handlebars known as 'ape hangers'. A long, crescent-shaped saddle reminiscent of a motorbike saddle enhanced the rider's comfort. This 'banana seat' was supported by an inverted U-shaped 'sissy bar', which, along with the front springer fork, reduced the impact of jolts. True to his design's coastal origins and rebel associations, Fritz named the Sting-Ray after a marine creature that is streamlined, flexible and somewhat menacing.

The bike's anti-conformist spirit sealed its success. According to Schwinn, though the Sting-Ray, at $50, was priced extravagantly for the time, 43,000 units were purchased in the first year alone. In 1964 Schwinn introduced the pastel Fair Lady model for girls, and in 1968 it launched the Krate, a model that conformed closest to its hot-rod progenitors by featuring a 16-inch front wheel, bigger and thicker back wheel, and even a gear stick for changing gears.

Competitors soon turned out imitation lowriders, including Raleigh's Malvern Star and Chopper bikes. The Sting-Ray's dominance wasn't eroded, however, until the subcultural forces that gave rise to its design further evolved. BMX, the sport of off-road bicycle racing and acrobatics, which originated with Sting-Rays, reached a frenzy in the late 1970s, and other manufacturers began producing bicycles that were better suited to its demands. In 2004, however, Schwinn, under new ownership, re-introduced the Sting-Ray to Americans who had become nostalgic for artefacts from their childhoods. **Julie Lasky**

Schwinn Sting-Ray (1963)
Bicycle
Al Fritz
Schwinn

Paperclip (1899)
Johan Vaaler
Various
Steel wire

Paperclip, 1899, by Johan Vaaler

The paperclip is the ideal shape, is effortlessly elegant and fulfils to perfection the function for which it was conceived. Easy to carry, produce and recycle, it is also cheap, available to everyone and a stranger to no one, regardless of social standing or age group. Moreover, its seductive power is nothing short of mystifying – it unsettles our hands and our minds, waiting for different uses and appealing to our creative behaviour.

One of the things that strikes me most about design is its potential for being a most democratic creative discipline. Although we are witnessing the rise of a design star system – fuelled by the media, the market and by our own insatiable consumerism – the essence of design is intimately connected to this perspective of democratizing knowledge, of making available to all each new achievement in humankind's unceasing struggle to master the environment in which we grow and develop.

Designers cater for others, serving needs and wants that are not necessarily their own, even when their work is more exploratory, conjectural or artistic. Service and responsibility are at the top of a designer's checklist. That is why, ultimately, designers are the architects of everyday comfort, ease and beauty – they work towards making our world a better and safer place in which to live. Our daily lives can be affected, for better or for worse, by a thousand little things. Designs need not be grandiloquent gestures but rather chirurgical interventions. Good designs are most often sensed and felt in the details.

Designer do not merely materialize other people's ideas. They are an integral part of the creative process, researching the roots of the problem trying solve as well as analysing the results of their proposed solution. Designers hold extraordinary tools to achieve this. Combining complementary perspectives such as aesthetics, ethics, emotion, economy, culture, functionality and sustainability into a overall view, designers search for the answers that provide the best solutions to the problems of contemporary society. By doing this, they reach new material solutions or processes for organizing information for widespread use. Design can be perceived as crossing all human activities, as a toolbox that heightens and intensifies the potential of other fields of work, and acts as a catalyst between them.

However, the meaning and importance of design now and in the future is also inextricably linked to three key elements: designers are trained to be versatile, quick and proactive. They are trained to be forward-thinking in their extreme capacity to adapt to new contexts, to respond to new scenarios efficiently, to speculate about the future. And most importantly of all, they now master the possibility of producing artefacts that have room for idiosyncrasies, for personality, for free, individual interpretation, with open source and customizable projects that can be mass-distributed. Design is emerging as the operational tool for the twenty-first century.

Maybe I have strayed a little too far from the paperclip. But what is there to be said about perfect things? Practically nothing. **Guta Moura Guedes**

Dilly Bag
Anon
Handcrafted Aboriginal product
Dyed and twined Pandanus palm

Dilly Bag, traditional Aboriginal design, by Anon

Australian design has come of age in recent years and many young Australian designers are now beginning to enjoy international recognition and success. As one of the youngest participants in the growing global design industry, it is interesting that Australia is also home to arguably the oldest continuing design culture in the world. This irony was at the forefront of my thoughts when I considered the problem of selecting a single 'design classic'. A classic is something that has withstood the test of time and yet remains significant today.

In the end I chose the twined conical basket or Dilly Bag made continuously for thousands of years by Aboriginal people in Arnhem Land, northern Australia. These distinctive baskets are depicted in ancient rock paintings more than 20,000 years old. Many exquisitely crafted examples of them were collected by anthropologists in the nineteenth and early twentieth centuries and are spread throughout the world's museums of natural history. Since the late 1980s others have been purchased by museums of contemporary art and design in Australia, Europe and North America.

Made from locally available plant materials, the basic blueprint for the design and manufacture of these baskets has been handed down over countless generations. Requiring significant technical and material knowledge, the lengthy process begins with collecting and carefully splitting the fronds of the Pandanus palm and then drying the strands in the sun. These are often dyed by being boiled with specific roots, barks and sometimes ash, before being laboriously twined together. The fronds spiral out from the centre of the base and each maker develops a subtly distinct personal style.

The baskets are used to collect and carry food and other possessions. More elaborate versions are made for ceremonial purposes and have a sacred cultural significance. Since the 1960s skilled Aboriginal women artisans have been making them specifically for the growing non-Aboriginal collector and tourism markets, providing valuable income to remote and regional communities. The baskets resonate with a sense of the place and culture from which they come and can now be seen in homes around the world, displayed as artworks, as souvenirs or used for practical purposes.

The Dilly Bag is in many ways a product of invention born of necessity, using the resources at hand. This same description could be applied, not necessarily to the products, but certainly to the approaches of the emerging wave of young Australian designers currently making an international impact. Not constrained by the weighty histories of design nations such as Italy, Sweden or the United Kingdom, Australian designers (such as those I have selected for this book) have invented their own way of doing things, turning the challenges of a small local market, a limited manufacturing sector and the tyranny of distance to advantage in a variety of unique, surprising and innovative ways. **Brian Parkes**

TP1 – T4 radio and P1
record player – (1959)
Portable radio and
portable vinyl disc player
Dieter Rams
Braun AG
Thermoplast

TP1 (T4 & P1), 1959, by Dieter Rams

My admiration for Dieter Rams' work has grown with time. I admire him far more today than when, as a student, I first saw his works. That was the 1980s when Rams' work was regarded with a certain ideological hostility, as if he were to blame for all the wrongs of functionalism. Rams was German, he came from Ulm and he designed simple objects – so simple they were not even worth imitating because it seemed as if anyone could make them. In reality, Rams is one of those designers whose work it is impossible to imitate – copies can't remotely come near to the mysterious beauty of the original.

I grew up at a time when everyone in Milan was competing to come up with the most original object possible. It had to be something no one had even dreamt of; it didn't matter if the objects were not useful. I remember a kettle by Philippe Starck that caused severe burns if you tried to hold it by what you assumed to be the handle; in fact, this was actually the spout through which the boiling water was poured. But such things did not matter: use was not a priority. It was a time of excess when everyone loved to be amazed. In the end, we were so heavily bombarded with abstruse, arbitrary objects that when I happened to spot a picture of an old Braun radio designed by Rams I drew a sigh of relief – as you do when you are lost in a crowd of strangers and finally glimpse a friendly face. At last a radio that you could recognize as a radio. Rams set the archetype for radios (and many other kinds of portable) when he designed his pocket-receivers. His T4 pocket radio (1959) could be fixed to the portable vinyl disc player called the P1, and together these two modular units formed the TP1. The T4's successor, the T41 (1959), with a larger circular frequency window, or the T52 (1961), used also as a car radio, were the most popular of Rams' radio designs, all of which influenced the appearance of portables radios for years to come.

During the 1950s and 1960s, miniaturization, as made possible by new transistor technology, presented an attractive challenge to many designers. In a sense, there were no precedents. Rams established a totally new solution of compact beauty in contrast to the cumbersome monumental objects full of rhetoric that were in fashion at that time.

Nowadays, Rams' works have become what Max Bill called 'objects in charge of a mental use'. Even though their technological usefulness has been surpassed, the abstract aesthetic appeal of his objects is such that today they are perceived as art pieces. His immobile platonic forms make collectors pay any price for displaying their crystalline fixity in their own living rooms.

When I was asked to select a design classic, this example came to mind. It's not clear to me what being a 'design classic' means, but I believe that the strength of an object lies in its ability to change with the passing of time and adapt in the viewer's eyes. So that at all stages of life, each of us can see something good in it. **Francesca Picchi**

Steel Shelving Unit, 1997, by Anon for MUJI

Steel Shelving Unit (1997)
Anon
MUJI
Stainless steel, ashen boards,
polypropylene

Companies that were establishing themselves in Japan during the 1980s had a strong tendency to believe that good products had to have a strong brand name. Around this time, MUJI appeared on the scene aiming to create essential daily goods with excellent style. The marketing concept was to become popular through rational designs and lower costs achieved by re-evaluating materials, avoiding complex production processes and minimizing packaging. In the early days, MUJI started with just over 40 items, including general goods and foodstuffs; this soon expanded into a wider field, including furniture and clothing; and today there are 306 shops in Japan and 62 overseas, selling about 6,000 items. Paradoxically, although the company's policy has not changed since its establishment, the MUJI no-brand has become a celebrated brand in itself.

Since the company's expansion into furniture in 1997, the Steel Shelving Unit (of which I am a proud owner) has been a big commercial success. The main feature of this product is a basic minimum frame that is designed to fit the dimensions of the Japanese standard house, but which can be tailored to individual styles using shelf and drawer variations. An easy-to-use object with good drawer storage, the Steel Shelving Unit's design is based on the modular A-series of paper sizes (A3, A4, etc.). Due to its success, a whole range of MUJI products – not just lights and radios, but also larger items such as refrigerators – were designed so that they could be accommodated easily within the Steel Shelving Unit. Despite being systematized, the design offers five height choices; seven shelf materials, including wood, steel and stainless steel; and four drawer types, including polypropylene and wire; and by adding different racks and doors, buyers can create their own customized versions.

Like other MUJI bestsellers, this system was not the work of a specific designer and features an anonymous design. When sales started in 1997, most rental accommodation in Japan had hard-to-use built-in closets, so MUJI focused on the demand for easy-to-assemble knockdown storage units that buyers could carry home with ease. Initially, Mr Kagaya (a member of GK design and an advisor to MUJI) and MUJI's merchandiser together sketched the rough designs, which were then fabricated in a Taiwan factory. Although the basic frame has not changed, some adaptations have been introduced to match altering tastes. For example, in 2002, MUJI's design director Naoto Fukasawa made a small but carefully considered colour change, going from stainless steel to light grey. Then, in 2004, rust-free stainless steel was chosen for kitchen components, and 2006 saw the launch of a desk design targeted at Soho businesses.

Even today MUJI boasts huge annual sales of the Steel Shelving Unit, selling over 40,000 every year. Although the Steel Shelving Unit continues to evolve with contemporary trends, it will always remain a unique personalized form of furniture. And as a result, it seems very likely to maintain its reputation as a design classic in the future. **Chieko Yoshiie**

Artists' Biographies

5.5 designers

www.cinqcinqdesigners.com
Vincent Baranger: born France, 1980;
Jean-Sébastien Blanc: born France, 1980;
Anthony Lebossé: born France, 1981;
Claire Renard: born France, 1980
All live and work in Paris

Education
Vincent Baranger: 2003 Diplôme Supérieur d'Arts Appliqués Design Produits, École Nationale Supérieure des Arts Appliqués et des Métiers d'Art, Paris; Jean-Sébastien Blanc: 2004 Diplôme d'Études Appronfondies Conception de Produits Nouveaux, École Nationale Supérieure d'Arts et Métiers, Paris; Anthony Lebossé: 2003 Diplôme Superieure d'Art Appliqués, Ecole Nationale Supérieur des Arts Appliqués et des Métiers d'Art, Paris; Claire Renard: 2004 Diplôme d'Études Appronfondies Conception de Produits Nouveaux, École Nationale Supérieure d'Arts et Métiers, Paris

Professional Background
2004 Established 5.5 designers

Selected Projects
2004 Sand Furniture; 2005 Réanim (dressings for furniture collection), Seating Prosthesis, Leg Prosthesis, Suture Kit, Graft Kit, 5.5 designers; Ouvriers-Designers (Workers-Designers), different pieces of hand-manufactured tableware, Bernardaud; Les Vices de la Déco, 5.5 designers; Les Objets Ordinaires (Ordinary Objects), Portecintre, Sugar Handle, Plug Lamp, La Corbeille Éditions; 2006 Furniture to Garden, chair, table, and bench, b-ton design

Selected Bibliography
2004 5.5 designers, *Sauvez les Meubles*, Paris; 2005 *Young European Designers*, DAAB; 2006 Joachim Fischer (ed.), *Living Trends*, Future Concept Lab, Milan; Christine Colin (ed.), Francesco Morace, *Design and Imitation*, Paris; Bénédicte Duhalde, 'Ouvriers-Designers', *Intramuros*, no.121, January, Paris; Karen Chung, 'Too Clever by Half', *Wallpaper**, March, London; Eve M. Khan, 'Playing Doctor', *I.D.*, September, New York

Selected Exhibitions
2004 'Observeur du Design 2005', Cité des Sciences et de l'Industrie, La Villette, Paris; 2005 'Observeur du Design 2006', Cité des Sciences et de l'Industrie, La Villette, Paris; 2006 'Tête à Tête' (Exhibition for Children), Beauborg, Centre Pompidou, Paris; 'Hôpital des Objets', Galerie Peyroulet et cie, Paris; 'Ouvriers-Designers', Superstudio Piu/Zona Tortona, Milan; 'Ressuciter', Commissaires, Montreal; 'Rehab', Cite Design, New York; 'Réanim, la Medicina de los Objectos', l'appartement, Barcelona; 'New French Designers', Droog Design, Amsterdam; 'Design Parade', Hyères, France; 'Design Reference, Paris Contemporary Design', Rotterdam, Budapest; 'European Ceramic Context', Börnholm, Denmark; 'Observeur du Design 2007', Cité des Sciences et de l'Industrie, La Villette, Paris; 'Labels VIA', VIA Agency, Paris; 'Eden ADN', Biennale Internationale de Design, St Etienne

Awards
2006 VIA, Label for 3 Products (Portecintre, La Corbeille Éditions; Furniture to Garden, b-ton design; Ouvriers-Designers, Bernardaud Foundation), France

ANTEEKSI

www.anteeksi.org
Selina Anttinen: born Finland, 1977;
Malin Blomqvist: born Finland, 1978;
Johanna Hyrkäs: born Finland, 1970;
Erika Kallasmaa: born Finland, 1969;
Jussi Kalliopuska: born Finland, 1973;
Maippi Ketola: born Finland, 1980;
Johannes Nieminen: born Finland, 1977;
Vesa Oiva: born Finland, 1973; Johan Olin: born Finland, 1974; Pauli Ojala: born Finland, 1980; Anna Oksanen: born Finland, 1972; Anne Peltola: born Finland, 1969; Tuomas Siitonen: born Finland, 1974; Aamu Song: born Korea, 1974; Janne Suhonen: born Finland, 1975; Mari Talka: born Finland, 1973; Tuomas Toivonen: born Finland, 1975; Nene Tsuboi: born Japan, 1976
All live and work in Helsinki

Education
Various design schools including: Institute of Design, Lahti Polytechnic, Lahti, Finland; University of Art and Design Helsinki, Helsinki; Helsinki University of Technology; University of Oulu, Oulu

Professional Background
All working together as ANTEEKSI and on individual projects

Selected Bibliography
2003 Nahoko Mori, 'Anteeksi', *Plus Eighty One*, Vol. 22, Tokyo; 2005 Barbo Larson, 'Anteeksi', *Forum*, 1, Stockholm; Katherine E. Nelson, 'I'm Sorry, Excuse Me', *Metropolis*, June, New York

Selected Exhibitions
2003 'Fäsäri I Fashion Show', at ANTEEKSI, Helsinki; 2004 'Milari I Furniture Fair', in the streets of Kallio, Helsinki; 'Fäsäri II Fashion Show', at ANTEEKSI; 'Blickfang Design Fair, Zürich' fair; 2005 'Milari II Furniture Fair', in the streets of Kallio, Helsinki; 'ICFF Furniture Fair', New York; 'Linelmia Show', Korjaamo, Helsinki; 'Helsinki International', Kojaamo, Helsinki; 'Ihmemaa, Habitare Furniture Fair', Helsinki; 'Fäsäri III Fashion Show', at ANTEEKSI; 2006 'Sai Sabai Temporary Art Museum', Bangkok; 'Milari III Furniture Fair', on the streets of Kallio, Helsinki; 'Remix – Exhibition', at Fiskars, Finland

Awards
2004 Golden Blickfang Preis, Blickfang Design Fair, Zurich

Antenna Design

www.antennadesign.com
Sigi Moeslinger: born Austria, 1968;
Masamichi Udagawa: born Japan, 1964
Both live and work in New York

Education
Sigi Moeslinger: 1988 Art Center College of Design, Vevey, Switzerland; 1991 Art Center College of Design, Pasadena, California; 1996 Interactive Telecommunications Program, New York University, Tisch School of the Arts, New York; Masamichi Udagawa: 1987 Industrial Design Department, Chiba University, School of Engineering, Chiba; 1991 Cranbrook Academy of Art, Bloomfield Hills, Michigan

Professional Background
Sigi Moeslinger: 1991–4 IDEO, San Francisco; 1996–7 New York University, Interval Research Fellow, New York; Masamichi Udagawa: 1987–9 Yamaha, Product Design Laboratory, Hamamatsu, Japan; 1992–5 Apple Computer, Cupertino, California; 1996–7 IDEO, New York; Both: 1997 Established Antenna Design

Selected Projects
2000 New York City Subway Cars: R142, R143, R160, Kawasaki, Alstom, Bombardier; 2003 MTA: MetroCard Ticket Vending Machine, Cubic, Scheidt & Bachman; Bloomberg Terminals and peripherals, Bloomberg LP; 2004 JetBlue self service check-in Kiosk, IBM; 2006 New York City Subway, Help Point Intercom, Siemens

Selected Bibliography
2001 Paola Antonelli, *Workspheres: Design and Contemporary Work Styles*, exhibition catalogue, New York; Steven Heller, *The Education of an E-Designer*, New York; 2003 Laetitia Wolff and Michael Porciello, *Graphis Product Design 3*, New York; Ellen Lupton, Donald Albrecht, Susan Yelavich and Mitchell Owens, *Inside Design Now*, London; 2004 Lauren Parker, *Interplay: Interactive Design*, exhibition catalogue, London; David Redhead, *Electric Dreams: Designing for the Digital Age*, exhibition catalogue, London; Lorraine B. Diehl, *Subways: The Tracks That Built New York City*, exhibition catalogue; *Subway Style: 100 Years of Architecture & Design in the New York City Subway*, New York; 2005 Paola Antonelli, *SAFE: Design Takes on Risk*, exhibition catalogue, New York; 2006 Janet Abrams and Peter Hall (eds.), *Else/Where: Mapping*, Minneapolis; Marta Serrats, *Sign Graphics*, New York; Dan Saffer, *Designing For Interaction*, Berkeley

Selected Exhibitions
2000 'Firefly', part of Creative Time Arts in Anchorage, Brooklyn Bridge Anchorage, New York; 2001 'The Emperor's New Clothes', Artists Space, Project Space, New York; 2002 'Power Flower', Windows at Bloomingdale's, New York; 'Blowing Gently...', Frederieke Taylor Gallery, New York; 2003 'Cherry Blossom', as part of Design Triennial, Cooper-Hewitt, National Design Museum, New York; 2005 'Nosy Parker', as part of Balance and Power exhibition, Krannert Art Museum, Champaign, Illinois; 2006 'Pattern Recognition', Frederieke Taylor Gallery, New York

Awards
2006 Masters of Design Awards, Rising Star, Fast Company, San Francisco; Finalist – Product Design, National Design Awards, Cooper-Hewitt, National Design Museum, New York

Atelier Oï

www.atelier-oi.ch
Aurel Aebi: born Switzerland, 1966;
Armand Louis: born Switzerland, 1966;
Patrick Reymond: born Switzerland, 1962
All live and work in La Neuville, Switzerland

Education
Aurel Aebi: 1989 Diploma in Architecture, École d'Architecture Athenaeum, Lausanne; Armand Louis: 1989 Naval Constructor Diploma, Lausanne; Patrick Reymond: 1989 Diploma in Architecture, École d'Architecture Athenaeum, Lausanne

Professional Background
1991 Established Atelier Oï

Selected Projects
1997 Torslanda, newspaper and clothes hanger, Ikea; 1998 Mesosfär, lamp, Ikea; 1999 Wogg 24, bed, Wogg; 2000 Skirt Light; 2002 Poche-Cul, redesign of the traditional milk chair, Milan; 2004, Pavillon Wogg 31, Wogg; 2005 Scenography in collaboration with Ronan and Erwan Bouroullec, Neue Räume, Zurich; Collection Lamps Tome, paper lamp, MIFF; Bank Plus, Roethlisberger; Scenography for 1000 Women for Peace, Nobel Peace Prize; 2005–7 Jewellery skills center, Swatch Group, Cormondrèche, Switzerland

Selected Bibliography
2002 'Atelier Oï/Free Spirit', *Frame*, Nov/Dec; 2004 'A Successful Troika', MD, Germany; 2006 *Small Beautiful Made in Switzerland*; Stefan Zwicky, 'Schweizer Möbel Lexikon', *Das Atelier Oï*, Zurich

Selected Exhibitions
2002 'Botte-Cul', MIFF; 2003 'Criss & Cross', São Paulo; 'Swiss design', traditionally progressive, New York; 2004 'La Quinzaine Internationale du design de jardin', Festival des jardins, Chaumont-Sur-Loire; 'Baguette chef d'orchestre', MIFF; 2005 'Gastraum', Neue Räume, Zurich; 'A Composition of Cords', Swiss Cultural Center, MIFF

Awards
2003 iF Design Award for Swivable Container, Hanover; 2004 Design Plus Light and Building/Messe Frankfurt for UniQ Lamp, Frankfurt

Maarten Baas

www.maartenbaas.com
Born in The Netherlands, 1978

Lives and works in Eindhoven

Education
2002 Design Academy, Eindhoven

Professional Background
2003 Established Studio Baas, Eindhoven; 2005 Established Studio Baas and den Herder, Eindhoven

Selected Projects
2003 Smoke Piano, Baas; 2004 Smoke ZigZag chair, Baas; 2005 'Hey Chair, Be a Bookshelf', bookshelf, Baas; Treasure Armchair, Baas; Treasure Dining Chair, Baas; 2006 Clay Chair, Baas and den Herder; Clay Bookshelf, Baas and den Herder

Selected Bibliography
2004 Kristi Cameron, 'Why Maarten Baas Burns Through History', *Metropolis*, May; Linda Hales, 'Maarten Baas's Claims to Flame', *Washington Post*, 22 May, Washington; 2005 Marcus Fairs, 'Designing with Fire', *icon*, February, London; Marcel Wanders, *International Design Yearbook*, London; Aad Krol, *Dutch Design Yearbook 2*, Rotterdam; Alice Rawsthorn, 'A World in Smoke and Clay', *International Herald Tribune*, 5 November

Selected Exhibitions
2004 'Where There's Smoke', Moss, New York; 'Nocturnal Emissions', Groninger Museum, Groningen; 'Brilliant', Victoria & Albert Museum, London; 2005 '05 Degrees of Separation', Moss, Art Basel Fair, Miami; 2006 'NLA', ACME, Los Angeles; 'Underdogs & Überprodukte', MU, Eindhoven; 'Clay & Smoke', Cibone, Tokyo; 'Crossovers', Contrasts Gallery, Shanghai; 'Clay Furniture', Design Museum, London

Suyel Bae
MILLIMETER MILLIGRAM. INC
www.mmmg.net
Born in the Republic of Korea, 1974
Lives and works in Seoul

Education
1999 BA Metalworks, Kookmin University; 2000 Metalworks at Kookmin University, Seoul

Professional Background
1999 Established MILLIMETER MILLIGRAM. INC

Selected Projects
2002 Communication balls, The Universe (prototype), Alessi Workshop; 2003 Do Show design project, Glass Pack, Do Art

Selected Bibliography
2002 Youngmi Kim, 'Suyel Bae, Craftman', *Monthly Design*, April, Seoul; 2003 Judy Zhu, 'Millimeter Milligram', *Vision*, April, Shanghai; Hirosh, 'Millimeter Milligram', *Dream Design*, No.9, May, Tokyo; 2005 Suyel Bae, *Designer Says, Designer*, Seoul; Fritz K. Park and Emil Goh, 'Creative Country', *IDN*, Volume 12, November, Hong Kong; 2006 En Keung Jun, 'Young Designer', *Dong-A Newspaper*, 13 February, Seoul

Selected Exhibitions
2001 'Good Design Festival', Coex, Seoul; 2003 'Seoul Design Festival', Seoul Art Centre; London Designers Block, London; Tokyo Designers Block, Idee, Tokyo; 2005 'de05 (Design Edge)', Suntec, Singapore

Awards
2002 Young Designer Award, Korean Design Award, Seoul

Manuel Bandeira
Ultradesign
www.manuelbandeira.art.br
Born in Brazil, 1973
Lives and works in Brazil

Education
1997 Architecture, Federal University of Bahia, Salvador Bahia, Brazil
1999 MA Industrial Design, Domus Academy, Milan

Selected Projects
1996 Cadeira Contorno, concept chair, Jorge Lima/Liceu; 1998 Viva, recycled umbrella lamp, Manuel Bandeira and Francisco Gomes Paz; 2001 Água Armchair, outdoor armchair, Tidelli; 2003 Sequinha®, underwear dryer, Ultradesign; Levita armchair, chair, Ultradesign; 2004 Escada chair, stair chair, Ultradesign; 2006 Euripedes Hanger, Kenison Furniture

Selected Bibliography
2001 Marlene Milan Acayba, *From 11th to 15th Design Prize of Museu da Casa Brasileira*, São Paulo; 2002 Alastair Fuad-Luke, *The Eco-Design Handbook*, London; 2002 Pedro Ariel Santana, *Design Brasil*, São Paulo; 2006 Ivens Fontoura, *Uma Visão do Design Moveleiro Latino Americano*, Bento Gonsalves

Selected Exhibitions
2002 'Uma Historia do Sentar', Novo Museu by Oscar Niemeyer, Curitiba, Brazil; 2003 '17th Design Prize of Museu da Casa Brasileira', Museu da Casa Brasileira, São Paulo; 2006 '1st Bienal Brasileira de Design', Oca, Ibirapuera, São Paulo

Awards
2003 17th Design Prize of Museu da Casa Brasileira, São Paulo; 2004 Young & Design, Milan; 2006 X Salão Movel Sul Prize of Design, Bento Gonçalves, Brazil

Yves Béhar
fuseproject
www.fuseproject.com
Born in Switzerland, 1967
Lives and works in San Francisco

Education
1991 Art Center College of Design, Vevey, Switzerland; 1993 Art Center College of Design, Pasadena, California

Professional Background
1992–5 Lunar Design, San Francisco; 1996–8 Frog Design, San Francisco; 1999–present fuseproject, San Francisco

Selected Projects
2004–5 MINI_motion, watch, luggage, apparel, MINI/BMW Group; 2006 Voyage, chandelier, Swarovski; DXL, protective helmet, protective bodywear, Pryde Group; KADA, table, stool, power-hub, Danese; Jawbone Air, headset, Aliph; Leaf, LED table light, Herman Miller; One Laptop Per Child (OLPC), aka $100 Laptop, OLPC

Selected Bibliography
2004 Joseph Rosa, *Yves Béhar fuseproject Design Series 2*, exhibition catalogue, San Francisco; Chantal Prod'hom, *Aaron Betsky and Steven Skov Holt*, Concept/ Commerce, Basel/Boston/Berlin; Carleen Hawn, 'Masters of Design', *Fast Company*, June, New York; 'Coolest Inventions', *Time*, November, New York; 2005 Chee Pearlman, 'Up Close and Personal', *Newsweek*, May, New York; Bruce Nussbaum 'Meet the Innovation Gurus', *Business Week*, August, New York; 2006 Jade Chang, 'All About Yves', *Metropolis*, June, New York; Bruce Nussbaum 'Winners: Best Product Design of 2006', *Business Week*, July, New York; Bradford McKee, '52nd Annual Design Review', *I.D.*, July, New York; Douglas McGray, 'The Laptop Crusade', *Wired*, August, San Francisco

Selected Exhibitions
2004 'Triennial 2003', Cooper-Hewitt, National Design Museum, Washington DC; 'California Design Biennial', Pasadena Museum of California Art; 'Yves Béhar', Musée de Design et d'Arts Appliqués Contemporains, Lausanne; 'Yves Béhar fuseproject/design series 2', San Francisco Museum of Modern Art; 2005 'Swarovski Crystal Palace', Art/Basel, Miami, Florida; 2006 'SAFE: Design Takes on Risk', Museum of Modern Art, New York; 'Second Skin', Ruhrmuseum, Essen

Awards
2004 National Design Award, Cooper-Hewitt, Smithsonian National Design Museum, Washington DC; 2005 *I.D.* Magazine Annual Design Competition, New York; 2006 Gold IDEA Award – IDSA & Business Week, Austin, Texas

Mathias Bengtsson
Bengtsson Design Ltd
www.bengtssondesign.com
Born in Denmark, 1971
Lives and works in London

Education
1994 Vehicle design, Art Centre College, Montreux, Switzerland; 1997 BA Furniture Design, Danish Design School, Copenhagen, Denmark; 1999 MA Industrial and Furniture Design, Royal College of Art, London

Professional Background
1999 Collaboration, Design Laboratory, London; 2001 Established own studio

Selected Projects
1999 Slice Chair, Bengtsson Design Ltd; 2000 Slice Chaise Longue, Bengtsson Design Ltd; 2002 Mac Couch, Bengtsson Design Ltd; 2003 Spun Carbon Fibre Chair, Bengtsson Design Ltd; 2004 Mac Chair, Bengtsson Design Ltd; 2005 Spun Carbon Fibre Bench, Bengtsson Design Ltd; 2006 Etched Jewellery, Bengtsson Design Ltd

Selected Exhibitions
2002 'Unexpected', Sotheby's, New York; 'The Adventures of Aluminium', Design Museum, London; 2003 'Great Brits', MIFF, Paul Smith HQ; 'Tokyo Designers Block, Great Brits', British Council; 2004 '20–21st Century Design Art', Philips, New York; 2005 'Mathias Bengtsson', Rohsska Museum, Gothenburg, Sweden; 2006 Slice Chair, Permanent Collection, Museum of Fine Arts, Houston, USA

Awards
2000 Finalist, *Blueprint*/ Ness Furniture, *Blueprint*, London; 2001 Finalist, Peugeot Design Award, London; 2004 1st Prize in outdoor furniture, Future City/St James Group, London

bernabeifreeman
www.bernabeifreeman.com.au
Rina Bernabei: born Australia, 1968; Kelly Freeman: born Australia, 1976
Both live and work in Sydney

Education
Rina Bernabei: 1991 Bachelor of Design, University of Technology, Sydney Kelly Freeman: 1998 Bachelor of Design, (Hons.), University of New South Wales, Sydney

Professional Background
Rina Bernabei: 1992–3 Lucci/Orlandini Design Studio, Milan; Dal Lago Studio, Milan; 1996–present Senior lecturer in industrial design, University of New South Wales, Sydney; Kelly Freeman: 1999–2000 Design Coalition, Sydney; 1999–present Lecturer in industrial design, University of New South Wales

Selected Projects
2003 Peony Chandelier, bernabeifreeman; Leaf Pendant, bernabeifreeman; Mirror Mirror, bernabeifreeman; Peony Floor Light, bernabeifreeman; 2005 Lace Pendant, bernabeifreeman; Exchange Hotel, chandelier and bar screens, bernabeifreeman; 2006 Cloth Table, bernabeifreeman; Brodie Table, bernabeifreeman

Selected Bibliography
2004 Simon Horauf, 'Idea 2004 Winner Product Design', *Inside*, no.34, November, Sydney; 2005 Ewan McEoin, 'The Promised Land', *Voi Tutti*, no.3, Sydney; 2006 Alex Gordon, 'The New Wave', *Vogue Living*, March/April, Sydney; Hannah Edwards, 'Stars in the Making', *The Sun Herald*, 26 March, Sydney

Selected Exhibitions
2004 'Sydney Style', Sydney Opera House; 'Design 2004', Melbourne Museum; 2005 'Bombay Sapphire Design Discovery', Space Furniture, Sydney, Melbourne and Brisbane; 2006 'The Presence of Things', Monash Gallery, Melbourne (Travelling Australia); 'Freestyle', Melbourne Museum

Awards
2004 Idea Award, *Inside* Magazine Annual, Melbourne; 2005 Australian Design Award, Powerhouse Museum, Sydney; 2006 Finalist, Bombay Sapphire Design Discovery Award, Australia

Mana Bernardes
www.mana.cx
Born in Brazil, 1981
Lives and works in Rio de Janeiro

Education
2001 Art therapy, Instituto de Artes Terapeutas do Brasil, Rio de Janeiro; 2003–present Industrial drawing, Pontificia Universidade Catolica, Rio de Janeiro

Professional Background
1996–2002 Professor of art, 'D+ da Conta', Museu da Republica, Rio de Janeiro; 1999 Artisan management, Centro de Assessoria do Movimento Popular, Rio de Janeiro; 1999–2001 Jewellery designer, R.Sobral, Rio de Janeiro; 2001–present, Jewellery designer, MANA, Rio de Janeiro

Selected Projects
2005 Magnomento, a metal latch made with transparent plastic, Thurleflex Ind.; Jóia Trabalhosa – Toilsome Jewellery Necklace, Juliana Lima; 2006 Urban Indian Bracelet, Tatiana Barbosa; Brinco Brinquedo – Toy Earring, Suelen Figueira

Selected Bibliography
2004 Juliana Motta, *Vogue Mistérios da Criação*, São Paulo; 2005 Bernardo Senna, *Momentum Design Contemporâneo do Rio de Janeiro*, Rio de Janeiro; Heloisa Marra, *Jornal o Globo*, 19 November, Rio de Janeiro; Maria Helena Estrada, *Arc Design Que Luxo*, December, São Paulo

Selected Exhibitions
2004 'In Partnership with her Mother, Rute Casoy', Daqui do Brasil, Rio de Janeiro; 'Metal and Lights', Metal and Lights in the Design Biennial, Saint Etienne; 'Design + Social', Sebrae Alagoas, Alagoas, Brazil; 2005 'Displays 4 pieces in the exhibition on Aluízio Magalhães', Centro Cultural dos Correios, Rio de Janeiro; 'Tim's stand exhibition', São Paulo Fashion Week, São Paulo; 'Conectar-se Pelo Cordão', *I.D.* Magazine presents 25 short films on the identity of video art, Chelsea Art Museum, New York; 'Video Art Conectar-se Pelo Cordão, 'J'en Rêve', Fondation Cartier, Paris; 2006 'Estampa de Cada Rosto' Paço Imperial Museum, Rio de Janeiro.

Big-game
www.big-game.ch
Grégoire Jeanmonod: born Switzerland, 1978; Elric Petit: born Belgium, 1978; Augustin Scott de Martinville: born France, 1980 Grégoire Jeanmonod, Augustin Scott de Martinville: live and

work in Lausanne; Elric Petit: lives and works in Brussels

Education
Grégoire Jeanmonod: 2003 BA Product and Industrial Design, École cantonale d'art de Lausanne; Elric Petit: 2001 BA Industrial Design, École Nationale Supérieure des Arts Visuels de la Cambre, Brussels; 2003 MA Industrial Design, École Nationale Supérieure des Arts Visuels de la Cambre, Brussels; Augustin Scott de Martinville: 2003 BA Product and Industrial Design, École cantonale d'art de Lausanne; 2005 MA Product and Industrial Design, École cantonale d'art de Lausanne

Professional Background
Grégoire Jeanmonod: 2004 – present: Designer, Pozzo Styling, Vevey, Switzerland; Elric Petit: 2005 Lecturer at École Nationale Supérieure des Arts Visuels de la Cambre, Brussels; 2006 Art Director, MXlight, Brussels; All: 2004 Established Big-game; Augustin Scott de Martinville: 2006 Professor at École cantonale d'art de Lausanne

Selected Projects
2005 HERITAGE IN PROGRESS: Moose, Vlaemsch; Treteau, Ligne Roset; Site Lamp, Hot by Mitralux; Le Marquis Reymond, coat peg; New Rich collection, accessories, +41; 2006 PACK, SWEET PACK: Tetra, seat; Box, stool, Fragile, vase; Styrene, lamp, prototype; PACK, SWEET PACK: Flatpack, rug, Galerie Kreo

Selected Bibliography
2005 Denis Laurent, 'Big-game sort le grand jeu', Upstreet, no.54, Paris; Luis Royal, 'Treteau (2005)', 100% – O Independente, 29 Apri, Lisbon; Urs Honegger, 'Grosswilddesign', Hochparterre, no.12, December, Zurich; 2006 Andrea Eschbach 'Trophies for Today', Frame, no.48, Jan/Feb, Amsterdam; Mareike Müller, 'Playboys made in Switzerland', Rondo, Der Standard, 8 September, Vienna; Alex Bagner, '50 people places things', Wallpaper*, October, London; Julie V.Iovine, 'Belgium's next generation of design stars?', I.D., November, New York; Françoise Foulon, 'Une imposture venielle', DITS, n°7, Autumn/Winter, Hornu, Belgium

Selected Exhibitions
2005 'HERITAGE IN PROGRESS', Salone Satellite, Milan; 'Label-design.be, Design in Belgium after 2000', Grand-Hornu Images, Hornu, Belgium; 'Design Brussels', Brussels Expo; 'Le jeune design bruxellois', Salon International du Design d'intérieur, Montreal; 2006 'Brussels design generation 2006', Belgisches Haus, Cologne, Germany; 'New Rich', Internationale Möbelmessse, Cologne; 'PACK, SWEET PACK', Salone Satellite, Milan

Awards
2005 Goldene Hase, First Prize for

design category, 'Die Besten' competition/Hochparterre – Zurich; Award of Excellence/SIDIM 2005, Montreal; 2006 Swiss Federal Design Grants, Bern

Steven Blaess
BLAESS™
www.blaess.com
Born in Australia, 1969
Lives and works in Dubai and Sydney

Education
1994 Bachelor of Design in Human Environments, University of South Australia, Adelaide; 1998 Business Certificate, Sydney

Professional Background
1996–present PINKUSBLAESS PTY LTD, Melbourne, Sydney and Dubai; 1999–present BLAESS™, Sydney and Dubai; 2001–present BLAESS TM architecture, interior, furniture, product projects; 2004 Alessi Workshop, Lago Maggiore, Italy

Selected Projects
1999 3pac Tampon Purse Pack Case for Johnson & Johnson, PINKUSBLAESS PTY LTD; 2000 Circa lighting, iGuzzini; 2001 Meditation Pod, seating, Edra; 2004 Corona lighting, pendant, wall mount, outdoor bollard lighting series, Artemide; 2005 Nursing Pad Case for Johnson & Johnson, PINKUSBLAESS PTY LTD; Marli Apribottiglie, bottle opener, Alessi

Selected Bibliography
2001 Petra Trefalt, 'Zwischen den Welten (Between Worlds)', design report, no. 7+8, Stuttgart; Michelle Hespe, 'A Taste of Blaess', Cream, Autumn issue, Sydney; 2004 Giovanni Odoni, 'Fuoriluogo', Casamica, March, Milan; 2004 Aldo Coloneti,'Un nuovo rapporto tra esperti del cibo e designer', Multipli di Cibo Centro Progetti Foodesign Guzzini, Bologna; Kate Bezar, 'Steven Blaess', Dumbo Feather, Pass It On, no.3, December, Sydney; Sean Topham, 'Taking the Trailer Park Uptown', Move House, Munich; 2005 Paul McGillick, 'Knocking On Doors', Indesign, no.24, February, Sydney; 2006 Marie Le Fort, 'Steven Blaess', AD, June, Paris

Selected Exhibitions
2001 'Edra', Salone del Mobile, Milan; 'Edra + Italian Design Showcase', Harrods, London; 2002 'Worldwide', Atelier Renault, Paris; 'Hybrid Objects', Melbourne Museum, Melbourne; 'Tokyo Designers Block, Hybrid Objects', Australian Embassy, Tokyo; 2004 'Food design guzzini – Fratelli Guzzini', Triennale, Milan; 2005 'Alessi', Salone del Mobile, Milan; Marli Apribottiglie bottle opener, Permanent Design Collection, Powerhouse Museum, Sydney; 2006 'Conversations of Things New – Italian Manufacturing and Australian Design', Melbourne and Sydney

Awards
2001 Finalist Torch Relay Design Competition for Sydney Olympics 2000

BLESS
www.bless-service.de
Desiree Heiss: born Austria ,1971;
Ines Kaag: born Germany, 1970
Desiree Heiss lives and works in Paris;
Ines Kaag lives and works in Berlin

Education:
Desiree Heiss: 1994 Academy for Applied Arts, Kunst, Vienna; Ines Kaag: 1995 University of Art and Design, Hannover

Professional Background:
1997 Established BLESS

Selected Projects
1999 BLESS No.06 Customizable Footwear for J. Colonna defilée; BLESS No.07 Chairwear B, Zucca defilée; BLESS Packaging System for CD-Edition 20 to 2000, Raster-noton; 2000 BLESS Advanced Earbags, designed by Tom Natvig; 2003 Sunshield in collaboration with Alan Mikli; Droog exhibition T-shirt; 2004 Mobil as shop interior for Addition Adelaide, Tokyo; Mobils and Handrail as museum-shop interior for GfzK Leipzig

Selected Bibliography
2001 Olivier Zahm, 'Bless, Recent History', Purple, spring; 2002 Renny Ramakers, Less+More: Design in Context, Amsterdam; 'Espace Critique de prod et de culture contemp', Catalogue Manifesta, May; Catherine Geel, 'Bless, l'art de pied en cap', BAM, no.221, October, Paris; 2003 Pierre Doze, 'Design Modeste', L'Officiel, April, Paris; Pierre Doze, Intramuros, no. 106, April/May, Paris; Guillaume Frauly, 'Destins Croisés', Les Inrock, June, Paris; 'Bless You', Neo 2, October; 2004 'Bless in Hotel Costes', Ryuko Tsushin, January, Tokyo; Marianne Wellershoff, Kultur Spiegel, June, Berlin

Selected Exhibitions
2000 'BLESS shop 8', Vivre sa Vie, Transmission Gallery, Glasgow; 'BLESS shop 2', Festival des Arts de la Mode, Hyères, France; 'BLESS shop 1, I love Fashion', Gallerie ffwd, Berlin; 2001 'Interpassive Shopping', Galerie des Galeries de Lafayette, Paris; 'Ars Viva 01/02 – Kunst und Design', Museum für Angewandte Kunst, Cologne; 2002 'Ars Viva 01/02 – Kunst und Design', GfzK, Leipzig and Neues Museeum, Weimar, Germany; 'Manifesta 4', European Biennial of Contemporary Art, Frankfurt; 'BLESS shop 12', Neues Musuem Nürnberg, Nuremburg; 2004 'Designmai', Berlin; 'BLESS Perpetual Home Motion Machine Transmitted Privacy', Gallery FR66, Paris; 'BLESS at home in Toronto', Art Metropole, Toronto; 2005 'BLESS – Design Relativators, The Vacuum Cleaner Cjaor', Townhouse Gallery, Cairo; 'Bless No. 26 cable jewellery', Galery Oona; 'Mo De', Hillside Terrace, Tokyo; 'Passagen', Galerie Reckmann, Cologne; 'Le cas du sac', Musée de la Mode et du Textile, Paris; 'Fashination', Moderna Museet, Stockholm; 2006 'BLESS', Tensta Kunsthall, Sweden

Awards
2001 Ars Viva, Berlin; 2004 Andam, Paris

Jörg Boner
Jörg Boner productdesign
www.jorgboner.ch
Born in Switzerland, 1968
Lives and works in Zurich

Education
1988 Diploma as carpenter/joiner; 1992 Diploma as draughtsman; 1996 Diploma with honours as product designer, Academy of Applied Art, Basel

Professional Background
1996–9 Designer, Berger & Stahl design studio, Basel; 1997 Co-founder of the Swiss design group N2 Design; 1999 Established Jörg Boner productdesign in Basel and since 2001 in Zurich

Selected Projects
2000 AJAX, desk, ClassiCon; 2002 Who's gonna drive you home tonight?, bench, Spazio Opos; 2004 BoCu, chair, Team by Wellis; 2005 DRESSCODE, wardrobe, Nils Holger Moormann; 2006 ALMA, floor lamp, Limited Edition

Selected Bibliography
2001 Lotte Schilder Bär and Norbert Wild, Designland Schweiz – Gebrauchsgüterkultur im 20. Jahrhundert, Zurich; Gabriele Lueg and Köbi Gantenbein, Swiss Made – Aktuelles Design aus der Schweiz, Zurich; Paola Antonelli, Workspheres – Design and Contemporary Workstyles, exhibition catalogue, New York; 2002 Christine Collin, Le Design Dans la Collection du Fond National d'Art Contemp., Paris; 2003 Valentina Croci, 'La luce ribalta l parametric', Ottagono, no.159, Bologna; 2004 Ralf Michel, 'Haute Cuisine', Frame, no.36; 2005 Renate Menzi, 'Ein Schrank aus starkem Stoff', Tages Anzeiger, 19 February, Zurich; Kristina Radarschad, 'Making of Dresscode', Form, no.203, Offenbach am Main; Maja Fueter, 'Portrait Jörg Boner, Raum und Wohnen, September, Zurich; 2006 Andrea Eschbach, 'Besuch bei Jörg Boner', design report, March, Leinfelden

Selected Exhibitions
2000 'submeet', Migrosmuseum für Gegenwartskunst, Zurich; 2001 'swiss made', Museum für angewandte Kunst Köln, Cologne; 'workspheres', Museum of Modern Art, New York; 2002 'Milan in a Van, Best of Milan International Furniture Fair 2002', Victoria & Albert Museum, London; 2006 'Swiss Design Now', Museum of Contemporary Art, Shanghai

Awards
2000 First Prize, Swiss Design Support Award, Zurich; 2005 Good Design Award, Chicago Athenaeum, Chicago; 2006 iF Gold Design Award, Hanover

Fernando Brízio
Born in Portugal, 1968
Lives and works in Lisbon

Education
1996 Equipment Design, Fine Arts Faculty, Universidade de Lisboa; 2003 Product Design, Architecture Faculty, Universidade de Técnica de Lisboa

Professional Background
1997 Proto Design, Lisbon; 1999 Established Fernando Brízio Studio; 2005 Visiting professor, École cantonale d'art de Lausanne, Lausanne; 2006 Head of the design department at School of Arts and Design, Caldas da Rainha, Portugal

Selected Projects
1999 Casa, ceramic lamp, Authentics; 2001 Tableshirt, tablecloth, Details; 2003 Mapping, Intramuros stand at IN'NOVA fair, Lisbon; Sound System #1, aluminium lamp, Fernando Brízio Studio; 2005 Salt and pepper dispenser, Cor Unum; 2006 Tromp D`Oeil, single drawer chest, Droog Design and Corian

Selected Bibliography
2003 Pierre Doze, 'Fernando Brízio: Figuration Libre', Intramuros, no.109, October/November, Paris; 2004 Brigette Fitoussi, 'Expérimental', Numéro, January, Paris; Antonella Boise, 'Fernando Brízio', Interni, April, Milan; Henrique Cayatte and Guta Moura Guedes, Portogallo 1990–2004, November, Milan; Frédéric Morel, Qu`est-ce que le design? (aujourd'hui), Paris; 2005 Liliana Albertazzi, 'Fernando Brízio', Intramuros, no.119, July/August, Paris; 2006 Jennifer Hudson, 1000 New Designs and Where to Find Them, London; Kate Franklin and Martin Raymond, 'Design Notebook', Viewpoint, no.20, October, London; Kieran Long, 'Fernando Brízio', Icon, no.031, January, London; Marcus Fairs, Twenty-First Century Design, October, London

Selected Exhibitions
2000 'Essere Ben Essere', Triennale di Milano, Milan; 2001 'Dry Kitchen', Droog Design/DMD, Milan; 2002 'Naydeikkuna – Contemporary Portuguese Design', Artek, Helsinki; 2004 'Shooting Stars of Europe', Designmai Forum, Berlin; 'Au doigt et à la baguette', Design Museum, London; 2005 'Plaques Sensibles', Beaubourg, Centre Pompidou, Paris; '97', Cristina Guerra Contemporary Art, Lisbon

Awards
1995 Young Designer ICEP Award for the Japanese market, Lisbon

Stephen Burks
Readymade Projects
www.readymadeprojects.com
Born in USA, 1969
Lives and works in New York

Education
1992 Department of Architecture & Institute of Design, Illinois Institute of Technology, Chicago, Illinois; 1995 Graduate School of Architecture, Planning & Design, Columbia University, New York

Professional Background
1993 Idee Co. Ltd., Tokyo; 1996 Swatch,

Milan; 1997–2001 Readymade, New York 2002 Established Readymade Projects, New York

Selected Projects
2000 Display, shelving system, Cappellini; 2003 Not So Soft Furniture, table and chair, Mogu; 2004 Missoni Patchwork, limited edition vases, Readymade Projects; Light Frame, pendant lamp, David design; Ambrogio & Gio, magazine rack and clothes dryer, Zanotta; 2006 Missoni Profumi, fragrance bottle and packaging, Estée Lauder; Modus Parallel, Cup Chair and Remnant, shelving system, chair and rug, Modus

Selected Bibliography
2004 Virginio Briatore, 'Stephen Burks', *Interni*, April, Milan; Chantal Hamaide, 'Stephen Burks Designer', *Intramuros*, no. 113, July/August, Paris; 2005 Pilar Viladas, 'Stephen Burks, Puff Dada', *New York Times*, Style Magazine Men's Fall Fashion, New York; 2006 Sonia Audoly, 'Born In The USA', *Elle Décor*, March, Milan

Selected Exhibitions
2003 'Cooper-Hewitt Design Triennial', Cooper-Hewitt, National Design Museum, New York; 2004 'Stephen Burks Designer', Intramuros/Audience & Creation, Paris; 2006 'Stephen Burks Readymade Projects', Knoxville Museum of Art Design Lab, Knoxville, Tennessee; 2007 'Designs for Life', Museum of Science & Industry, Chicago

Sam Buxton
www.sambuxton.com
Born in UK, 1972
Lives and works in London

Education
1997 Middlesex University, London; 1999 Royal College of Art, London

Professional Background
1999–2000 Established studio At The Third Stroke, London; 2000–2 Established Design Laboratory with Mathias Bengtsson, London; 2002 Established own studio, London

Selected Projects
2000 Business card, miniature fold-up sculpture; 2001 KENZO Parfums Paris, worldwide display system, KENZO; 2002 MIKRO-House, fold-up sculpture, Worldwide Co.; 2003 Hub Commissioning Centre, information point, SB Studio; MIKRO-City, miniature stainless-steel city installation, Design Museum, London; 2005 Vauxhall Skate City, urban skating obstacle, collaboration with pro-skateboarder Pete King; SIEMENS CBI Exhibition Stand, Laser Shape

Selected Exhibitions
2001 'Design Now London', Design Museum, London; 2003 'Great Brits at the Milan Furniture Fair', Paul Smith HQ, Milan; 'Great Brits at Tokyo Design Week', United Nations University, Tokyo; 2004 'Designer of the Year 2004', Design Museum, London; 'Great Brits in Brazil', House

Museum, São Paulo; 'MIKROs and Clone Bodies', Object Gallery, Sydney; 2005 'Twinkle Twinkle – British Council', Istanbul

Awards
2004 Shortlisted for the Designer of the Year Award, Design Museum, London

Elio Caccavale
Elio Caccavale Design Studio
www.eliocaccavale.com
Born in Italy, 1975
Lives and works in London

Education
1997 Facoltà di Architettura, Università degli Studi di Napoli Federico II, Naples; 2001 BA (Hons) Product Design, Glasgow School of Art; 2003 MA Product Design, Royal College of Art, London

Professional Background
2003–present Associate lecturer, BA Product Design, University of the Arts London Central Saint Martins; 2004 Design consultant Branson Coates Architecture, London; 2004–present Visiting lecturer, MA Industrial Design, University of the Arts London Central Saint Martins; 2004–2005 Visiting lecturer, Intermediate Unit 9, Architectural Association School of Architecture, London; 2005–present Elio Caccavale Design Studio, London; Visiting lecturer, MA Design Interactions, Royal College of Art, London

Selected Projects
2004 Utility Pets, Smoke Eater, Toy Communicator, Comforting Device, Low-resolution Pig TV, Memento Service, self-production; A Cupboard Full of Mothers, Science Drama, Wellcome Trust & Nesta; 2005 Hybrids: Towards a New Typology of Beings and Animal Products, Wellcome Trust; MyBio Dolls range: MyBio Boy, MyBio Pig, MyBio Spider Goat, MyBio Bioreactor, MyBio Glowing Rabbit, MyBio Jellyfish, Batch Production Science Learning Centre; 2006 De_Coded, First Stage Research Project, PEALS Policy, Ethics and Life Sciences Research Institute

Selected Bibliography
2003 Louise Schouwenberg, 'Learning in London', *Frame*, no.27, July, Amsterdam; Elio Caccavale, 'Room Service', *Wallpaper**, December, London; 2004 Stefano Mirti and Walter Aprile, 'A Man and His Pig', *Cluster*, no.004, October, Turin; 2005 Alice Wang, 'Utility Pets', *Art World*, no.003, January, Shanghai; Shumon Basar, 'Animal Magic', *Tank*, no.003, December, London; 2006 Elio Caccavale, 'A Pig Saved My Life', *Etcetera*, no.100, March, Brussels; Anthony Dunne, 'Utility Pets', *Domus*, no.889, February, Milan; Robert Klanten & Matthias Hübner, *Dot Dot Dash – Designer Toys, Action Figures and Character Art*, October, Berlin; Walter Aprile, Britta Bolan & Stefano Mirti, *Interaction Design Primer (Vol.1)*,

September, Milan

Selected Exhibitions
2004 'Transmediale. Fly Utopia!', Haus der Kulturen der Welt, Berlin; 2005 'Subtle Technologies', Innis Town Hall, University of Toronto, Toronto; 'MyBio', Science Learning Centre, London; 2006 'PopNoir', The Israel Museum, Jerusalem; 'ZeroOne San Jose: A Global Festival of Art on the Edge (ISAE 2006)', San Jose, California; 'Bêtes de style', MUDAC, Lausanne; 'Biennale Internationale Design 2006 Saint-Etienne', France; 2007 'The New Italian Design', La Triennale di Milano, Milan

Awards
2003 Kenny Yip Medical Award, Royal College of Art, London; 2005 Wellcome Trust Sciart Award, London

Louise Campbell
Louise Campbell Studio
www.louisecampbell.com
Born in Denmark, 1970
Lives and works in Copenhagen

Education
1992 BSc, London College of Furniture, London; 1995 MA, Danmarks Designskole, Copenhagen

Professional Background
1996 Established own studio

Selected Projects
1998 A Retreat Funnel, chair, self-production; 1999 Casual Cupboards, storage units, out of production; Honesty, chair, one-off made by Jacob Trolle; Bless You, chair, self-production; 2000 Folda, sofa and armchair, out of production; 2002 Waiting Rooms: Leave your Mark, Entertainment, Interaction; Series of three furniture designs for waiting rooms, one-offs made by Jacob Thau and Erik Jørgensens Møbelfabrik; 2003 Bille goes Zen, one-off, Lars Bille Christensen; 2004 Fatso, Radiator, LP Balls, Campbell Pendant, 14 one-off lamps, in collaboration with Louis Poulsen; 2005 Collage, pendant and floor lamp, Louis Poulsen; Prince Chair, Hay; 2006, Home Work, table top accessories, Stelton; Spider, stacking chair, Hay; Veryround, chair, Zanotta

Selected Bibliography
2003 Mel Byars, *The Design Encyclopedia*, New York; Katherine E. Nelson, 'Against the grain', *I.D.*, vol.50, number 6, October, Cincinnati; 2005 Julie Müller Stahl, *Dish*, New York; 'Natural High', *Casa International*, no.36, December, Shanghai; 2006 Henrik Most, 'Corridor of Power', *Frame*, no.48, January/February, Amsterdam; Tina Jørstian, 'Laser Cut, but', *Louisiana*, no.20, February, Copenhagen; Palle Schmidt, Den nye Generation, Copenhagen; Hirokuni Kanki, 'Complaint', *Axis*, no.119, February, Tokyo; David Sokol, 'Ministry of Color', *Interior Design*, no.77, March, New York; Ellen Himelfarb, 'Story of', *Azure*, May, Ontario

Selected Exhibitions
2002 'quick, quick, slow', Danish Design Centre, Copenhagen; 'Waiting Rooms', Trapholt Museum of Art and Design, Kolding; The Generation X', travelling exhibition; 2003 'Derfor/curator and designer of exhibition', Paustian, Copenhagen; 2004 'Stuff/Curator of exhibition', Harbourfront centre, Toronto; 'studies with light' with Louis Poulsen', travelling exhibition; 2005 'Solo exhibition', SimSum, Madrid; 'Nordic Cool, Hot Women Designers', National Museum of Women in the Arts, Washington D.C.

Awards
2004 The Finn Juhl Architectural Award, Copenhagen; 2005 iF Gold medal Award, Hanover; 2006 *Wallpaper** Interiors Maverick Award, London; EDIDA *ELLE DECO* International Design Award; Home Beautiful Award, Sydney

Leo Capote
LCapote Arquitetura & Design
Born in Brazil, 1981
Lives and works in São Paulo

Education
2002 Universidade Paulista, São Paulo

Professional Background
2001–02 Campana Design (Fernando and Humberto Campana), São Paulo

Selected Projects
2001 Spoon Simple, picture holder, Leo Capote; Spider, picture holder, Leo Capote; Spoon Chair, Leo Capote; 2002 Luminária Passe Bem, lamp, Leo Capote; 2003 Jar Lamp, Leo Capote; Disk Chair, Leo Capote; Disk Stool, Leo Capote; Spoon Chair 2, Leo Capote

Selected Bibliography
2002 *XVI Exibição Brasileira de Decoração, Casa Cor Guide*, (annual), Brazil; 'Gente Nova na Praça', *Casa Cláudia*, no.10, May, São Paulo; 2002 Designers, *Casa & Jardim*, no.572, September; Ricardo D'angelo, 'Luz que Aquece' – Viver Bem, no.10, October, São Paulo; 'Cotidiano Revisitado', *Viver Bem*, no.5, October, São Paulo; 2003 Kiki Romero, 'As Boas Compras', *Veja*, April, São Paulo; Simone Quintas, 'Cobiça, Jarra de Luz', *Casa & Jardim*, no.581, June; 'Acensa suas Noites com Luminárias Especiais', *VIP*, no.11, November; Vania Silva, 'Utilidades Domésticas Revisitadas', no.117, December

Jennifer Carpenter
TRUCK Product Architecture.
www.the-truck.com
Born in USA, 1970
Lives and works in New York

Education
1992 BA, Yale, New Haven, Connecticut 1998 M Arch, Columbia, New York

Professional Background
1992–5 Designer, Cesar Pelli & Associates, New Haven; 1999–2000

Architect, Rogers Marvel Architects, New York; 2000–present Partner, TRUCK Product Architecture, New York

Selected Projects
2003 Studio Nova by TRUCK, line of dinnerware, drinkware and accessories, Studio Nova; Kate Spade home fixtures, store/retail fixtures, Lenox; 2004 Built by Me Collection, children's furniture, Offi; Tambour Table, storage table, TRUCK; 2005 STRETCH Fence, Design Trust for Public Space; 2006 TRUCK for Steeltek, toothbrush holder and other bath accessories, Steeltek/Sitlax; TRUCK for Nurseryworks, children's furniture, Nurseryworks

Selected Bibliography
2004 Elizabeth Weil, 'The Modernist Nursery', *New York Times Magazine*, no.28, November, New York; Jane Margolies, 'Getting Tough About Kid's Stuff', *I.D.*, March/April, New York; 2005 Julie Müller Stahl, *Dish: Internat'l Design for the Home*, February, New York; Stacy Perman, 'The Total Taxi Rethink', *Business Week*, 21 July; Photo essay by Dan Forbes, 'A Touch of Mass: Against the Grain', *Time*, Summer, New York; Marianne Rohrlich, 'For Tiny Tasks, Big League Design', *New York Times*, 29 September, New York; Mairi Beautyman, 'Taxi Upgrades', *Wired*, November; Mireille Hyde, 'Place Setting: Product', *Metropolis*, December, New York; Cara Brower, Rachel Mallory and Zachary Ohlman, *Experimental Eco-Design*, Mies, Switzerland; 2006 David Anger, 'Recipe for Comfort', *Better Homes and Gardens*, January, Des Moines, Iowa; Kelly McMasters, 'Brooklyn Block Party', *Metropolis*, June, New York

Selected Exhibitions
2002 Tokyo Designers Block with Idee, Aoyama Mansion, Tokyo; 'Future Furniture' by Interior Design magazine, New York D&D Building, New York; 2003 '15 in 15' by *Metropolis*, ICFF, New York; 'Transformations: New International Design', Parsons Gallery, The New School, New York; 2005 'Designing the Taxi' by the Design Trust for Public Space, Parsons Gallery, The New School, New York; 2006 'Blockparty: Art and Design in the New Urban Home', 267 A State Street, with BKLYN Designs, New York

Awards
2001 and 2002 Future Furniture Award, *Interior Design* magazine, New York; 2003 Good Design Awards for Built by Me Collection and U-Turn Tables, Chicago Athenaeum, Chicago; 2006 Technological Advancement Award, History Channel City of the Future Competition, New York

PIERRE CHARPIN
www.pierrecharpin.com
Born in France, 1962
Lives and works in Ivry Sur Seine

Education

1984 École Nationale Des Beaux-Arts, Bourges, France

Selected Projects

1993 Stacking Chair, prototype; 2002 Stands Collection, stand, Design Gallery Milano; Triplo Vase, Venini; 2003/05 Ceram X Vases, Craft; 2004 Basket-Up, coffee table, Montina; Carpet, Post Design; Carafe Eau de Paris, glass carafe, Eau de Paris; 2005 Ceramica 2 and Lampada 27, Oggetti Lenti Collection, objects and furniture, Design Gallery Milano & Haute Définition Paris; 2006 Platform Collection, coffee table and wall mirror, Galerie Kreo

Selected Bibliography

2000 Marco Romanelli, 'Ceramica e Vetro: Martin Szekely, Pierre Charpin, Jasper Morrison', *Abitare*, No.399, October, Milan; 2002 Laurence Salmon, 'Pierre Charpin en État de Grâce', *Intramuros*, no.101, June/July, Paris; Christian Poirier, 'Pierre Charpin Maxims words on stands', *Inview*, no.20, September, Paris; 2004 Albert Hill, 'Stands and deliver', *Wallpaper**, no.247, May, London; 2005 Alex Bagner, 'High scorers', *Wallpaper**, no.80, July/August, London; Claire Fayolle, 'Pierre Charpin: Objets d'Utilité Publique', *Beaux-Arts*, no.247, December, Paris; 2006 Christian Simenc, 'Les Particules Élémentaires', *Journal des Arts*, no.233, March, Paris

Selected Exhibitions

2000 'Solo Show', Galerie Kreo, Paris; 2001 'Torno Subito', Musée des Arts Décoratifs, Paris; 2002 'Stands', Design Gallery Milano, Milan; 2005 'Oggetti Lenti', Design Gallery Milano, Milan; 'Pierre Charpin 2005', Grand Hornu Images, Hornu, Belgium; 2006 'Platforms', Galerie Kreo, Paris

Awards

2005 Designer de l'Année du Salon du Meuble de Paris, Paris

Lucas Chirnside

Lucas Chirnside studio
Born in New Zealand, 1976
Lives and works in Melbourne

Education

2000 Gerrit Rietveld Academie (Student Exchange), Amsterdam; 2002 BA Architecture (Hons), Royal Melbourne Institute of Technology, Melbourne

Professional Background

2000 S333 Architecture and Urbanism, Amsterdam; 2002 Watson Architecture and Design, Grimshaw Architects, Melbourne; Established own practice; 2005 Jackson Clements Burrows Architects, Melbourne

Selected Projects

2004 Polytope Seating, prototype, Edag Future/Lucas Chirnside studio; 2005 HUB circumference wristwatch, concept; PI World Time Clock, Lucas Chirnside studio; 2006 smlwrld Desk Clock,

limited edition, Lucas Chirnside studio; Shuriken lamp, concept; Ta Ke, floor lamp, prototype

Selected Bibliography

2002 Michael Trudgeon, 'Design Cultures', *Object*, no.40, October, Sydney; Simon Braun, 'Orient Express', *Blueprint*, no.202, December, London; Ross McLeod, 'Hybrid Objects', *(Inside)*, no.25, December, Melbourne; 2005 Juliana Engberg, *A Molecular History of Everything*, January, Melbourne; Ebony Bizys, 'Raw Like Sushi', *Vogue Living*, January/February, Melbourne; Idea Awards, 'Idea Awards', *(Inside)*, no.37, June/July, Melbourne; 2006 Megan Michelson, 'Go List', *Outside*, February, New York; Penelope Barker, 'Around the World', *Cream*, February/March, Sydney; Leta Keens, 'Clocking On', *Belle*, February/March, Melbourne; Kelly Holt, 'Time for Design', *Poloxygen*, no.16, March, Sydney; Ying-Lan Dan, 'Time Travel', *Monument*, no.72, April/May, Sydney; *Freestyle, New Australian Design for Living*, October, Sydney

Selected Exhibitions

2002 'Hybrid Objects', Tokyo Designers Block, Australian Embassy, Tokyo; 2004 'Tyranny of Distance', Tokyo Designers Block, Gallery Piazza Harajuku, Tokyo; 'Tyranny of Distance', Seoul Design Festival, Hangaram Art Museum, Seoul; 2005 'A Molecular History of Everything' (with Bianca Looney), Australian Centre for Contemporary Art, Melbourne; 'Printing Matters' (with Bianca Looney), Witte de With Centre for Contemporary Art, Rotterdam; 'Salone Satellite (Melbourne Movement)', Salone del Mobile, Milan; 2006 Bombay Sapphire Guest Designer Salone del Mobile, Superstudio 'PIU, Milan; 'Freestyle, New Australian Design for Living' Melbourne Museum, Melbourne

Awards

2004 Australia Council for the Arts, Milan Studio Residency, Milan; 2005 Concept Design Award, Furnitex 'Vivid', Melbourne; Design Discovery Award, Bombay Sapphire, Sydney

Clemens Weisshaar & Reed Kram

Kram/Weisshaar AB
www.kramweisshaar.com
Reed Kram: born USA, 1971; Clemens Weisshaar: born Germany, 1977
Both live and work in Munich and Stockholm

Education

Reed Kram: 1998 MIT, Cambridge, Massachusetts; Clemens Weisshaar: 1996 Central Saint Martins, London; 1998 Royal College of Art, London

Professional Background

Reed Kram: Media Design, Copenhagen and Stockholm; Clemens Weisshaar: 1996–9 Konstantin Grcic Industrial

Design, Munich; 1999–2002 Industrial Design, Clemens Weisshaar, Munich and London; Reed Kram: 1996–8 Co-founder of the Aesthetics & Computing group at MIT, Cambridge, Massachusetts; 1999–2002; Both: 2002 Established Kram/Weisshaar AB

Selected Projects

2001 Media & Installation Design, PRADA Epicenter Store New York, PRADA; 2002 Living 2030, proposal for media installation, Beaubourg, Centre Pompidou; 2004 Media & installation design, PRADA Epicenter Store Los Angeles, PRADA; 2005 Countach, Moroso; 2006 The Design Annual, Exhibition Architecture Messe Frankfurt

Selected Bibliography

2001 Jens Hommert and Rem Koolhaas, *Prada Projects 1*, Fondazione Prada, Milan; Leo Gullbring, 'Graphic Spaces', *Frame*, March, Amsterdam; 2002 Anne Pascual, 'Flagshipstore und Epicenter', *De:Bug*, no.59, May, Berlin; 2003 Mike Toth, *Fashion Icon*, New York; 2004 Nils Forsberg, 'Technologue', *Forum*, January, Stockholm; Hanae Komachi, 'Up and Coming Creators', *Axis*, no.110, August, Tokyo; 2005 Francesca Picchi (ed.) Interview by Stefano Mirti and Walter Aprile, 'Table Genetics', *Domus*, no.879, March, Milan; Ronald Jones, 'A Breed Apart', *I.D.*, no.505, May, New York; 2006 Florian Böhm (ed.), *Breeding Tables*, Barcelona

Selected Exhibitions

2001 'PRADA Works In Progress with OMA', Fondazione Prada, Milan; 'Stealing Eyeballs', K/Haus & Museum in Progress, Vienna; 2004 'Architecture, Media, Industrial Design for PRADA', Ospedale degli Innocenti, Florence; 2005 'La Raccolta', Spazio Krizia, Milan; 2006 'Breeding Tables', Design Museum, London

Awards

2005 Breeding Tables selected for permanent collection, Vitra Design Museum, Weil am Rhein; 'La Raccolta' selected for permanent collection, Beaubourg, Centre Pompidou, Paris

Kenneth Cobonpue

www.kennethcobonpue.com
Born in The Philippines, 1968
Lives and works in Cebu

Education

1991 Pratt Institute, New York; 1996 Akademie Baden, Württemburg, Germany; 1996 Holz, Berufsgenossenschaft, Munich

Professional Background

1990 Designer & craftsman, Centro Azur, Florence; 1991–4, 1996–present Designer and general manager, Interior Crafts of the Islands, Inc., Cebu; 1995 Firma Decker & Reposa; 1996 Woodworking & cabinet-making apprenticeship, Schreinerei, Munich; 2001–present Director, Kenneth

Cobonpue USA, Washington D.C.

Selected Bibliography

2000 Elizabeth Reyes, *Tropical Living*, Hong Kong; 2002 Elizabeth Reyes, *Tropical Interiors*, Hong Kong; Ross Lovegrove, London; 2004 Tom Dixon, *International Design Yearbook*, London; 2005 Marcel Wanders, *International Design Yearbook*, London

Selected Exhibitions

2003–4 'Life Symphony', Felissimo Design House, New York; 'Design 21, Theme: Why the World Needs Love', New York; 2004 'Design 21, Theme: Why the World Needs Love', Barcelona; 2004–5 'Kenneth Cobonpue: Idea + Form + Craft', Material Connexion Gallery, New York; 'Design 21, Theme: Why the World Needs Love', Kobe, Japan; 2005 'Design 21, Theme: Why the World Needs Love', Tokyo, New York, Paris

Awards

2002 Presidential Citation for Embodying the Ideas of Asian Design, Malacanan Palace, Philippines; 2003 Japan Good Design Award, Tokyo; 2005 Design Excellence Award, Hospitality Design Product Competition, New York; Design For Asia Award 2005, Hong Kong

Paul Cocksedge

Paul Cocksedge Studio
www.paulcocksedge.com
Born in UK, 1978
Lives and works in London

Education

2000 Sheffield Hallam University, Sheffield, UK; 2002 Royal College of Art, London

Professional Background

2003 Co-founded Paul Cocksedge Studio, London

Selected Projects

2003 Watt?, light, Paul Cocksedge Studio; NeON, lights, Paul Cocksedge Studio; Bombay Sapphire Light, Paul Cocksedge Studio; 2004 Styrenissimo, large scale installation of Styrene, London, Paul Cocksedge Studio; 2005 Crystallize, chandelier for Swarovski Crystal Palace Collection, Paul Cocksedge Studio; 2006 On Reflection, customized bin for Vipp, Paul Cocksedge Studio; Light As Air, lighting sculptures, Paul Cocksedge Studio

Selected Bibliography

2002 Fiona Rattray, 'Class of 2002', *Independent* magazine, June; 2003 Lidz-Ama Applah, 'Paul Cocksedge', CNN – Design 360; Sara Manuelli, 'Light Heavyweight', *Design Week*, October, UK; Lesley Jackson, 'Light Fantastic', *Elle Décor*, November, UK; 2004 Richard Parkin, 'Designer of the Year', *BBC – Arts*; Max Fraser, 'Max Fraser Review', *Pen*, no.140, Japan; Tamsin Blanchard, 'At the Flick of a Switch', *Observer* magazine, no.11, January; Helen Kirwin-Taylor, 'Struck by Lighting', *Telegraph* magazine,

no.7, February; Karen Klages, 'Designer Makes Magic With Light', *Chicago Tribune*, April, USA; Pilar Viladas and David Farber, 'Design Language', *New York Times*, 9 May, New York; Stephen Cook, 'Leading Light in Free Thinking', *Guardian*, 29 June, London; Kyoko Nakajima, 'Up and Coming Creators', *Axis*, August, Japan; 2005 Julian MacDonnell, documentary for 'Designers Under Pressure Series', Channel 4, Pumpkin TV; Tamsin Blanchard, *Love Your Home*, UK; Marcel Wanders, *International Design Yearbook*, London; Charlotte J. and Peter M. Fiell, *1001 Lights*, UK; Helen Kirwin-Taylor, 'Meltdown Grows into Bright Idea', *Wall Street Journal*, 8 April, Europe Edition; Kieran Long, 'Milan Top 10', *Icon*, June, London; 2006 Frank Lukeit, 'Light Magic – Paul Cocksedge', *Deutsche Welle-Euromaxx*; Nicole Swengley, 'Lighting to Blow You Away', *The Times*, 12 May, London; Janice Blackburn, 'A Strangely Un-English Education', *Financial Times*, 25 March, London

Selected Exhibitions

2001 '6 Visions', Issey Miyake Gallery, Tokyo; 2003 'Ingo Maurer presents Paul Cocksedge', Spazio Krizia, Milan; 2004 'Brilliant', Victoria & Albert Museum, London; 'Designer of the Year', Design Museum, London; 'Open Borders', Curated by Droog Design for Lille, European Capital of Culture, Lille; 'Bombay Sapphire Blue Room', Powerhouse Museum, Sydney; 'Twinkle Twinkle', Laforet Musuem, Tokyo; 2005 'Twinkle Twinkle', 100% Design, Istanbul; 'Twinkle Twinkle', 100% Design, Moscow; 'Swarovski Crystal Palace Collection', La Segheria, Milan; 2006 'Light As Air', Rabih Hage Gallery, London

Awards

2003 1st Prize Glass Prize Awards, Bombay Sapphire Foundation, London; 2004 nominated for Designer of the Year, Design Museum, London

D-BROS

Draft Co., Ltd.
www.d-bros.jp
Ryosuke Uehara: born Japan, 1972; Yoshie Watanabe: born Japan, 1961
Both live and work in Tokyo

Education

Ryosuke Uehara: 1997 Tama Art University, Tokyo; Yoshie Watanabe: 1983 Yamaguchi University, Yamaguchi, Japan

Professional Background

Ryosuke Uehara: 1997 D-BROS, Draft Co., Ltd., Tokyo; Yoshie Watanabe: 1986 D-BROS, Draft Co., Ltd., Tokyo

Selected Projects

2003 Hope Forever Blossoming, vase, D-BROS; Rivulet of the Heart, cup and saucer, D-BROS; 2004 Seconds Tick Away, wall clock, D-BROS; Live Together, glassware, D-BROS; 2005 Time Paper, wall clock, D-BROS; Method of Drinking Fairy Tale, glassware, D-BROS; Waltz, cup and saucer, D-BROS; 2006 Hotel Butterfly, light, glass, letterhead, luncheon mat and bookmark, D-BROS

Selected Bibliography
2004 *Axis*, no.112, December, Tokyo; *Graphic*, no.20 July, Tokyo; 2005 *idea*, no.309, May, Tokyo; *Casa Brutus*, no.68, November, Tokyo

Awards
2000 Tokyo Art Director's Club (ADC) for packing tape: A Path to the Future, Tokyo; 2002 Tokyo ADC for vase: Hope Forever Blossoming and greeting card, Tokyo; Hara Hiromu prize, Tokyo for 'Bouch' book; New York ADC Award for 'Bouch' book, New York; 2004 New York ADC GOLD award for 'Piece and Peace' calender, New York

DEMAKERSVAN
www.demakersvan.com
Judith de Graauw: born The Netherlands, 1976; Jeroen Verhoeven: born The Netherlands, 1976; Joep Verhoeven: born The Netherlands, 1976
All live and work in Rotterdam

Education
2004 All graduated, Product Design, Design Academy, Eindhoven

Selected Projects
Judith de Graauw: 2004 Lost&Found, stool and lamps, DEMAKERSVAN; Jeroen Verhoeven: 2004 Cinderella Table, DEMAKERSVAN; Joep Verhoeven: 2004 How to Plant a Fence, fence, DEMAKERSVAN and IDFence

Selected Exhibitions
2004 'Graduation Show', Design Academy, Eindhoven; 2005 'Dutch Village', ICFF, New York; 'Salone del Mobile', Palazzo Trusardi, Milan; Jeroen Verhoeven: 2006 'Modern Craft', Museum of Modern Art, New York

Awards
Joep and Jeroen Verhoeven: 2005 *Arhend & Items Magazine*, Eindhoven

Stefan Diez
www.stefan-diez.com
Born in Germany, 1971
Lives and works in Munich

Education
1994 Course in Cabinet Making, Ursula Maier studio, Stuttgart
2002 BA Industrial Design, Academy of Fine Arts, Stuttgart

Professional Background
1998 Assistant to Richard Sapper; 1999–2002 Assistant to Konstantin Grcic, Munich; 2003 Established own studio

Selected Projects
Tema, cutlery, Rosenthal/Thomas; Friday, chair; Buoy, soap dispenser, Authentics; Big Bin, multi-purpose storage system, Authentics; 2005 GENIO, cooking pots, Rosenthal-Thomas; 2006 Shuttle, storing

container, serving container, orange juicer, pepper and herb mill, Rosenthal/Thomas; Kuvert, bags, Authentics; UPON Wall, coat hanger, Schönbuch; Bent, tables and chairs (in collaboration with Christophe de la Fontaine), Moroso

Selected Bibliography
2002 Klaus Meyer, 'Grips und Gefühle', *design report*, July/August, Stuttgart; 2005 Jan van Rossem, 'Mann der Tat', *Architektur & Wohnen*, no.3/05, June/July, Frankfurt; 2006 Francesca Picchi, 'Networked generation', *Domus*, no.890, March, Milan; Klaus Meyer, 'Besuch bei Stefan Diez', *design report*, no.5/06, May, Stuttgart; Oliver Herwig, 'Nahtlos schön – A Bag of Tricks', *form*, no.208, May/June, Frankfurt; Liliania Albertazzi, 'L'ingéniosité de Stefan Diez', *Intramuros*, 124, May/June, Paris; Hanae Komachi, 'Up and coming creators in the eyes of prominent designers 3', *Axis*, 122, July, Tokyo

Selected Exhibitions
2004 'Designparcours Munich, Quality Control for Nymphenburg Porzellanmanufaktur', Munich; 'Designparcours Munich, exhibition for Rosenthal Porzellan', Bayerisches Nationalmuseum, Munich; 'Rosenthal Design Award', exhibition design, Pinakothek der Moderne, Munich; 2006 'Ideal House Cologne 06', IMM Cologne

Awards
2003 Best of the Best Award for Instant Lounge (in collaboration with Christophe de la Fontaine), IMM Cologne; 2004 Winner of the Fokus, Gold for GENIO and Silver for Tema, Stuttgart; 2005 iF-Design Award for GENIO and Tema, Hanover; A&W Mentor award, selected by Richard Sapper (A&W Designer of the Year 2005), Hamburg; Best of the Best Award for Couch, IMM Cologne; Red Dot Award for Couch, Essen; 2006 Silver for GENIO, Designpreis der Bundesrepublik Deutschland, Frankfurt

Florence Doléac
www.doleac.net
Born in France, 1968
Lives and works in Paris

Education
1994 Diplôme ENSCI les ateliers, École Nationale Supérieure de Design Industrielle, Paris

Professional Background
1993–2003 Member of the Group Radi Designers, Paris; 2000–present Professor, École cantonale d'art de Lausanne; 2003 Established own studio; 2006–present Professor, École Nationale des Arts Décoratifs, Paris

Selected Projects
2001 Robot, Porte-fruits et légumes en céramique, Florence Doléac/Radi Designers, Commande Publique du Ministère de la Culture et de la Communication; Poterie d'Amélie,

Andreas Bergmann/Vallauris and Calude Aello/Vallauris; 2002 La Chaise Mise à Nu, Housse de Chaise-Tapis, Édition Galerie Aline Vidal; 2004 Chamarande, Modules d'Accueil pour les médiatrices du Centre d'Art, Domaine Départemental de Chamarande; 2005 L'Espace Créateurs des Galeries Lafayette, Conception du Nouvel Espace Créateurs du magasin principal, Galeries Lafayette 40 BD Haussmann; 2006 Patapouf, Siège de Jardin, Édition Fermob

Selected Bibliography
2003 Pierre Doze 'Intramuros Portrait Florence Doléac', 'Le principe d'incertitude', *Intramuros*, no.106, April–May, Paris; Françoise Ayxendri 'Portrait "La Fée du Design"', *Elle*, no.3048, June, Paris; 2006 Cécilia Bezzan and Chantal Prod'hom, 'Tac Tic', *In Folio*, June, Lausanne; Alexandra Midal 'Design & Designer Florence Doléac', *Pyramyd*, September, Paris

Selected Exhibitions
2003 'Divagations', Villa Noailles, Hyères; 2004 'Flogistique', Avec l'Aide à la Première Exposition de la D.A.P., Direction des Arts Plastiques, Ministère de la Culture, Galerie Aline Vidal, Paris; 'Exposition Personnelle', Toolsgalerie, Paris; 'Vague à l'Âme, Tapis Réalisé en Iran, Parsua', Espace Chevallier, Paris; 2006 'Tac Tic, Carte Blanche', MUDAC, Lausanne; 'Own Invites Florence Doléac', Boutique Own à Bruxelles, Brussels; 'Exposition Personnelle', Galerie Philippe Jousse & Galerie Aline Vidal, Paris

Awards
2001 Villa Kujoyama, Afaa, Pour le Développement du Revêtement de Sol 'Stimulisol', Kyoto; 2003 Aide à la Première Exposition, D.A.P. avec la Galerie Aline Vidal, Paris; 2005 Critt Chimie, Centre Régional d'Innovation et de Transfert de Technologie Pour le Projet 'Poignée Molle', Paris; 2006 Bourse AGORA du Ministère de la Culture et de la Communication, pour un projet de recherche 'Le Corps Aborbe', Paris; Les Globes de Cristal, Lauréate du Prix de la Presse et la Culture, Paris

Doshi Levien
Doshi Levien design office
www.doshilevien.com
Nipa Doshi: born India, 1971; Jonathan Levien: born Scotland, 1972
Both live and work in London

Education
Nipa Doshi: 1989 National Institute of Design, Ahmedabad; 1995 Royal College of Art, London; Jonathan Levien: 1989 Rycotewood College, Thame, Oxon, UK; 1991 Buckinghamshire College of Brunel University, High Wycome, UK; 1995 Royal College of Art, London

Professional Background
Nipa Doshi: 1998 Architect, David Chipperfield, London; Jonathan Levien:

1997 Ross Lovegrove (Studio X), London; Both: 2000 Established Doshi Levien design office

Selected Projects
2003 Mosaic range for Tefal, cooking pans, Tefal; 2005 Doctor's Case and Stethoscope, Wellcome Trust installation, Fulcrum Model makers; 2006 My World: the new subjectivity in design, Fulcrum Model Makers; Fan, prototype, Doshi Levien; Matlo, drinking vessel, Doshi Levien; My World: the new subjectivity in design, marble table, Texxus

Selected Bibliography
2003 Henrietta Thompson, 'Opposites Attract', *Blueprint*, October, London; Jenny Dalton, 'Young British Designers', *Elle Décor*, October, London; 2004 Fiona Rattray, 'Bright Young Things', *Independent* magazine, January, London; Henrietta Thompson, 'Healthy Dose of Sparkle', *Blueprint*, November, London; 2005 Virginio Briatore, 'United Kingdom of India', *Interni*, April, Italy; Albert Hill, 'Function and Fantasy', *World of Interiors*, September, London; Marcus Fairs, 'London issue', *icon*, no.028, October, London; Ilse Crawford, *Home is where the heart is*, New York; 2006 Nipa Doshi, 'Made for India', *Wallpaper**, March, London

Selected Exhibitions
2004 'Global Local', India Habitat Centre, Delhi; 2006 'My World: the new subjectivity in design', Lisbon Design Biennale, Lisbon; 'My World: the new subjectivity in design', Design Museum, London

Awards
2003 Winner of Best Furnishing or Accessory for Residential Interiors for Mosaic range for Tefal, FX Design Award, London; 2005 Voted *Blueprint* Sessions Product Designer of the Year, London; Award to produce work for the British Council exhibition 'My World', Arts Council of Great Britain, London

david dubois
www.davidduboisproduct.com
Born in France, 1971
Lives and works in Paris

Education
2003 École Nationale Supérieure de Création Industrielle, Paris

Selected Projects
2004 Clothes-Press, armoire mobile (limited edition), Galerie FR66; 2005 ABCD Bags, (limited edition), Galerie FR66, Adaptable Strap, T-shirt Bag, Bag Net, Leather Purse Bracelet, davidduboisproduct; Sink, DMA; 2006 System Bag, Chêne-à-Vélo, Water-Bench, MUDAM; Easystool, Galerie Kreo; Christian Lacroix Shop, new concept for CL-shops; 2007 Kronenbourg, beer glasses for the Kronenbourg brasserie, ARC

Selected Bibliography
2004 Anne-Marie Fèvre, 'La lampe, c'est l'hallu', *Libération*, 27 July, Paris; 2005 Anne-Marie Fèvre, 'Dubois, us et costumes', *Libération*, 18 March, Paris; Anne-Marie Fèvre, 'Le Var avant-garde', *Libération*, 8 August, Paris; Bénédicte Duhalde, 'Les Tuteurs de David Dubois', *Intramuros*, no.120, September/October, Paris; 2006 David Dubois and Christine Colin, *Design & Imitation*, Paris; Romina Calo, Björn Dahlström, Clément Minighetti, Claude Moyen and Monica Portillo, *Eldorado*, Luxembourg; Claire Fayolle, 'Un musée très design', *Beaux-Arts Magazine*, no.265, July, Paris

Selected Exhibitions
2004 'DEBUT', VILLA NOAILLES, Hyères; 2005 'LUMIÈRES BLANCHES', Reckermann Gallery, Cologne; 'TUTORING', Centre d'art/Domaine de Baudouvin, La Valette du Var; 'ABCD BAGS', Galerie FR66, Paris; 'DMA-EXPERIMENTA LISBOA', Museu Da Cidade, Lisbon; 2006 'ELDORADO', MUDAM, Luxembourg; 'STOOL', Galerie Kreo, Paris

Piet Hein Eek
Eek en Ruijgrok
www.pietheineek.nl
Born in The Netherlands, 1967
Lives and works in Geldrop

Education
1986 VWO, Amsterdam; 1990 Academy of Industrial Design, Eindhoven

Professional Background
1992 Established own studio, Geldrop

Selected Projects
2004 Scrap Wood Cabinet, Eek en Ruijgrok; 2006 Crisis Tube Chair, Eek en Ruijgrok; Football Lamp, Eek en Ruijgrok; Kanteen Table in Scrapwood, Eek en Ruijgrok

Selected Bibliography
2006 Max Fraser, *Boek Piet Hein Eek*, 23 September, De Boekenmakers

Selected Exhibitions
1996 'Door Cupboards', Stedelijk Museum, Amsterdam; 1998 'Garden Houses', Groninger Museum, Groningen

Martino d'Esposito
www.despositogaillard.com
Born in Cyprus, 1976
Lives and works in Lausanne

Education
1998 École supérieure d'arts appliqués de Vevey, Switzerland; 2002 BA Industrial Design, École cantonale d'art de Lausanne; 2004 MA Industrial Design, École cantonale d'art de Lausanne

Professional Background
2002–06 Professor of product design, École cantonale d'art de Lausanne; 2004 Professor of design, Swiss Federal Institute of Technology, Lausanne; Established own studio; 2005 Professor

of design, École d'Ingénieurs d'Yverdon, Switzerland

Selected Projects
2000 L6300 and L6350, watches, Lacoste; 2003 Profil (vase), Grow Bag (planter) and Drink'Tray, Ligne Roset; 2004 Lumin'Air, and Triple Lustre, lights, Ligne Roset; 2005 MartiniChéri, martini glass, Ligne Roset

Selected Bibliography
2001 Marion Godau, 'Rational und humorvoll-Junges Schweizer design', *design report*, no.9, September, Frankfurt; 2002 A.-M. F., 'Remue-Ménage, *Libération*, 4 February, Paris; rédaction, 'Fumer ou dormir', *Intramuros*, no.100, April, Paris; C.H., 'In Reih und Glied', *design report*, 6, June, Frankfurt; 2003 École cantonale d'art de Lausanne, 'Ecal', *Industrial Design*, Lausanne; 2004 John D. Harris, *Inventions 2004*, Neuilly sur Seine, Michel Lafon; rédaction, '1000 Modèles Design', *L'Officiel*, no.2, summer, Paris; Albert Hill 'Outside the Box', *Wallpaper**, no.70, August; 2005 Julie Lasky, 'Count It! One Good Reason We Produced This Issue', *I.D.*, no.40, January, New York; rédaction, '1000 Modèles Design', *L'Officiel*, no.3, summer, Paris; Paola Antonelli, *SAFE: Design Takes on Risk*, September, exhibition catalogue, New York; Pierre Keller, *Swiss Design Now*, September, Shanghai; Julie Lasky & I.D. editors, 'The Way You Move Me', *The New Yorker*, Art & Architecture issue, 17 October, New York; 2006 'Inout Designers', *Inout*, February, Lausanne

Selected Exhibitions
2001 'Earthquake Tables' exhibited for École cantonale d'art de Lausanne's showroom, Salone Satellite, Milan; 2002 'Milking stool', École cantonale d'art de Lausanne showroom, Swiss Cultural Center, Milan; 'Earthquake Tables' Swiss Cultural Center, Paris; 2003 'Milking stool' Binnen Gallery, Amsterdam; 'Earthquake-tables' in 'From Sweeden With Love', Museum für Gestaltung, Zurich; 2004 'Conductor's Baton', École cantonale d'art de Lausanne showroom, Scala and Grand Hotel et de Milan; 'Conductor's Baton', Design Museum Tank, London; 2005 'Drink'Tray' at 'Swiss Design Now', MoCA, Shanghai; 'Drink'Tray' in 'Signes Quotidiens', Swiss Cultural Center, Paris; 2005–6 'Earthquake Tables' in 'SAFE: Design Takes on Risk', MoMA, New York

Awards
2000 Ebel award 'Autour de la main', La Chaux-de-Fonds, Switzerland; 2001 1st prize for École cantonale d'art de Lausanne showroom, Salone Satellite, Milan

FRONT
www.frontdesign.se
Sofia Lagerkuist: born Sweden; Charlotte von der Lancken: born Sweden; Anna Lindgren: born Sweden; Katja Säuström: born Sweden
All live and work in Stockholm

Education
2004 All: Masters Degree Industrial Design, Konstfack University College of Arts Craft and Design

Selected Projects
2003 Wallpaper by Rats, Front; Table by Insects, Droog Design; Lamp by Rabbit, Moooi; 2004 Reflection Vase, Droog Design; Design by Explosion, soft lounge chair, prototype; Walking Table, prototype; Branch Table, Barry Friedman Gallery and Droog Design; Falling Vase; 2005 Sketch Furniture, Barry Friedman Gallery; 2006 Animal Thing, Horse Lamp, Pig Tray and Rabbit Lamp, Moooi; Office Collection – Bin, Hanger, Peg, Screen, Materia

Selected Bibliography
2004 Lesley Jackson, *icon*, February; Ronald Jones, *I.D.*, September/October, New York; 2005 Ross Lovegrove, *The International Design Yearbook 2005*, London; Anneke Bokern, *Frame*, July/August; Kieran Long, *icon*, September

Selected Exhibitions
2004–5 'Sallone Satellite', Salone del Mobile; 2005 'Droog meets Front', Droog @ Home Gallery, Amsterdam; 'Design 05', Barry Friedman Gallery (New York), Art Basel Miami; Tokyo Design Week, Tokyo; 'Koncept Design', National Museum, Stockholm; 2006 Interior at Stockholm Furniture Fair, 'Interior – Bar by Front', Stockholm Furniture Fair, Stockholm

Awards
2004 Future Design Days, Stockholm; Best Young Designer of the Year, *Wallpaper**, UK; 2006 Design Prima Award for Bin 'Best in Show' London

FUCHS + FUNKE
www.fuchs-funke.de
Wilm Fuchs: born Germany, 1972; Kai Funke: born Germany, 1970
Both live and work in Berlin

Education
2001 Diploma Industrial Design, Institute for Product and Process Design, Universität der Künste Berlin

Professional Background
2001 Established FUCHS + FUNKE

Selected Projects
1999 (Pillow) Chair, folding chair, Habitat; 2000 (Reclining) Chair, transforming chaise longue, FUCHS + FUNKE; 2000–2002 (Lounge) Chair, transforming easy chair, FUCHS + FUNKE; 2002 Bauhaus-Archiv Shop furniture, FUCHS + FUNKE; Polygon Chair, easy chair, FUCHS + FUNKE; 2003 Waist, lamp, Habitat; Bonanza, cantilever chair, FUCHS + FUNKE; 2004 Berlin Mitte Chairs, talk-show easy chairs, FUCHS + FUNKE; 2005 Plastic Papton, public seating, FUCHS +

FUNKE; 2006 869g Chair, experimental lightweight chair, FUCHS + FUNKE

Selected Bibliography
2001 Chantal Hamaide, 'Attitudes atyp-iques', *Intramuros*, no.10+11, Issy-Les-Moulineaux; 2002 'Sitzen und Liegen', *md*, no.1/02, Leinsfelden-Echterdingen; Klaus Meyer, 'Raffinesse mit Gebrauchswer', *design report*, no.7+8, Stuttgart; Cecilia Fabiani, '20 progetti in cerca di azienda', *ddn*, no.98, Milan; 2003 Patricia Holz, 'Mod rockers', *Azure*, March/April, Toronto; Mateo Kries, Britt Angelis, *Design Berlin – New Projects for a Changing City*, Berlin; Almerico de Angelis, 'Design a Berlino', *MODO*, May/June, Milan; Carolin Kurz, *Berliner Style*, London; 2004 'Magic of Chairs', *DesigN*, 2; *International Design Yearbook 2004*, London; 2005 Oliver Jahn, 'Die Wiege des Designs', *monopol*, no.5, Berlin; 2006 Blaine Brownell, *Transmaterial*, New York

Selected Exhibitions
2000 'Salone Satellite', Salone del Mobile, Milan; 2001 'Talents', Tendence, Frankfurt; 2002 'Salone Satellite', Salone del Mobile, Milan; 'New Fairy Tales', Interieur, Kortrijk; 2003 'Tube', Designmai, Berlin; 'Design Berlin', Vitra Design Museum, Berlin; 'Eine neue deutsche Welle', Galerie Einzigart, Zurich; 'Design Berlin', Cube Gallery, Manchester; '100% Folding Chairs', 100% Design, London; 2004 'Inspired by Cologne', IMM, Cologne; 'Wohnmunition', Designmai, Berlin; 'Materialica Design Show', Materialica, Munich; 'Design Berlin', Museum of Architecture, Moscow; 2005 'Neue Lieferung', Designmai, Berlin; 'Jung + Deutsch', Edison Höfe, Berlin; 'Jung + Deutsch', Art Front Gallery, Tokyo; BtoB – from Berlin to Brussels, Designbrussels, Brussels; 2006 'Intercity: Berlin–Prague', Galerie Manes, Prague

Awards
2002 Special Mention, *design report* Award, Milan; Bronze Leaf, International Furniture Design Award, Asahikawa, Japan; Outdoor Award, Design for Europe Kortrijk, Kortrijk, Belgium; 2003 Selection 100% Folding Chairs, London; 2004 Materialica Design Award, Munich

Martino Gamper
www.gampermartino.com
Born in Italy, 1971
Lives and works in London

Education
1994 Design, University of Applied Art Vienna; Sculpture, Academy of Fine Art Vienna; 1998 Product Design, Royal College of Art, London

Professional Background
1996–98 Designer for Studio Matteo Thun, Milan; 1998 Furniture designer for Studio Barrazuol, Treviso, Italy; 2000 Established own design studio

Selected Bibliography
2000 Martino Gamper, 'Corners', *a place to be contained within*, July, London, self-published; 2004 Martino Gamper, *What Martino Gamper did between 2000–2004*, September London, self-published

Awards
2003 OXO Tower Design Award in Furniture Category, London

Alexis Georgacopoulos
www.georgacopoulos.com
Born in Greece, 1976
Lives and works in Lausanne

Education
1999 BA Industrial and Product Design, École cantonale d'art de Lausanne

Professional Background
2000–present Head of the Industrial and Product Design Department, École cantonale d'art de Lausanne; 2003–present Lecturer, École Polytechnique Federale Lausanne

Selected Projects
2000 Inflatable Bottle Cooler, ECAL; 2002 D/side furniture, chair, table and bench, Attitudes Art Gallery; Milk from Production to Consumption, milking stool (prototype), ECAL; 2003 Rain-mat, A door mat and an umbrella stand, DIM; Faster pussycat, chair (prototype); 2004 Newton, fruit bowl, own production; 2005 Stando, blown glass vase, self-production; 2006 CMYK, trays, self-production; Ultragrip, football (prototype)

Selected Bibliography
2002 'Le botte-cul', exhibition catalogue, Lausanne; Memos Filippidis, 'The Farm', *Vimagasino*, Vimagasino; 2003 Valentina Crocci, 'Winning with the leaders', *Ottagono*, no.159, April, Milan; Jeroen Van Roojien, 'Moving up the world', *Frame*, no.31, April, Amsterdam; Memos Filippidis, 'What's new', *Figaro Maison*, July, Athens; 2004 'Mundo Low Tech', *Arc Design*, no.35, February, São Paulo; 'The Conductor's Stick', Exhibition catalogue, April, Lausanne; Memos Filippidis, 'Musical design', *Vimagasino*, no.187, May, Athens; Marrizio Vitta, 'Design Maturo', *L'Arca*, no.193, June, Milan; Albert Hill, 'Outside the Box', *Wallpaper**, July-August, London; 'Signale', *design report*, no.9, September; Christos Xanthakis, 'Portrait', *Homme*, no.19, November, Athens; 2005 Ariella Risch, 'Illywords-coopetition', *Illywords*, no.9, April, Trieste; Volker Albus, 'Somewhat different', exhibition catalogue, November, Stuttgart; 'Swiss Design Now', exhibition catalogue, November, Shanghai; Norbert Wild, 'Take Away', exhibition catalogue, December, Zurich; 2006 Meret Ernst and Christian Eggenberger, *Design Suisse*, December, Zurich

Selected Exhibitions
2000 'Are en Forme, MUDAC, Lausanne;

2001 'Swiss Made', Museum für Angewandte Kunst, Cologne; 2003 'Swiss Peaks', Totem Gallery, New York; 'The Milking Stool', Design Museum, London; 2004 'The Conductor's Stick', Grand Hotel et de Milan, MIFF, Milan; 'Shooting-Stars of Europe', DesignMai 2004, Berlin; 'Criss-Cross/the best of Swiss design', DesignMai 2004, Berlin; 2005 'Swiss Design Now', Museum of Contemporary Art, Shanghai; 'Take Away', Museum Für Gestaltung, Zurich; 'Somewhat different', The Townhouse Gallery, Cairo; 'Brave New Worlds', DesignMai 2005, Berlin; 2006 'Football Fever', Design Museum, London

Adam Goodrum
Born in Australia, 1972
Lives and works in Sydney

Education
1989 Churchlands High School, Perth; 1993 University of Technology, Sydney

Professional Background
1996 Designer, Parker Furniture, Sydney; 2000 Lecturer, University of New South Wales, Sydney; 2002 Cabinet Maker, William Russell Architects, London; 2003–6 Lecturer, University of Technology, Sydney

Selected Projects
1998 Folding Series, Loop Chair, Folding Table, Folding and Baby's Highchair, Adam Goodrum; 2001 Spaghetti Installation, sculptural artistic installation, Adam Goodrum; 2005 Spa Furniture Range, White Flax; Temporary Dwellings, folding house, Adam Goodrum; 'and the serpent tempted Eve', Tsunami Bowl, Lissasous Table and Eve, Adam Goodrum; 2006 Armour Screen, prototype, Adam Goodrum

Selected Bibliography
2004 Bonnie Malkin, 'Way to Go', *The Sydney Magazine*, 16, August, Sydney; Ben Wyld, 'Adam + Eve', *Domain, Sydney Morning Herald*, November, Sydney; Prue Pascoe, 'Shaken Not Stirred', *Inside*, no.34, Melbourne; 2005 Chelsea Hunter, 'All About Eve', *Black + White*, no.76, January, Sydney; Nicole Bearman, 'Design Tonic', *Monument*, no.65, February/March, Sydney; Julica Jungehulsing, 'Der Glamourose', *Schoner Wohnen Decoration*, no, July/August, Berlin; Susan Yelavich, 'I.D. Annual Design Review 2005, Category: Concepts, Award: Honourable Mention', *I.D. Annual Design Review*, vol. 52, no. 5, July/August, New York; Tracey Clement, 'Timepieces', *Driven, Peugeot*, October, Melbourne; 2006 Geoff Piggott, 'Form & Function', *LINO*, no.13, Autumn, Sydney

Selected Exhibitions
2004 'Workshopped', Strand Arcade, Sydney; 'Australian Design', Tokyo Design Week, Tokyo; 2005 '…and the serpent tempted Eve', Object Gallery, Sydney; 'Bombay Sapphire, Superstudio', Superstudio, Milan; 'Canberra Biennale',

Parliament House, Canberra; 2006 '20/20 Vision', NSW Parliament House, Sydney; 'Australian Freestyle', Melbourne Museum, Melbourne and Sydney Opera House, Sydney

Awards

2004 Design Award, Bombay Sapphire, Sydney; 2005 Stitch Chair selected, 2004 *I.D.* International Design Review, New York; Australian Council for the Arts Grant, Sydney

graf:decorative mode no.3

www.graf-d3.com
Hiroto Aranishi (furniture maker): born Japan, 1970; Shigeki Hattori (designer, representative): born Japan, 1970; Takashi Matsui (designer): born Japan, 1970; Yuji Nozawa (carpenter): born Japan, 1971; Hideki Toyoshima (artist): born Japan, 1971
All live and work in Osaka

Selected Projects

2000 graf building opening, graf : decorative mode no.3; 2001 'analogue from digital', furniture series, SOGO furniture Co. Ltd.; 2002 Yayoi Kusama Furniture by graf, collaborate furniture series; TABIO KINGS ROAD, sock shop, graf : decorative mode no.3; 2004 SUNAO, cutlery, Tsubame Shinko Co. Ltd.; 2006 'morozoff grand', confectionery brand, Morozoff Co. Ltd. and Studio Cultivate

Selected Bibliography

2000 'Design_Hot Sheet', *Wallpaper**, September, London; 2003 Tomoko Ishiguro, 'The Next Reality in Japan Design – A Creative Unit Offering Possibilities as Medium; graf', *Axis*, no.103, May/June, Tokyo; 2005 Toshiro Sato, 'Design Group-Craftmanship' *Nikkei Marketing Journal*, 29 July, Tokyo; 'A Yen for Simplicity, How to Spend It' *Financial Times, Interiors Edition*, October, London; 'The Style Files', *Wallpaper**, November, London; Yoshitomo Nara + graf, *Studio Voice*, vol.358, October, Tokyo; Takashi Matsui of graf, 'The Artist as an Artisan', *Hinism*, vol.5, October, Tokyo

Selected Exhibitions

2002 'Nakanoshima Sketch – structure for living', La Galerie des Nakamura, Tokyo; 2003 'Structure for living', Club 360, Milan; 'S.M.L', graf media gm, Osaka; 2005 'Yayoi Kusama Furniture by graf', Galleria Carla Sozzani, Milan; 'Yoshitomo Nara + graf', International Triennale of Contemporary Art Yokohama 2005, Yokohama; 2006 'Talents 2006 – EDIDA "winners as talents", Ambiente 2006, Frankfurt

Awards

2005 Sakuya Konohana Award, Osaka; 2006 Elle Deco International Design Award, Tokyo; Asahi Beer Art Prize, Tokyo

Tal Gur

Tal GuR design
www.talgur.com
Born in Israel, 1962
Lives and works in Tel Aviv, Jerusalem and Kibbutz Gilgal, Israel

Education

1996 BDES, Bezalel Academy of Art and Design, Jerusalem; 2000 Mino Paper Art Village, Mino, Japan; 2004 Tajimi City Pottery Design and Technical Center, Tajimi, Japan

Professional Background

1996 Established Tal GuR design studio; 1997–present Lecturer, Bezalel Academy for Art and Design, Jerusalem

Selected Projects

1999 Gelem – Og light fixture, Vashty bowl, Zer vase, Sarah light fixture, Yam Yabasha illuminated end table, Amos lampshade, Tal GuR design; 2000 Eash, light, Hofit; 2001 Tubes – Pepe stool, Doo lampshade, Orba light fixture, Anana light fixture, Romold for Tal GuR design; 2002 Simple Gestures – Shlonsky light fixture, Yabasha light fixture, Luba lampshade, Yam wall light fixture, Vazit vaze, Zaga chandelier, Tal GuR design; 2003 Sturdy Straws – Sturdy Straw chair, ottoman, lampshade and partition, Tal GuR design; 2004 Tajimi Project – Left Turn vase, Serve-Bowl, fruit bowl, Tal GuR design; 2005 Gwangju Biennale Light into Life – Gwangju chair, Flo Flow Flowers lamp shade, Tal GuR design; A Matter of Taste, table and rocking chairs, Tal GuR design

Selected Bibliography

2000 Virginio Briatore, 'Piccole Idee', *Interni*, no.502, June, Milan; 2001 Kyoko Nakajima, 'Why are Japanese designers flocking to London?', *Axis*, vol.89, Jan/Feb, Tokyo; Katsutoshi Ishibashi, 'Creators' Work', *Axis*, vol.90, Mar/Apr, Tokyo; Nirith Nelson, 'Letter From' – 'Israeli design', *New Design*, Sep–Oct, London; 2002 Billy Nolan, 'products', *Frame*, no.27, Jul–Aug, Amsterdam; 2003 Nirith Nelson, 'Fusion Design: Contemporary Israeli Design', *Product Design*, May, China; 2004 Jiyoung Choi, 'Designer's Tal Gur', *Maru*, vol.3, March, Korea; 2005 Alan Fennell, 'Tal Gur takes part at TDB 2004', *Designing Ways*, no.61, January, South Africa; Lee Soon-jong, Rhi Joomyung, Eun Byungsoo and Park Inn-seak, *Light Into Life*, October, Seoul; Shimazaki Makoto, *We Love Chairs*, November, Tokyo; 2006 Mel Byars, *New Chairs*, London; Ellen Lupton and Erzi Tarazi, *New Design from Israel*, exhibition catalogue, January, New York

Selected Exhibitions

2000 'Yalla design', Gallery Le Bureau des Esprits, Milan; 2001–2 'Industrious Designers: Israeli Design', Abitare il Tempo, Verona; 2002 'Simple Gestures', Periscope Gallery for Contemporary Design, Tel Aviv; 2002–5 'Domains –

Contemporary Israeli Design', Tokyo Designers Block 02, International Design Center, Nagoya; Jintai Museum, Beijing, International Expo Center, Shanghai, The Hangaram Museum, Seoul, The Museum of Fine Art, Taipei, UTS gallery, Sydney; RMIT gallery, Melbourne, Indira Ghandi gallery, New Delhi; 2003 'Sturdy straws', The Heder Gallery for Contemporary Art, Tel Aviv; 2005 'Promisedesign', Triennale, Milan, Designmai, Edison Hofe, Berlin; 'The Gwangju First International Design Biennale', Gwangju Design Center, Gwangju, Korea; 2006 'Solos: new design from Israel', Cooper-Hewitt, National Design Museum, New York

Awards

2001 Crate & Barrel, The Israel Museum, Jerusalem; 2001 Design Award, Ministry of Culture, Israel; 2004–5 Israel Cultural Excellence Foundation, IcExcellence, Israel

Ineke Hans

INEKEHANS/ARNHEM
www.inekehans.com
Born in The Netherlands, 1966
Lives and works in Arnhem

Education

1991 3-D Design, Hogeschool voor de Kunsten Arnhem, Arnhem; 1995 MA Furniture Design, Royal College of Art, London

Professional Background

1989–90 Studio Sipek, Amsterdam; 1997 Habitat UK Ltd, London; 1998 Established INEKEHANS/ARNHEM studio, Arnhem

Selected Projects

1997–2005 Ordinary Furniture – Tête-à-Tête Set, Beer Set, Deluxe Set, Relax Set, table and chairs, INEKEHANS/ARNHEM for INEKEHANS©OLLECTION; 2000–01 Black Beauties, Office Set, chair and desk, Happy Horse for children, Crash Car, INEKEHANS/ARNHEM for INEKE-HANS©OLLECTION; 2002 Black Gold, Five-armed Candleholder, Erik Jan Kwakkel for INEKEHANS©OLLECTION; Black Magic, Laser Chair, INEKEHANS/ARNHEM for INEKE-HANS©OLLECTION; 2005 Jolly Jubilee, easy chair, Arco; Garlic Crusher, Bowls & Spoons, Royal VKB; Forest for the Trees, coat stand, Lensvelt; 2006 Swarovski Crystal Palace, Pixelated Peacock, Dazzling Dahlia, Leaf Light, Swarovski

Selected Bibliography

2003 Ineke Hans & Ed Van Hinte, *Black Bazaar – Design Dilemmas*, no.010, Rotterdam

Awards

2005 Red Dot Award for Garlic Crusher, Germany; 2006 Design Plus Award for Garlic Crusher, Germany; Red Dot Award for Bowls & Spoons, Germany

Jaime Hayon

Hayonstudio
www.hayonstudio.com
Born in Spain, 1974
Lives and works in Barcelona

Education

1996 Instituto Europeo di Design, Madrid; 1997 École Nationale Supérieure des Arts Décoratifs, Paris

Professional Background

1997–2004 Fabrica, Treviso, Italy; 2000 Established Hayonstudio

Selected Projects

2003–6 AQHayon Collection, bathroom collection, Artquitect Editions; 2004 Josephine, lamp, Metalarte; 2006 Funghi, lamps, Metalarte; Showtime, furniture collection, BD; Kubokube, cabinet and Halo, mirror, Pallucco

Selected Bibliography

2003 Suzanne Wales, 'Club Med', *Surface*, no.43, New York; Suzanne Wales, 'Object Lesson', *Pol Oxygen*, no.4, Australia; 2005 Cover, *Interoi*, no.10, October, Verona; Rayon Ubeda, 'La Vida de un Diseno', *El Pais*, May, Madrid; 2006 'The Furniture Fiend', *Wallpaper**, February, London; Eva Schapp, 'Clash and Combine', *Frame*, no.49, March/April; 'La Maigo Interni', *Interni*, no.4, April, Verona; 'Elle Decoration "Aujourd'hui Talent"', *Elle Décor*, no.155, June, France

Selected Exhibitions

2003 'Mediterranean Digital Baroque', David Gill, London; 2005 'AQHayon', Design Museum, London; 2005–2006 'Mon Cirque', Iguapop Gallery, Barcelona, Oxop Gallery, Minneapolis, Stylepark, Cologne; 'Ready to Read', Lisbon Experimenta Georgetown, Lisbon

Awards

2006 *Wallpaper**, London; Best Bathroom 2006, *Elle Déco International*

Simon Heijdens

www.simonheijdens.com
Born in The Netherlands, 1978
Lives and works in London and Rotterdam

Education

1999 Hochschule der Künste Berlin; 2002 Cum Laude, Design Academy, Eindhoven

Professional Background

2001 Established own practice

Selected Projects

2002 Moving Wallpaper, self-production/Droog Design; 2003 Your Choice, 50 differently priced bags, Sugarcubes no.1–200, Droog Design; 2004 Tree, location-sensitive outdoor light installation, self-production; Broken White, ceramics, Droog Design; 2006 Lightweeds, indoor light installation, self-production; Reeds, light element, self-production; Rising Slowly, chandelier, Swarovski

Selected Bibliography

2004 Francesca Picchi, 'Simon Heijdens – The Biography of objects', *Domus*, no.876, December, Milan; Renny Ramakers, *Droog 10+1*, March, Amsterdam; 2005 Eva Schaap, 'New Dutch on the Block', *Frame*, no.46, September/October, Amsterdam; Liat Stanislavsky, 'Simon Says', *Binyan Vediur*, no.101, November, Tel Aviv; 2006 Filippo Romeo, 'Simon Heijdens', *Case da Abitare*, April, Milan; Die Gestalten, *Into the Nature*, June, Berlin

Selected Exhibitions

2004–6 'Tree', light installation for public space, London, Milan, Eindhoven, Amsterdam, Sheffield, Berlin, Cologne, Nottingham, Hasselt, 'Simply Droog', Munich, Den Haag, São Paulo, New York; 2005 '4 New Dutch Designers', Droog Design, Amsterdam; 'European Design Show', Design Museum, London; 'Grazic dei Fior', Haunting dogs, Milan; 2006 'NLA' Netherlands to Los Angeles, Acme Gallery, Los Angeles; 'Swarovski Crystal Palace', Swarovski, Milan; 'Contrasts and Contradictions', Contrast Gallery, Shanghai

Awards

2002 Best Graduation Project, Rene Smeets prize, Design Academy, Eindhoven

Jackson Hong

Jackson Hong Design
Born in South Korea, 1971
Lives and works in Seoul

Education

1995 BFA, Department of Visual & Industrial Design, Seoul National University; 1999 MFA, Department of Visual & Industrial Design, Seoul National University; 2002 MFA, Department of 3-D Design, Cranbrook Academy of Art, Bloomfield Hills, Michigan

Professional Background

1997–9 Designer, Technical Centre, Samsung Mobile Inc, Seoul; 2003–4 Product Designer, ECCO Design Inc, New York; 2005–6 Product Design Director, NP TF Team, NCsoft Corp, Seoul; 2005–present Part-time lecturer, department of industrial design, Ewha Womans University, Seoul

Selected Projects

2002 Hannstar LCD TV Project (IDEO), Hannstar; 2003 Bic Smart Pen Project (ECCO Design), Bic; KIRIN Kodak Digital Camera Project (ECCO Design), Kodak; Pure Digital One-time-use Digital Camera Project (IDEO), Pure Digital; 2005 External Intestine X, Jackson Hong Design; PE (Personal Entertainment) Project, NCsoft Corp.; 2006 Masked Citizen X Working Prototype, version 2.0, Jackson Hong Design

Selected Bibliography

2004 Chung Woo Lee, 'The King of Dark Interface: Jackson Hong', *Haute Magazine*,

November, Seoul; 2005 Chung Woo Lee, 'Jackson Hong and Hyungkoo Lee', *DT 01*, no.01, April, Seoul; Fritz K. Park and Emil Goh, 'How Korea Stays Ahead of the Crowd', *IdN Magazine*, Vol.12, no.4, September, Hong Kong; Iris Moon, 'Exhibition Review: Mechanic Beasts and Fire Wars – Jackson Hong's solo show', *Art Asia Pacific*, no.46, Fall, New York

Selected Exhibitions
2001 'In Progress', Network Gallery, Cranbrook Academy of Art, Bloomfield Hills; 2002 'Seoul Design Festival', Hangaram Design Museum, Seoul; 2005 'Shanghai_Cool', Shanghai Duolun Museum of Art, Shanghai; 'Mechanic Beasts and Fire Wars', Kimjinhye Gallery, Seoul; '17X17', Total Museum of Contemporary Art, Seoul; 'Upset', One and J Gallery, Seoul

Awards
2001 Gersin Design Insight, Cranbrook Academy of Art, Bloomfield Hills

Junya Ishigami
Born in Japan, 1974
Lives and works in Tokyo

Education
2000 Tokyo National University of Fine Arts and Music, Tokyo

Professional Background
2000–4 Kazuyo Sejima & Associates, Tokyo; 2004–present Junya Ishigami & Associates, Tokyo

Selected Projects
2003 Tables for restaurant noel, Nichinan Tekko, Ube, Yamaguchi, Japan; 2004 Round Table, space design for LEXUS exhibition, DOMECO Ltd; 2005 Table, Kirin Art project 2005, Akahoshi; Low Chair, Mituisi

Selected Bibliography
2003 Junya Ishigami, *Shinkenchiku*, October, Tokyo; 2004 Yoshiyuki Ishii *Domus*, no.871, June, Milan; 2005 Junya Ishigami, *SD Review 2005*, December, Tokyo; Francesca Picchi, *Domus*, no.881, May, Milan; Arnaldo Farnese, *www.Scoop*, August, New York; *AXIS*, October, Tokyo; 'brain', *ONE TEN GALLERY TALK*, no.45, December, Tokyo; 2006 'JA Winter', *Yearbook 2005*, January, Tokyo; Junya Ishigami, 'GA Houses 2006', *Project 2006*, no.92, April, Tokyo; Yoshiyuki Ishii, 'bijutsu techo', *Creators Edge*, April, Tokyo

Selected Exhibitions
2004 'LEXUS exhibition', Salone de Mobile, Milan; 'row house', *SD Review*, HILLSIDE TERRACE F building, Tokyo and Osaka; 2005 'table' Kirin Art project 2005, Kirin Art project 2005, Tokyo and Osaka; 2006 'table' Basel Art Unlimited, Basel Art Unlimited, Basel

Awards
2004 SD Prize, *SD Review 2005*, Tokyo; 2005 Kirin Prize, Kirin Art Project, Tokyo; 2006 1st Prize, Housing Proposal for Tokyo Electric Power Company, Tokyo

Ichiro Iwasaki
Iwasaki Design Studio
http://www.iwasaki-design-studio.net/
Born in Japan, 1965
Lives and works in Tokyo

Education
1986 Industrial Design, Salesian Polytechnic, Tokyo

Professional Background
1986–91 Sony Design Center, Tokyo; 1991–2 some design studios, Milan; 1995–Iwasaki Design Studio, Tokyo; 2005–Lecturer, Product Design Course, Tama Art University, Tokyo; Lecturer, Department of Design, Tokyo National University of Fine Arts and Music, Tokyo

Selected Projects
2001–6 various advanced designs, Sony, Panasonic, Canon, Ricoh, au by KDDI, JVC; 2002–5 Opus-1, spinning reel, Evergreen International; 2003 Takata 04-neocar, Child car seat, Takata; 2004 Telephone 510, MUTECH; cellular phone, A5405SA (Sanyo), A1403K, (Kyocera), au by KDDI; LOUV/METEO, personal chair and side table, AIDEC; 2006 Stainless Steel Kettle, PYPEX

Awards
2001–6 iF design award, Germany, and Good Design Award, Japan

ixi
www.ixilab.com
Izumi Kohama: born Japan, 1968; Xavier Moulin: born France, 1969
Both live and work in Okinawa

Education
Izumi Kohama: 1991 Musashiro Art University, Tokyo; Xavier Moulin: 1994 University of Industrial Design, Paris

Professional Background
Izumi Kohama: 1991 Interior Design, TACK Design, Tokyo, Japan; Xavier Moulin: 1994 Took part in 'Black radish' (Urban Mobile Design Agency), Projects for Homeless People and Gypsies, Paris; Izumi Kohama: 1995 Interior Design, Aldo Cibic and Partners, Milan; Xavier Moulin: 1996 Art Director, Aldo Cibic and Partners, Milan; Izumi Kohama: 1998 Interior Design, Antonio Citterio and Terry Dwan, Milan; Both: 1999 Established ixi, Paris, Milan, Tokyo; Xavier Moulin: 2006 established Okinawa Design Centre, ZERO in partnership with Ryoko Oshiro, Okinawa

Selected Projects
1998, 2006 Smart Home Fitness, Up, ladder and shelf, Kenji Nakachi; 2000 Interstice, Balladeuse, light, ixi; 2001 Homewear – Mangastrap, ixi; Homewear – Pady Bag, ixi; 2002 Homewear – Stoolpants, ixi; 2004 Homewear – High Tension, suit, ixi; 2005 ixi land, office, ixi

Selected Bibliography
2000 Jerome Sans, *My Home is Yours, Your Home is Mine*, November, Seoul; 2001 Charlotte and Peter Fiell, *Designing* the 21st Century, London; Stephanie Comer, 'Special Issue: Design', *Wired*, January, New York; Mami Kataoka, *My Home is Yours, Your Home is Mine*, July, Tokyo; 2002 Ellen Lupton, *Skin*, New York; 2004 Sang-kyu Kim, *Good Chair*, December, Seoul; 2005 'Good Chair', *Interni and Deco*, Spring, Seoul; Andrea Branzi, 'Un riformismo colorato', *Interni*, December, Milan; 2006 'ixi land', *Visita*, March, Tokyo

Selected Exhibitions
2000 'My Home is Yours, Your Home is Mine', Samsung Foundation Museum, Seoul; 2001 'Blow up shaped air in Design, Architecture, Fashion and Art', Vitra Design Museum, Weil am Rhein, Germany; 'My Home is Yours, Your Home is Mine', Tokyo City Opera Gallery, Tokyo; 2002 'Skin', Cooper-Hewitt, National Design Museum, New York; 2004 'Good chair' ('Homewear', wearable furniture), Seoul Arts Center, Hangaram Design Museum, Seoul; 2006 'Second Skin', Vitra Design Museum, Weil am Rhein

Awards
2000 Grant, French Ministry of Foreign Affairs + 'Villa Kujoyama' of Kyoto, Paris

Trent Jansen
Trent Jansen Studio
www.trentjansen.com
Born in Australia, 1981
Lives and works in Sydney

Education
2002 Industrial Design, University of Alberta, Edmonton, Canada; 2004 BA Design, College of Fine Arts, University of New South Wales, Sydney

Professional Background
2004 Intern, Marcel Wanders Studio, Amsterdam; Established Trent Jansen Studio, Sydney; Graphic and Object Designer, The Study, Sydney; 2006 Design tutor, School of Design, College of Fine Arts, University of New South Wales, Sydney

Selected Projects
2004 Sign Stool, Trent Jansen Studio; 2005 Red x storage system, prototype; Pregnant Chair, prototype; Sign Project, exhibition, Trent Jansen Studio; 2006 Topple Lamp, ISM Objects

Selected Bibliography
2005 Penelope Barker, 'Around the World', *Cream*, February/March, Sydney; Ewan McEoin, 'Design Discovery Award', *Voi Tutti*, no.3, Sydney; Whimsy Candle, 'Metropolis', *Japan Today*, November, Tokyo; 2006 'New Wave', *Vogue Living*, March/April, Sydney; Caroline Casey, 'Domain', *Sydney Morning Herald*, 17 November, Sydney

Selected Exhibitions
2004 'New Design "National Graduate Exhibition"', Object Gallery, Sydney; 2005 'Workshopped', Strand Arcade, Sydney; 'Sydney Sign Project – Sydney Design Week 2005', Space Furniture, Sydney; 'Bombay Sapphire Design Discovery', Space Furniture, Sydney; 'Tokyo Sign Project – Tokyo Design Tide 2005', Lazy Hazy Planet, Tokyo; 'Evian Lighting Project', Space Furniture, Sydney; 2006 'Vogue Living (The New Wave)', Object Gallery, Sydney

Awards
2004 Object New Design National Graduate Award, Sydney; Australia Council New Work Award, Sydney; 2006 Spiral Rendez-vous Award, Tokyo

Patrick Jouin
www.patrickjouin.com
Born in France, 1967
Lives and works in Paris

Education
1986 Baccalauréat D., Loquidy, Nantes, France; 1992 ENSCI-les Ateliers Diploma, Paris

Professional Background
1992 Designer, Compagnie des Wagons-lits, Paris; 1993–4 Designer, Tim Thom, Thomson Multimedia, Paris; 1995–9 Designer, Philippe Starck studio, Paris

Selected Projects
1995 Boos Dolby Surround Speaker (in collaboration with Philippe Starck), Thomson multimedia; 55" Television set (in collaboration with Philippe Starck), Saba; Don O stereo radio and tape player (in collaboration with Philippe Starck), Thomson multimedia; 1998 FACTO, table and chair, Fermob; 1999 MORPHEE, sofa-bed, Ligne Roset; AI shelving, Proto Design; Wave table set, Gien earthenware, Alain Ducasse; 2000 Cute Cut, sofa, Moderno; 2001 Fold, chair, Xo; Concept car, Renault; Audiolab, module for electro-sound designers listening in museums; 2002 Tajine Dishes, Vallauris potteries; Fagoë, toy collection, Fagoë; 2004 d.i.Vox, radio, d.i.Time, clock, d.i.Maths, calculator, Lexon; Desk for private client, RINCK; Solid, stereolitography furniture collection, Materialise.MGX; 2006 Ether, light fixture, Murano Due; Oneshot, foldable stool, Materialise.MGX; Optic Cube, Kartell; EVOL earthenware collection, Gien earthenware

Selected Bibliography
2000 Jennifer Hudson, *Design Year Book*, London; 2004 Terence Conran and Max Fraser, *Design/Designers*, London; Patrick Jouin, *Design & Designer*, Paris; Mel Byars, *The Design Encyclopedia*, exhibition catalogue, New York; 2005 Hervé Descottes, *Ultimate Lighting Design*; 2006 *Architecture Now! Volume IV*, London; Jennifer Hudson, *1000 new designs [and where to find them]*, London

Selected Exhibitions
2003 'Maison & Objet Show', Fair, Paris; 2004 'Launch-exhibition of the SOLID collection', Maisonneuve Gallery, Paris; 2005 'Maison & Objet Show', Fair, Paris; 'European Design Show', Design Museum, London; 'Art Basel I Miami', Moss Gallery, New York; 2006 'Simply Droog 10+3 years of Creative Innovation and Discussion', Grand Hornu

Awards
2003 Designer of the Year award, 'Maison & Objet Show', Paris; 2006 Travel + Leisure Design Award for Best Restaurant, Mix, Las Vegas, New York

Chris Kabel
www.chriskabel.com
Born in The Netherlands, 1975
Lives and works in Rotterdam

Education
1996 Technical University Delft, The Netherlands; 2001 Design Academy, Eindhoven

Selected Projects
2001 Sticky Lamp, Droog Design; 2002 ltotree, lamp, Chris Kabel Studio; 2003 Flames, candle, Moooi; 2004 Shadylace, parasol, Droog Design; 2005 Tailor Chair, Chris Kabel Studio; Black Money Vase (BMV), Chris Kabel Studio; Platinum Money Vase (PMV), Toolsgalerie Ltd; Pin Light, Chris Kabel Studio; XYZ table, Chris Kabel Studio

Selected Bibliography
2004 Aaron Betsky (ed.), *Droog 10+1*, Amsterdam, Droog; Francesca Picchi, 'Viaggio in Olanda', *Domus*, no.872, July/August; Philippo Romeo, 'A Me Gli Occhi, Profilo di Chris Kabel', *Case da Abitare*, no.85, March; 2005 Ingeborg de Roode, *Nest, Designs For the Interior*, Rotterdam; Kersting Finger, *Tape, an Excursion Through the World of Adhesive Tapes*, Berlin; 'New Dutch on the Block', *Frame*, no.46, September/October; Ed van Hinte, *Under Cover, Evolution of upholstered furniture*, Rotterdam; Marcel Wanders, *International Design Yearbook*, London; 2006 'Bernards World', *I.D.*, LI/VI, July; Anne-Marie Fèvre, 'Beau et Usage de Beau', *Libération*, 12 May, Paris

Selected Exhibitions
2003 'Design Prize Rotterdam', Boijmans van Beuningen, Rotterdam; 2004 'Brilliant!', London; 2005 'Design Made', Hangaram Design Museum, Seoul; 'Nest, Designs for the Interior', Amsterdam; 'New Dutch Designers', Droog Gallery, Amsterdam; 'Project Rotterdam', Boijmans van Beuningen, Rotterdam; 2006 'Black Money Vase and other useful objects', Toolsgalerie, Paris

Awards
2003 Nominated for Design Prize Rotterdam; 2005 First prize for Souvenir for Bruges

Lambert Kamps
www.lambertkamps.com
Born in The Netherlands, 1974
Lives and works in Groningen

Education
1994 MDGO Fashion School, Groningen;
1999 School of Arts, Minerva, Groningen

Professional Background
1999 Established own studio

Selected Projects
1999 M3, inflatable furniture, Lambert
Kamps; 2000 Matrix, inflatable construction, Lambert Kamps; 2001 Ferry Go
Round, inflatable ferry, Lambert Kamps
and Tjeerd Veen Hoven; Air Bridge,
Lambert Kamps; 2002 Space Maker, inter-
active inflatable tent, Lambert Kamps;
Boxing, interactive boxing game, Lambert
Kamps; Balloon tent, air balloon tent,
Lambert Kamps; 2003 Box tent, inflatable
tent construction, Lambert Kamps;
2004–5 Light-Emitting Roof Tiles,
Lambert Kamps; 2005 War of furniture,
autonomous moving furniture, Lambert
Kamps; Magnet Clock, Lambert Kamps;
2006 Turning Tree, moving dead tree,
Lambert Kamps; Crop Circle Machine,
machine to make crop circles in a field,
Lambert Kamps; Big Ball, artificial moon,
Lambert Kamps

Selected Bibliography
2005 Marijke Kuypers, 'Communiceren
met Lichtpan', *Silk + Silkscreen*, March,
The Netherlands; Roelien Plaatsmaat,
'Times Square inde bolder', *Items*,
July/August, Amsterdam; Julie Lasky and
Anneloes van Galen, 'Puff Baddy (How
Lambert Kamps…)', *I.D.*,
September/October, New York; Anneloes
van Galen, 'Blow Up', *Cosmos*, October,
Australia; 2006 WIM van der Beek,
'Review (Lambert Kamps)', *Kunstbeeld*,
February, The Netherlands

Selected Exhibitions
2001 'Ferry-Go-Round', Festival
Noorderton, Groningen; 'Lucht-Brug/
Air-bridge', Bleu Moon, Groningen; 2002
'Space Maker', Festival Noorderzon,
Groningen; 2004 'Stekel (light emitting
roof tiles)', Stekel, Diever, The
Netherlands; 'Split screan', Kunst Vlaai 5,
Amsterdam; 'Boxing', Low Lands,
Biddinghuizan, The Netherlands; 2005
'Duch Pot', Fabio Novembre, Milan;
'Boxing/Matrix', Festival de Beschaving,
Utrecht, The Netherlands; 'Pneu Foria',
75 + NAP, The Hague; 'Logo Tech',
Kunstenlab, Deventer, The Netherlands

Awards
2002 Hendrik de Vries Sti pendium,
Groningen; 2005 Uterechtse Design Prijs,
Utrecht; Groningse Nieuwe, Groningen

Yun-je Kang
Samsung Electronics Co., LTD.
www.samsung.com
Born in Korea, 1968
Lives and works in Seoul

Education
1994 BA Industrial Design Department,
Chung-Ju University, Chung-Ju, Korea

Professional Background
1997 Visual Display Department,
Samsung Electronics, Seoul; 1998–9
Concept Design Department, Samsung
Electronics, Seoul; 2000–present
Creative director, Visual Display
Department, Samsung Electronics, Seoul

Awards
2000 iF design award, Yepp YP-D MP3
player, Germany; 2001 Bronze Prize,
Concept Design for Modular TV, IDEA;
G mark, Yepp YP-64S, MP3 player,
Japan; 2004 Silver Prize, Product Design
for Dlp Projection TV L7., IDEA,
USA; 2006 Silver Prize, Product Design
for Pocket Imager, IDEA, USA; iF design
award, Pocket Imager, Germany

Meriç Kara
www.merickara.com
Born in Turkey, 1977
Lives and works in Istanbul

Education
2001 Industrial Design, Middle East
Technical University, Ankara; 2002 Master
in Design, Domus Academy, Milan

Professional Background
2003–5 Fabrica, Research and
Communication Center, Benetton Design
Department, Treviso, Italy; 2006–present
creative designer, Republica Advertising,
Istanbul; 2006–present Designer, Santral
Istanbul, Istanbul; Freelance designer

Selected Projects
2005 B-Sides, more than 30 designs in
the B-Sides collection for Fabrica;
Ceramic Candle Holder, Paola C.; Rolling
Stone, Fabrica

Selected Bibliography
2004 *Maison Française, Turkey*, no.105,
February, Istanbul; 2005 Andrew Luke
(Sinboy), *Fab Magazine*, Summer, Treviso;
2006 Tuce Yasak, *XXI*, no.42, February,
Istanbul; Julien Martinez Calmettes, *Pure
Design Objects of Desire*, April,
Barcelona; Frederico Duarte, *Icon*,
no.036, June, London; Josep Maria
Minguet, *sitdowndesignnow!*, August,
Barcelona; Esin Eskinat, *icon – Turkey*, 01,
December

Selected Exhibitions
2003 'Random Drink', Inside Design
Amsterdam, Amsterdam; 'The Loft
Couch', Coroflot Members Show, ICFF,
New York and 100% Design, London;
2004 'Christmas Trees – Fabrica', Street
Festival, Ni_anta_i (Benetton & Sisley),
Istanbul; 2005 'A Perfect Weekend',
Fabrica with Tommaso Cora Space
Design, Design Week, Istanbul; 'Digit –
Paola C. – Fabtab', MIFF; 'Mogu Funny
Functions – Fabrica', MIFF; 'Rolling Stone
– Fabrica with Keren Rosen', Milan
Streets, MIFF

Bosung Kim
www.bosungkim.com
Born in Korea, 1974
Lives and works in California

Education
2000 BA Industrial Design, Kook-Min
University, Seoul; 2005 MA Product
Design, Royal College of Art, London

Professional Background
2000–2002 KODAS, Seoul; 2005–
present IDEO, Palo Alto

Selected Projects
2000 Todacosa, retail cosmetic shop,
TODACOSA; 2001 Samil Pharm,
Headquarters, Canopy Architecture and
Interior, Samil Pharm Ltd.; Muritzy, digital
car amplifier, Muritzy; 2002 Kusshand,
kitchen appliances, KUSSHAND; Lluon
Modular PC, TG; 2004 Consuming
Monster, conceptual design for GM Love;
2005 Nomadic Eye Chandelier, self
production; Metamorphosis Chandelier,
self- production

Selected Bibliography
2005 Corinne Julius. 'Summer's
Madness', *Evening Standard*, 25 May,
London; Liz Brown, 'Lighting by British
Student', *Frame*, no.47, November,
Amsterdam

Selected Exhibitions
1999 'Accident', French Cultural Centre,
Helsinki; 'SDM Exhibition', Samsung
Design Membership, Seoul; 2001 'Design
for the Public', Design Museum, Seoul

Awards
2002 CES Innovation Awards, Las Vegas

André Klauser
www.andreklauser.com
Born in Germany, 1972
Lives and works in London

Education
1995 Fachhochschule Münster, Münster;
2002 Product Design, Royal College of
Art, London

Professional Background
1999–2001 Jasper Morrison Design
Office, London; 2002 Established own
studio

Selected Projects
2003 Solitaire Olive Dish, ceramic snack
dish, Thorsten van Elten; 2005 Meccano
Chair, self-production; 2006 Mandi Chair,
Adal; 2007 Cone Tables, Elmar Flötotto;
World Clock, Authentics

Selected Bibliography
2002 RCA Gang, *Popeye*, no.639, Tokyo;
2004 Max Fraser, 'Manufacturing
Renaissance', *icon*, February, London;
Gerrit Terstiege, 'Impulses from
Cologne', *Form*, March/April, Frankfurt;
2005 Damian Barr, 'The Wizzards of
odd', *The Times*, 14 January, London;
Brendan I. Koerner, 'The Goods', *New
York Times*, 20 November, New York;
Marco Romanelli, 'A Future for Wooden
chairs', *Abitare*, December, Milan;
Dominic Lutyens, 'Sharp edges', *Art
Review*, April, London; Despina Curtis,
'Hall the Rage', *Wallpaper**, June/July,
London.

Selected Exhibitions
2001 'Best Before', Galeria Internos,
Milan; 2004 '5F 69', Hanbury Gallery,
London; 'Consume This', Victoria &
Albert Museum, London; 2005
'Prosomedia International Design
Competition', Prosomedia, Udine, Italy;
'Design Mart', Designboom at ICFF, New
York; 'Upstairs in the Woodlands',
Truman Brewery, London; 2006 'Talents',
Ambiente, Frankfurt

Awards
2006 Good Design Award for Mandi
Chair, Tokyo

Tetê Knecht
www.teteknecht.com
Born in Brazil, 1971
Lives and works in Lausanne

Education
1998 Industrial Design, Belas Artes School,
São Paulo; 2005 Master Industrial Design,
École cantonale d'art de Lausanne

Professional Background
1995–6 Lila's Rensins Assistant in devel-
opment and assembly of scenarios, São
Paulo; 1996 wooden furniture, Montage
Industrial Components, São Paulo; 2000
Organised Notech Design Exposition,
São Paulo

Selected Projects
2000 Goma, vase, self-production; 2001
Victor, lamp, self-production; 2004 Rust,
fruit bowl, self-production; 2005 Sabot,
shoe, self-production; Grand Pouf de
Paille, self-production; Oil, vase, self-production; 2006 Georgia, sex toy, Exquise
Design

Selected Bibliography
2001 Maria Helena Estrada, 'Design
Brasil', *Arc Design*, no.18,
February–March, São Paulo; Diana Sung,
Case da Abitare, no.51, October, Milan;
Maria Helena Estrada, 'Design Brasil
2000+1', *Arc Design*, no.22,
November–December, São Paulo; 2002
Jan van Rossem, 'Was zählt ist die Idee',
Architektur und Wohnen, no.03, March,
Hamburg; Helio Hara, 'Olhar Marciano',
'Luxe?', *S/n*, no.01, July, São Paulo; Vanni
Pasca, 'Young Design', *Interni*, no.523,
July–August, Milan; 2003 Maria Helena
Estrada, 'Design Brasil 2000+1'
'Campanas', no.1, June, São Paulo; 2005
Marco Romanelli, 'Glue + Straw =
Design' *Abitare*, no.456, December,
Milan; 2006 'News' section, 'Misturas',
Arc Design, no.46, February, São Paolo;
Laura Bossi, 'Latin American', *Domus*,
no.890, March, Milan; Tiffany Tripet,
'L'Experimentation de la Matière',
Espaces Contemporains, no.02,
May–June, Puidoux

Selected Exhibitions
2000 'Abitare il tempo', Fair-Verona,
Verona; 2001 'Notech Design', Brazilian
Museum of Sculpture, São Paulo; 'New
Alchemists', Rio Centro, Rio de Janeiro;
2002 'Singular e Plural', Institut Tomie
Ohtake, São Paulo; 2005 'au milieu du
chemin', La Placette, Lausanne; 'Design
Industriel', Galerie Kreo, Paris; 2006
'Promosedia, Salone Internazionale della
Sedia, Sezione ad Invito', Udine, Italy;
'Eidgenössische Förderpreise für Design',
Museum Für Gestaltung, Zurich

Awards
2005 nomination for Lucky Strike
Designer Award, Geneva

korban/flaubert
www.korbanflaubert.com.au
Janos Korban: born Australia, 1961;
Stefanie Flaubert: born England, 1965
Both live and work in Australia

Education
Janos Korban: 1988 BA, University of
Adelaide; Stefanie Flaubert: 1988 BA
Architecture, University of Adelaide

Professional Background
Janos Korban: 1991–5 k m hardwork
metalworkshop, Stuttgart; Stefanie
Flaubert: 1991–4 Behnisch and Partner,
Architects, Stuttgart; 1995–2001 Bates
Smart Sydney, Architects, Sydney; Both:
1993 Established korban/flaubert,
Stuttgart; 1995 relocated korban/flaubert
to Sydney

Selected Projects
2000 Membrane Chaise, korban/
flaubert; 2001 Arterial foyer sculpture,
korban/flaubert; 2003 Tetra-five foyer
sculpture, korban/flaubert; Cellscreen,
korban/flaubert; Swaylamp,
korban/flaubert; 2005 Adventure,
installation at Object gallery,
korban/flaubert; Array Screen, korban/
flaubert; Weblight, korban/flaubert; 2006
Burstlight, korban/flaubert, Prism Screen,
korban/flaubert

Selected Bibliography
2002 Cover image, *Independent*,
15 September, London; 2005 Joachim
Fischer, Young Asian Designers,
Hamburg; Rhana Devenport,
'korban/flaubert and the Adventure
Laboratory', *Object Magazine*, no.46,
June, Sydney; Masahiro Kamijyo,
'korban/flaubert', *Axis*, no.10, October,
Tokyo; Penny Craswell, 'Tangled in Light',
Frame, no.47, November, Amsterdam;
2006 Blaine Brownell, *Transmaterial*, New
York; Donald Williams, *Studio: Design
at Work*, Sydney

Selected Exhibitions
'Membrane, Wrenchlamps' in permanent
design collection, Pinakothek der moderne,
Munich; 'Membrane, swaylamp' in
permanent design collection, Powerhouse
Museum, Sydney; 2002 'Mood River',
Wexner Center for the Arts, Columbus,
Ohio; 'Sway With Me' Kinetic Installation
Designers Block, Selfridges Oxford Street
window, London; 2004 'The Tyranny of
Distance' Tokyo Designers Block, Spiral
Building; 2005 'Adventure', Object
Gallery, Sydney; 2006 'Freestyle',
Melbourne Museum

Joris Laarman

Joris Laarman Studio
www.jorislaarman.com
Born in The Netherlands, 1979
Lives and works in Rotterdam

Education
2003 Design Academy, Eindhoven

Professional Background
2004 Established Joris Laarman Studio, Rotterdam

Selected Projects
2003 Heatwave, radiator, Droog Design/Jaga; 2004 Ivy, wall decoration, JL Laboratory; 2005 Arco 12 – Freedom of beech, chair, JL Laboratory; What's Luxury Credit Branch, Chi ha Paura; Value for Money – Stackanov, dishware, Droog BV

Selected Bibliography
2003 Louise Schouwenberg, 'Reinventing functionality', Archis, no.5; 2004 Francesca Picchi, 'Wall lingerie', Domus, no.868, March, Milan; Joris Laarman, 'Joris' choice', Domus, no.870, May, Milan; Louise Schouwenberg, Simply droog; Alex Wiltshire, 'Joris Laarman', Icon, no.046, October; 2005 Robert Thiemann, 'Ideal House', Frame, no.50, May; Eva Schaap, 'New Dutch on the Block', Frame, no.46; Shumon Bazar, Sweet sexcess, Tank, no.9; Michèle Naturel and Frederic Bodet, Le corps L'atelier Le paysage, Musées de Châteauroux; 2006 Aaron Betsky, 'Nederlanden', experimenta, no.54, January

Selected Exhibitions
2004 'Concrete poetry', Design Academy, Eindhoven, Milan; 'Improvisare', Chelsea Hotel, New York; Design Museum tank, London; 'Just what it is', Frac Nord-Pas de Calais, Lille, France; 2005 'solo', DSM, Heerlen, The Netherlands; 'European design show', Design Museum, London; 'over de mensen en de dingen', Z33 Hasselt, Belgium; 'Human touch', Gallery bund 18 & Droog Design, Shanghai; 'Simply Droog 10 Years of Droog Design', MUDAC, Lausanne; 'Biennale de Ceramique Contemporaine de Châteauroux', Musées de Châteauroux, Châteauroux; 2006 'European design show', Millennium Galleries, Sheffield; 'GEBORDUURD', Textiel museum, Tilburg, The Netherlands; 'Ideal House Exhibition', International Furniture Fair, IMM Cologne'

Awards
2004 Interior Innovation Award, IMM Cologne; Best Young Designer, Wallpaper*, London; 2006 Red Dot Design Award, Frankfurt

Nicolas Le Moigne

www.nicolaslemoigne.com
Born in France, 1979
Lives and works in Lausanne

Education
2005 BA Industrial Design, École cantonale d'art de Lausanne; 2007 MA Industrial Design, École cantonale d'art de Lausanne

Professional Background
2006–present Teaching assistant for design courses, École cantonale d'art de Lausanne

Selected Projects
2003 Pot-au-mur, flowerpot, Serralunga; 2004 Rings to twiddle thumbs, Viceversa Gallery; 2005 Public clock for the City of Geneva, Omega Electronics; 2006 Verso Diverso, watering can, Viceversa; Cigar case, Christofle; Fountain mirror, prototype; Side tables, self-production

Selected Bibliography
2003 Giulia Ber Tacchni, The Official Point of View, May, Milan; Pierre Doze, '1000 modèles', L'Officiel, Hors serie, no.1, June, Paris; Benedicte Duhalde, 'Beautiful plastic', Intramuros, no.107, June, Paris; 2005 Stephen Milioti, 'Style in a Bottle', The Word of Exterior, no.1, February, New York; Emmanuelle Niklas, 'Un Arrosoir qui Gagne', Tendance Déco, no.2, April, Lausanne; Tachy Mora, 'Sed de Deseño', Neo2, no.45, July, Madrid; Tina Schneider, 'Froh-Natur', Elle Décor, no.82, November, Munich; 2006 Thierry Mertenant, 'L'Horloge de la Place Neuve', La Tribune de Genève, no.41–7, 18 February, Geneva; Julien Martinez Calmettes, Pure Design/Objects of Desire, April, Barcelona; Frederika Lokholm, 1000 New Designs, June, London

Selected Exhibitions
2003 'ECAL – Serralunga (Milan International Furniture Fair)', La Posteria Gallery, Milan; 2005 'Swiss Design Now', Museum of Contemporary Art, Shanghai; 'Tokyo Designer's Week'; 'Signes du quotidien', Centre Culturel Suisse de Paris; 'IFF', Jacob K. Javits Center, New York; 2006 'ECAL – Christofle (Milan International Furniture Fair)', Mauro Brucoli antiquario, Milan; 'INOUT – Config.01', Musée de Design et d'Arts Appliqués Contemporains, Lausanne

Awards
2004 1st Prize, Public Clock for the City of Geneva, Geneva; 2005 1st prize, 'Re-think/Re-cycle', Designboom, Milan; 3rd prize, Berner Design Award, Bern

Simone LeAmon

Born in Australia, 1971
Lives and works in Melbourne and Milan

Education
1989 Advanced Certificate of Art and Design, Box Hill Institute of Tertiary and Further Education, Melbourne; 1992 BA Sculpture, Victorian College of the Arts, University of Melbourne; 1999 Advanced Certificate of Art and Design, Box Hill Institute of TAFE, Melbourne; 2004 MA Design, School of Architecture and Design (Industrial), Royal Institute of Technology, Melbourne

Professional Background
1996 Project assistant, Workshop 3000, Director Susan Cohn, Melbourne; 1996–2006 Lecturer, Box Hill Institute of TAFE, Media, Arts & Design, Melbourne; 1997–9 Program assistant, Craft Victoria, Melbourne; 1999–2003 Partner, n+1 equals Interdisciplinary Studio, Melbourne; 2003–6 Lecturer, Swinburne University, Visual Arts and New Media, Melbourne; 2004–6 Design lecturer, Industrial Design, RMIT University, Melbourne; 2006 Director, O.S. Initiative Interdisciplinary Studio, Melbourne

Selected Projects
1999 Bowling Arm, fashion accessory, a n+1 equals and O.S Initiative; 2002 Plaza Shirt, a project hybridizing small architectural dwellings and garment structure, n+1 equals, Australia; Pointy><Pointy, A garment exercise exploring the concept of elbow-room, n+1 equals, Australia; 2003 Bodywork, computerised racing suit, Simone LeAmon with Daryl Munton and Dave Morison Performance for Gertrude Contemporary Art Spaces, Melbourne; Supersystem Play, a furniture system with options for chair, sofa, bed and screen, Dulop Pty Ltd.; Helmet Heads: Graphic design for motorcycle helmets, Dainese s.p.a; 2004–6 Kissing MOTO, video works, Simone LeAmon for 'Anytime Soon', viafarini; 2005–7 Baileys Beads: A celestial necklace for my Italian love, lighting products, Oluce S.r.l.; 2006 Super Sofas, design propositions for body play and rest, Australian Centre for Craft and Design and Object Gallery

Selected Bibliography
2002 Fleur Watson, 'Profile', Monument, no.46, February, Sydney; Mimi Xiu, 'Profile', Australian Style, no.60, March, Sydney; 2003 Ashley Crawford, 'Get Your Motor Running', The AGE, no.30 August, Melbourne; Lucinda Strachen, 'All Curves and Danger', The AGE, no.4 September, Melbourne; (ed.), 'Collisioni Silenziose', L'ARCA, no.186, October, Milan; 2004 Helen Kaiser, 'Form and Content', Melbourne Magazine, no.24, November, Melbourne; Paul Andrew, 'Motorcycle baby', Inpress, November, Melbourne; Patrizia Coggiola, 'Moto Design', Mood, no.59/60, December, Milan; 2005 Din Heagney, 'Profile: Simone LeAmon', Desktop, no.203 March, Melbourne; (ed.), 'Design: From Brasil to China', Gulliver, no.4, April, Milan; Lucas Chirnside, 'Salone del Mobile 2005', Inside, no.37, August, Sydney; 2006 Belinda Stenning, 'Forms of Desire and Comfort', Curve, no.14, March, Melbourne; Stephen Crafti, 'Profile: Simone LeAmon', Indesign, no.25, May, Sydney

Selected Exhibitions
2001 'Conceptual Models', Salone del Mobile, Milan; 'Spacecraft', Monash University Gallery, Melbourne; 'Doubletake: Recycling in Australian Craft', Australian national touring exhibition, Perth, Brisbane, Sydney; 2002 'Hybrid Objects: Designers Block Tokyo', Australian Embassy, Tokyo; 2003 'Quiet Collision Australian Propositions in Art and Design', Association Viafarini, Milan; 'MOTO Showroom: The Return of Desire', Gertrude Contemporary Art Spaces, Melbourne; 2004 'Australian Digital Design Biennale', Melbourne Musuem; 2005 'Anytime Soon. Australi Fuorisalone', Salone del Mobile, 1000 Eventi Gallery, curated by Association Viafarini, Milan; 'Beyond Fashion', National Wool Museum, Geelong, Australia; 'Tales of the City', Australian Galleries, Melbourne & Sydney; 2007 'Freestyle: New Australian Design for Living', Sydney Opera House and Object Gallery, Sydney

Awards
2001 New Work Australia Council for the Arts, Sydney; 2003 Australia Council for the Arts, Sydney; 2004 2nd Prize, Lab 3000 Australian Digital Design Biennale, Melbourne

Sang-jin Lee

DIWILL/Sungshin Women's University
Born in Republic of Korea, 1965
Lives and works in Seoul

Education
1998 Seoul National University, Seoul; 2002 Domus Academy, Milan

Professional Background
1991–6 LG Electronic Design Center, Seoul; 2000–present DIWILL, Seoul; 2004–present Sungshin Women's University, Seoul

Selected Projects
1999 Billiard table design, Kangaroo Billiard; 2001 UPS design, Powervally; 2005 Object for X-Mas, 3M; Mania Festival, Seotaiji Company; Bluetooth headset design, Dreantek; 2006 Easy Chair, Bentek

Selected Bibliography
2001 Minyong Yoon, 'Idea man', Kyunghyang, 2 July, Seoul; 2004 Mira Kim, 'Young designer', Chosun, 22 October, Seoul; 2006 Hanna Lee, 'Next design leader's work', Maeil Business, 21 April, Seoul

Selected Exhibitions
2003 'Seoul Design Festival', Seoul Art Center; 2004 'Designers Block', London; 'Usual, Unusual', Gallery MOA, Heri Art Valley, Paju; 2005 'Sydney Esquisse', Launch Pad, Sydney; 'Shanghai Cool', Shanghai Duolun Museum of Art, Shanghai; 'Gwangju Design Biennale 2005 Special Exhibition 5 New wave in Design', Kimdaejung Convention Center, Gwangju; 2006 'Salone Satellite', Salone del Mobile, Milan

Awards
2004 Good Design Award, Korean Foreign Trade Association, Korea; 2006 Sponsored by the government of Korea

Mathieu Lehanneur

www.mathieulehanneur.com
Born in France, 1974
Lives and works in Paris

Education
2001 École Nationale Supérieure de Création Industrielle, Paris

Professional Background
2001 Established own design office; 2004 Post Graduate Design Manager, École des Beaux-Arts Saint-Etienne

Selected Projects
2001 Therapeutic Objects, designs applied to medicine, self-production; 2003 Bangkok Museum, furniture and lamps for the museum restaurant and bookshop, B-Mu; Useful Mouldings, modular units, FR 66; 2004 House 213.6, accommodation units for stray cats, self-production; 2005 J'en Rêve, scenography, Fondation Cartier; 2006 Restaurant Flood©, interior design, FLOOD©; Elements Collection, 5 products for healthy living, VIA

Selected Bibliography
2003 Laurence Salmon, 'Design sur Ordonnance', Intramuros, no.106, April–May, Paris; 2004 Pierre-Olivier Martin Vivier, 'Mathieu Lehanneur, Designer Non-Standard', L'Oeil, no.554, January, Paris; 'Corrupted Biotopes', Design Document Series, no.05, February, Seoul; 2005 Clara Young, 'Medicine Man', Modern Painters, no.9, September, London; Paola Antonelli, SAFE: Design Takes on Risk, exhibition catalogue, New York; 2006 Anne-Marie Fèvre, 'L'Habitat n'est pas une Enveloppe Morte', Libération, no.7677, 13 January, Paris; Claire Fayolle, 'Design Exploratoire', Beaux-Arts Magazine, no.260, February, Paris; Sophie Djerlal, 'Le Beau Qui Fait du Bien', Vogue, no.866, April, Paris; Wouter Wiels, 'Beautiful Designs for the Good of All', Beyond Beauty, no.7, April, Paris; Yves Mirande, 'Mathieu LeHanneur', Artravel, no.9, April–May, Paris; Julie Boukobza, 'Mathieu LeHanneur, Designer', French, no.8, Spring/Summer, Paris

Selected Exhibitions
2003 'Mosquito Bottleneck', with R&Sie, Gallery Henri Unbach, New York; 'Architectures Non Standard', Centre Pompidou, Paris; 2004 'Moulures Utiles + Sacco + Ral 9006', Centre d'Art Contemporain Bretigny, France; 'Domesticity', À Suivre…Lieu d'Art, Bordeaux, France; 2005 'Slim Retrospective', Villa Nevrand, Lyon; 'SAFE: Design Takes on Risk', MoMA, New York; 2006 'L'invenzione del quotidiano', Università la Sapienza, Rome

Awards
2000 Anvar, Paris; 2006 Carte Blanche VIA, Paris; Grand Prix de la Création, Paris

Arik Levy

Ldesign
www.ldesign.fr/www.ariklevy.fr
Born in Israel, 1963
Lives and works in Paris

Education
1991 Art Centre Europe, Switzerland

Professional Background
1991 Seiko Epson Inc., Japan; 1997 Established LDesign, Paris, with Pippo Lionni

Selected Projects
2003 Umbilical, sculpture lamps, Ldesign; 2004 4 to 8, dining table, Desalto; 2005 Rocks, coffee tables, Mouvements Modernes; Cloud Fractal, sculpture lamp, Mouvements Modernes/Ldesign; Mistic, vase/candle holder, Gaia & Gino; Black Honey, fruit bowl, Materialise.MGX; 2006 Minishake, suspension lamp, Materialise.MGX

Selected Bibliography
2005 Esther Henwood, 'Maison Imaginaire', AD, no.50, June, Paris; Helen Kirwan-Taylor, 'Design: The New Who's Who', Wall Street Journal, 11 November, New York; 2006 Chantal Hamaide, 'Arik Levy, L Dorado', Intramuros, no.122, January, Paris; Paul Makovsky, 'Breaking the mould', Metropolis, February, New York; Brigitte Fitoussi, 'The Transformer', Numéro, April, Paris

Selected Exhibitions
2004 'Love counts', Park Ryusook Gallery, Seoul; 2005 'SAFE: Design Takes on Risk', MoMA, New York; 'D Day – from primitive to virtual', Beaubourg, Centre Pompidou, Paris; 'Feel before you see', Garanti Gallery, Istanbul; 2006 'Republique Libre du Design', Centre Culturel Français, Milan; '24 H', Forum Diffusion, Paris 'ShimmerGlimmerTwinkleSparkleShine!', Moss Gallery, New York

Awards
2002 Label Via for Sofa Arik, Paris; 2003 Best of Best for Liko table, Interior Innovation Award, Cologne; 2005 Best Detail for 4 to 8 table, Interior Innovation Award, Cologne; 2006 Design Plus Award for Mistic vase, Ambiente Fair, Frankfurt

Julia Lohmann

www.julialohmann.co.uk
Born in Germany, 1977
Lives and works in London

Education
2001 Surrey Institute of Art and Design, Epsom, Surrey, UK; 2004 Royal College of Art, London

Professional Background
2001–2 Creative BaByDO Think Tank, Abbott Mead Vickers BBDO (AMV BBDO), London; 2003–6 Sessional lecturer in design, Epsom and Rochester, UK and Reykjavik; 2005 Established own studio

Selected Projects
2001 PET-Pick Bottle Carrier, Zeltec; 2004, Glove, pencil holder, prototype; Flock prototype, light installation, prototype; Ruminant Bloom, lamp, Julia Lohmann; 2006 Cow Benches, Julia Lohmann in-house production; Fruit bowl 1990, prototype

Selected Bibliography
2004 Kirsty Robinson and Sacha Spencer Trace, 'New Talent: Julia Lohmann', Marmalade, no.5, May, London; Marcus Fairs, 'Design Courses & London Came of Age this Year', icon, no.016/018, October/December, London; 2005 Garry Mason, 'Young and Dangerous', idFX, no.75, January, London; Kerstin Kühn and Emily Campbell, 'Talent Spot – Ones to Watch', Design Week, vol.20, no.3, January, London; Tom Dyckhoff, 'The Sofa That Gas a Bone to Pick', The Times, 29 March, London; Julie V. Iovine, 'House & Home', New York Times, 21 April, New York; Martin Raymond and Kate Franklin, 'Design Notebook: Zoomorphia', Viewpoint, no.18, Spring/Summer, Amsterdam; Masaichi Tamaoki (ed.), 'New Designer – Great Brits The New Alchemists', Elle Décor Japan, no.78, June; Alex Bagner, 'Salone 2005 – The Graduates', Wallpaper*, no.80, July/August, London; Martin Zentner, 'Eigene Vision', design report, September, Frankfurt; Maggie Ma (ed.), 'Four Legs Good', Casa International, no.34, October, Beijing; Vicky Richardson, 'The Seven Ages of Design', Blueprint, no.236, November, London; 2006 Mika Kakegawa, 'Garden Love – Elements Space', Casa Brutus, no.586, February, Tokyo; Josh Sims, 'A Beastly Business', Independent, 21 September, London

Selected Exhibitions
2004 'Design Mart', Design Museum, London; 'Biennale International Design', Saint-Etienne; 2005 'Great Brits – The New Alchemists', Paul Smith Headquarters, Milan; 'European Design Show', Design Museum, London; 'Twinkle Twinkle, 100% Design Moscow'; 'Great Brits – The New Alchemists', Tokyo; 2006 'Thirty-Seven', Galerie Kreo, Paris; 'New British Designers', Droog Design, Amsterdam

Awards
2001 Product Development Student Design Award, D&AD, London; Ecology and Best of Category, iF design award, Hanover; John Gillard Award, London; 2004 Esmée Fairbairn Foundation, London

Alexander Lotersztain

www.derlot.com
Born in Argentina, 1977
Lives and works in Brisbane and Barcelona

Education
2000 Griffith University, Queensland College of Art, Brisbane

Professional Background
2000 GK Planning and Design, Tokyo; 2001 Established own design studio

Selected Projects
2001 Soft Sofa, street furniture, Idee; 2004 DEK range, park seating, Street and Garden Furniture Co.; 2005 Stelios Papas, hair salon interior design, Alexander Lotersztain Studio; 2006 Espejos Collection, mirrors, Artquitect

Selected Bibliography
2001 Akari Matswra, Japan Design to the New Generation, Japan; 2002 Ricardo Rinietti, The Official Point of View, Italy; 2003 Sean Topham, Where is My Space Age?, UK; Madalena Padovani, 'New Plastics', Interni, no.537, Italy; 2005 Margie Fraser, 'Satellite Out of Africa', In Design, vol.20, Australia; 2006 Anne Misako, 'Success Story', MAP, February, Australia; David Clark, 'Australian Design', Vogue Living, March/April, Australia

Selected Exhibitions
2002 'Designers Block', St. Pancreas Chambers, London; 2003 'Popcork', Salone del Mobile, Milan; 'Sputnik St.', Stockholm Design Week, Stockholm; 2004 'Sputniks' Sputnik Pad, Tokyo; 2005 'Plastic Fantastic', Craft Queensland, Brisbane; 'Coral Light + Sound', Space, Furniture, Brisbane

Awards
1999 Student Award, Design Institute of Australia; 2000 Opus Award, Japan

Xavier Lust

ELIXIR sprl
xavierlust.com
Born in Belgium, 1969
Lives and works in Brussels

Education
1992 Interior Design, Institut St. Luc, Brussels

Professional Background
2000 Start of collaboration with MDF Italia, Milan; 2001 Start of collaboration with De Padova, Milan; 2002 Collaboration with EXTREMIS outdoor furniture, Belgium; 2003–present Collaboration with Driade, Milan

Selected Projects
2000 Le Banc, bench, MDF Italia; 2001 Finder, wall storage unit, DePadova, Zoltan; 2002 La Grande Table, MDF Italia; PicNik, picnic table and bench, EXTREMIS; Crédence, chest, De Padova; 2003–4 Paso Doble, umbrella stand, Driade Kosmo; 2004 Extra Chair, Driade Atlantide; Turner, candlesticks, Driade Kosmo; 2004–6 Banc-vélo, bike stand/bench, Elixir; Abri-vélo, bike stand, Elixir; 2005 T42 & T43, trays, Driade Kosmo; 2006 Abri-voyageur, bus shelter, Elixir; Bwb Chair, Zeritalia

Selected Bibliography
2001 Lise Coirier, A to Z, Design Made in Belgium, Brussels; 2003 Marco Romanelli, 'Il Design Belga Visto da Occhi Italiani', Abitare, no.428, May; Terence Conran and Max Fraser, Designers on design, London; 2005 Lise Coirier, Label-Design in Belgium, Design in Belgium after 2000, Brussels; Marco Bravo, Art Direction and Editing, I.Dot Points of View/What's Moving in Italian Design Today, Milan; Bénédicte Duhale, 'Xavier Lust, l'Esprit cCourbe', Intramuros 20 ans International design, no.117, March/April, France; Noel Montrucchio, 'Lust for Life', Spaces, no.13, UK; 2006 Oscar Asensio, D! DES!GNDES!GN Meubles et Lampes, Barcelona; Marie Pok, 'Xavier LUST Su Design à l'Architecture d'Intérieur', La libre ESSENTIELLE immo, no.39, April, Belgium

Selected Exhibitions
2003, VIZO Gallery, Brussels; 2004–5 'Imperfect by Design, 4th Triennale of Design', Musée d'Art et d'Histoire (Cinquantenaire), Brussels; 'I.Dot – Italian Design on tour', Chelsea Art Museum, New York, Neocon, Chicago; 'Back to School', Fad-Foment de las Arts Decoratives, Barcelona, Bultaup Center, St Petersburg, Designmay Forum, Berlin, China Design Alliance, Beijing; 2005 'EDIFICE présente Xavier Lust', EDIFICE, Paris; 'Guest of Honor', SIDIM, Montreal; 2005–6 'Label-design.be – Design in Belgium After 2000', Grand-HornuImages, ProMateria, DesignVlaanderen, Hornu; 2006 'Spotting', Stedelijk Museum CS, Amsterdam

Awards
2002 Best Furniture Design for Le Banc, 13th Design Week Awards, London; 2003 Young Talent Award, Henry van de Velde Award, Brussels; 2004 Good Design Award for PicNik, Chicago Athenaeum, Museum of Architecture and Design, Chicago; Special Mention for La Grande Table, Xst Compasso d'Oro Awards, Milan

Cecilie Manz

www.ceciliemanz.com
Born in Denmark, 1972
Lives and works in Copenhagen

Education
1995 University of Art and Design, Helsinki; 1997 Denmark's Design School, Copenhagen

Professional Background
1997 KYW Design, Professor Yrjö Wiherheimo, Helsinki; 1998 Established own studio office

Selected Projects
2000 Clothes Tree, prototype by PP Moebler; 2001 Fixed Seats prototype; 2003 Micado, side table, Fredericia Furniture; 2004 One of a Kind, chair, one-off; 2005 Mondrian and Caravaggio, lamp, Lightyears; 2006 Minima, glass-ware, Holmegaard

Selected Bibliography
2004 Kim Flyvbjerg, Designmatters, no.6, June, Copenhagen; 2005 Henrik Most, Cecilie Manz, Status, March, Copenhagen; Palle Schmidt, Den Nye Generation – Dansk Møbeldesign 1990–2005, Copenhagen; Libbie Fjeldstrup, 'Super', Politiken, 3 April, Copenhagen; 2006 Kraks Blaa Bog, Copenhagen

Selected Exhibitions
2000–4 'SE – Cabinetmakers Autumn Exhibition', Danish Museum of Art and Design, Copenhagen; 2001 'The Third Design for Architecture Symposium Exhibition', Alvar Aalto Academy, Jyväskyla, Finland; 'Bodil and Cecilie Manz in Conversation', Centrum Goed Wek, Belgium; 2002 'Living in Motion', Vitra Design Museum, Basel; 2004 'Northern Lights', MDS-G/Issey Miyake Gallery, Tokyo; 2005 'Cecilie Manz – Taking Stock', Danish Design Centre, Copenhagen; 'anders als immer/somewhat different', Institut fur Auslandsbeziehungen, Stuttgart

Awards
2003 Award of Honour, Inga & Ejvind Kold Christensens Foundation, Copenhagen; 2004 Danish Design Prize, Danish Design Centre, Copenhagen; 2006 Forum + 1 Award, Stockholm

Tatsuya Matsui

www.flower-robotics.com
Lives and works in Tokyo

Professional Background
2001 Established own studio

Matthias Aron Megyeri

Megyeri & Partners Ltd
www.MegyeriAndPartners.com
Born in Stuttgart, 1973
Lives and works in London

Education
2001 Diploma in Graphic Design/Visual Communication, State College of Design Karlsruhe, Germany; 2003 MA Design Products, Royal College of Art, London

Professional Background
2000–2 Art director, werk, bauen + wohnen, Zurich; 2003–present Co-founder and director, Designstudio Dear Dad, London; 2004–present Lectures, talks and workshops, Berlin, Eindhoven, London, New York, Ogaki, São Paulo, Singapore, Tel Aviv, Tokyo 2005–present Founder and Director, Megyeri & Partners Ltd

Selected Projects
2003 Sweet Dreams Security™ – Mr. Smish & Madame Buttly, razor wire, Megyeri & Partners Ltd; Sweet Dreams Security™ – R. Bunnit, Peter Pin and Didoo, Victorian railings, Whitton Castings Ltd.; Sweet Dreams Security™ – Landscape, glass objects, V. Seile GmBH; Sweet Dreams Security™ – Billy B., old English padlock, HY Squire & Sons Ltd.; Sweet Dreams Security™ – Daisy T., ADT alarmbox attachment, Megyeri & Partners

Ltd; Sweet Dreams Security™ – CityCatTV, CCTV attachment, Megyeri & Partners Ltd; 2004 Sweet Dreams Security™ – Heart to Heart, chain, Megyeri & Partners Ltd; 2005 Sweet Dreams Security™ – Snoopy D., window hanger, Plauener Spitze, Modespitze Plauen GmBH; Sweet Dreams Security™ – Mrs Welcome, Nottingham lace curtain, Lace Dimensions; 2006 Sweet Dreams Security™ – Mr. Welcome, Georgian cast wire glass, V. Seile GmBH

Selected Bibliography

2005 Paola Antonelli, *SAFE: Design Takes on Risk*, October, exhibition catalogue, New York; 2006 Sarah Simpkin, 'Soft Security – Fuzzy Frontiers', *Art Review*, April, London; Alison Clarke, 'SweetDreamsSecurity™', *Home Cultures*, May, Vienna

Selected Exhibitions

2003–4 'Hometime by the British Council, Gerrard O'Carroll, Tian He Stadium, Guangzhou, China Millennium Monument, Beijing, Multi-function Hall, Shanghai, Super Brand Mall, Pudong, Chongqing Stadium, Chongqing, China; 2004 'Design Mart', Design Museum, London; 2005 'Pop Noir – Critical Designs Selected by Dunne & Raby', The Israel Museum, Jerusalem; 'SAFE: Design Takes on Risk', MoMA, New York; 2005–7 'Great Brits', by the British Council, Pantheon, Athens, RMIT Gallery, Melbourne, UTS Gallery, Sydney, ICA Institute of Contemporary Arts, Singapore, 100% Design, Tokyo, HQ Paul Smith, Milan; 2006 'Safety Nest' SESC Pinheiros, São Paulo; 'UK Jack, OK!', Isetan, Tokyo, Comme de Garçons, Dover Street Market, London, and Colette, Paris

Awards

2004 Esmée Fairbairn Foundation bursary for Sweet Dreams Security™, London; 2005 Creative Pioneer Business Award for Sweet Dreams Security™, Nesta, London

Marcelo Moletta

www.marcelomoletta.com
Born in Brazil, 1976
Lives and works in Rio de Janeiro

Education

2003 Product Design, Pontifícia Universidade Catolica do Rio, Rio de Janeiro

Professional Background

2001 Sculptor, Unidos da Tijuca and Mocidade Independente de Padre Miguel, Rio de Janeiro; 2004 Recycling Materials, Rio de Janeiro; 2005 Product designer, Indústria nacional design, Rio de Janeiro; 2006 Project manager, Ernesto Neto – Atelie Nave, Rio de Janeiro

Selected Projects

1997 Carioca Chair, prototype; 1999 Cama Sutra, bed, prototype; 2000

Envelope Chair, Marcelo Moletta and Itapeva; 2002 Lixeira Seleta, trash can, prototype; 2003 Solar Battery Charger, prototype

Selected Bibliography

2001 Ethel Leon, 'Meu Design', *Valor Economico*, 3rd May, Rio de Janeiro/ São Paulo; Simone Raitzik, 'A Nova Cara do Design', *Revista Casa Claudia*, no.11, November, Rio de Janeiro/ São Paulo

Selected Exhibitions

2001 'Cria Design', PUC – PR, Rio de Janeiro; 'Mostra de Novos Talentos em Design', Shopping D&D, São Paulo

Nada Se Leva

www.nadaseleva.com.br
André Bastos: born Brazil, 1963;
Guilherme Ribeiro: born Brazil, 1968;
Both live and work in São Paulo

Education

André Bastos: 1984 Pre-Med, Universidade Federal UFRGS, Porto Alegre, Brazil; Guilherme Ribeiro: 1991 BA, Fordham University New York

Professional Background

André Bastos: 1994–2002 Owner, Villa Due Store, São Paulo; 2004–2005 co-owner/interior designer, OBRA, São Paulo; Guilherme Ribeiro: 1992–95 Senior designer, Schell & Mullaney, New York; 1995–96 Graphic designer, Bianco & Cucco, Milan; 1996–2000 Co-owner, creative designer, 2PG, Rio de Janeiro; 2000–02 Creative director, Promon, São Paulo; Both: 2005 Established Nada Se Leva

Selected Projects

2005 Print Collection – Antique Table, Firma Casa; Ligero Mirror, Firma Casa; 2006 Ligero Small Side Tables, Acriresinas; Print Collection – Bull's Eye Table, Tergo print; Ishi Collection – Wax Stool, Fiorini

Selected Bibliography

2006 Fabrizio Rollo, 'Tecno Décor', *Casa Vogue*, no.247, May, Carta, São Paulo; Laura Wie, 'Baroque Complements Modern Design', *Go Where*, May, United, São Paulo; Brunete Fraccaroli, 'Free, Light + Loose', *ABD Interiors*, no.2, May, ABD, São Paulo; Antonio Togo, 'Irreverence Togo', *Motomag*, no.14, July, Wide, São Paulo; Clarissa Schneider, 'What's New' *Casa Vogue*, no.251, July, Carta, São Paulo; Ronaldo Ferreira, 'What's News Kaza', *Kaza*, no.38, July, Agão, São Paulo

Awards

Nada Se Leva: Prêmio do Museu da Casa Brasileira; Sideboard Ligero, finalist; Guilherme Ribeiro: 2001 Cyber Lions Gold, Cannes Lions, Cannes; 2001–2 Best Brasil, São Paulo

Khashayar Naimanan

www.khashayar-naimanan.com
Born in the UK, 1976
Lives and works in London and Munich

Education

1999 BA Central Saint Martins, London
2003 MA Royal College of Art, London

Professional Background

2005–present Kram/Weisshaar, Munich

Selected Projects

2001–3 Collection 1 – Designstars Project, Khashayar Naimanan and Chris Fitzgerald; 2002 Clothing for Concealment – Fat Pockets, Khashayar Naimanan; Robin Barcodes, Robin of Fulham; 2003 Clothing for Concealment – Sleeving Jacket, Khashayar Naimanan; Hidden Wealth Project – Incognito, Porzellan-Manufaktur Nymphenburg; 2003–4 Hidden Wealth Project – Ça Ne Vaut Pas un Clou, Khashayar Naimanan; 2004 'Robin Board', Robin of Fulham, Khashayar Naimanan; 2006 Rugged Art – Tag

Selected Bibliography

2002 'News-Vitra Visions', *Blueprint*, no.211, September, UK; 2003 'Degree Shows', *Creative Review*, September, UK; 'Hopeful Monsters', *Blueprint*, no.211, September, UK; Gerrit Terstiege, 'Projects-Big Designers on a Small Scale', *Form*, no.192, November/December, Germany; 2004 'What's New', *Vogue*, January, UK; 2005 Paola Antonelli, *SAFE: Design Takes on Risk*, exhibition catalogue, New York; 2006 Helen Kirwan-Taylor, 'My Dinnerware, Myself: the Expressive Table', *Wall Street Journal*, no.239, January, Europe; Laura Bossi and Francesca Picchi, 'Geo Design 2: Schools, nomadic designers and migrating ideas', *Domus*, no.889, February, Italy; Thomas Eldelmann, 'Urbane Trendsetter', *A&W Architektur & Wohnen*, August/September, Germany; 'It's Not Worth a Nail', *Tally-Ho*, no.1, October, New York

Selected Exhibitions

2004 'Some of Us', Messe Centrum, Bremen, Germany; 'Evolution of Letterforms', Alliance Française, Harare, Zimbabwe; 'Inspired', Apart Gallery, London; 2005 'The Master's Selection', Ausstellungshalle Zeitgenössische Kunst, Münster; 2005–6 'Pop Noir: Design for Thought', Israel Museum, Israel; 'SAFE: Design Takes on Risk', MoMA, New York; 2006 'StoryTellers – StylePark in Residence', Konrad-Adenauer-Ufer, Cologne; 'Khashayar Naimanan vs. Arne Quinze – The Design Annual-Urban', StylePark, Frankfurt

Awards

2002 Winner, *Blueprint*, Vitra Design Museum Competition, UK

nendo

www.nendo.jp
Oki Sato: born Canada, 1977
Lives and works Tokyo

Education

2000 BA Architecture, Waseda University, Tokyo; 2002 MA Architecture, Waseda University, Tokyo

Professional Background

2002 Established own studio

Selected Projects

2004 Wind, stool, Swedese; 2005 Chab-table, side table, De Padova; Sorane, lamp/speaker system, Oluce; Snow, low table, Swedese; 2006 Hanabi, lamp, prototype; Polar, nesting table, Swedese; Yuki, screen, Cappellini; Sasae, coat stand, Gebrüder Thonet; Yab, coat hanger/screen, Blå Station

Selected Bibliography

2005 Marco Romanelli, 'Luce Light 2005', *Abitare*, no.449, April, Milan; Marco Romanelli, 'Milan 2005 Special Pieces', *Abitare*, no.451, June, Milan; Fiona Wilson, 'Newspaper', *Wallpaper**, June, London; 2006 Didi Gnocchi, è *De Padova 50 Years of Design: Institutions, Passions, Encounters*, Milan; Robert Thiemann, 'Unfulfilled Expectations', *Frame*, no.49, March, Amsterdam; nendo, 'Designer', *Interni International*, April, Milan; Marco Romanelli, 'Design 6 dai saloni', *Abitare*, no.460, April, Milan; Vicky Richardson, 'Upwards and Onwards to Milan', *Blueprint*, no.241, April, London; Kelly Rude, 'Designer Indentikit', *Azure*, May, Toronto; Rosie Spencer, 'Milan', *icon*, 035, May, London

Selected Exhibitions

2005 'AMAYADO-LOUNGE/Design Tide Tokyo', Space Intart, Tokyo; 2006 'bloomroom/Milan design week', Zona Tortona, Milan

Awards

2002 *SD Review*, Award Winner, Tokyo; Gold Award Koizumi International Lighting Design Competition for Students, Tokyo; 2004 Elle Déco International Design Award, Japan Nomination, Tokyo; Good Design Award, Tokyo; 2005 JID Award Biennial, Grand Prix + Interior Product Award, Tokyo; JCD Design Award, Award for Excellence & Special Mention, Tokyo

Nido Campolongo

Galeria do Papel Gráfica Editora LTDA
www.nidocampolongo.com.br
Born in Brazil, 1954
Lives and works in São Paulo

Education

1976 College of São Paulo Engineering; 1983 Drawing with Dalton de Luca, São Paulo; 1984 Engraving in Metal, FAAP, São Paulo

Professional Background

1985 Established Nido Campolongo Design Gráfica LTDA; 1999 Established Galeria do Papel Gráfica LTDA

Selected Projects

1998 Packaging, Natura (Supporting, 'Crer para Ver' (To believe to see) assisting Brazilian children); 2000 Trio Lamp, Galeria do Papel Gráfica Editora LTDA; 2001 Avenstruz Lamp, Galeria do Papel Gráfica Editora LTDA; 2002 Cone Table,

Galeria do Papel Gráfica Editora LTDA; 2004 'Vestir o Boracéa' (In collaboration with Adriana Yasbek), a project to promote education for homeless people in São Paulo; 2005 Woven fabric of paper yarn Lamps, Galeria do Papel Gráfica Editora LTDA; 2006 Caixa-estante, box-bookshelf, Natura

Selected Bibliography

2004 Éric Amouroux, 'Papier humaniste', *Ateliers d'Art*, no.52, July–August, Paris; 2005 Hubert Dubois, 'S'abriter en recyclant', *D' Architectures d'A*, no.143, January–February; Tatewaki Nio, 'International Design – São Paulo', *Pen*, no.156, July, Tokyo; 2006 Ailton Pimentel, 'O papel do designer', *Casa Vogue*, no.203, May, São Paulo

Selected Exhibitions

2000 '100 designers brasileiros', D&D Decoração & Design Center, São Paulo; 2001 'Escala Porto/CRAIT', Centro Regional de Artes – Regional Centre of Traditional Arts, Porto, Portugal; 'Singular Plural...quase...últimos 50 anos de design brasileiro', Fundação do Centro Cultural Tomie Otake, São Paulo; 2002 'Bricadeirasde papel – Paper entertainment', SESC SP, São Paulo; 'Oca', SESC SP, São Paulo; 'Papel na casa – Paperat Home', Museu da Casa Brasileira, São Paulo; Biennale du Design Saint-Etienne; 2003 ICFF, New York; 2004 'D&D Design & Natureza Responsabilidade Socioambiental', D&D Decoração e Design, São Paulo; 2005 'Brazil, Brazil', Design Centre of Marseille; 2006 'Permanet instalation at Syndicat mixte pour la Gestion et l'Incinération des Déchets Urbanais de la Région de Sarcelles', Sigidurs, Sarcelle, France; 'Brazil Design: The Intimacy of Extremes', Designmai, Berlin; 'Nido Campolongo and Pur'suco', Paris

Nódesign

www.nodesign.com.br
Flavio Barão Di Sarno: born Brazil, 1978;
Marcio Giannelli: born Brazil, 1977;
Leonardo Massarelli: born Brazil, 1979
All live and work in São Paulo

Education

Flavio Barão Di Sarno: 2002 Industrial Design Course with qualification in product development, University Presbiteriana Mackenzie, São Paulo; Marcio Giannelli: 2001 Industrial Design course with qualification in product development, Fundação Armando Álvares Penteado, São Paulo; Leonardo Massarelli: 2001 Industrial Design course with qualification product development, Fundação Armando Álvares Penteado, São Paulo

Professional Background

2001 Established Nódesign, São Paulo

Selected Projects

2002 Jelly Fish Lamp, Nódesign; 2003 Black Hole, storage, Nódesign; Trachea Lamp, Nódesign; 2004 Tangram Chair,

Nódesign; Exfoliating Soap Dish, Natura; Drip Soap Dish, Natura; 2006 Blob, mobile phone, Easy Track

Selected Bibliography
2003, Maria Helena Estrada, 'Design Brazil 2000 + 3', *Arc Design*, no.33, November, São Paulo; Lucia Pietroni, 'Use and Reuse in Young Design', *DIID – Disegno Industriale*, no.5, April, Italy; Eris Program, 'Abitare a Saint Etienne', *Plastdesign*, January, Milan; 2006 Wang Zhou, 'Chair: original design from Brazil', *Furniture and Interior Design*, no. 84, February, Hong Kong; Mel Byars, *New Chairs: Innovations in Design, Technology and Materials*, London; Logos; Mel Byars (ed.), *The Design Encyclopedia*, London and New York; 'Brazil and India, a new current in design', *AXIS*, no.121, 6th June, Tokyo

Selected Exhibitions
2002 'Brazil Faz Design', IBRIT, São Paulo and Milan; 'Design&Natureza', Shopping D&D, São Paulo; 'Prêmio Museu da Casa Brasileira', Brazilian House Museum, São Paulo; Biennal de Design de Saint-Etienne; 'Expo Eco-Design', Campinas, Brazil; 2003 'Expoeco Design 2003', Shopping Metrópole, São Paulo; 'Design&Natureza'. Shopping D&D, São Paulo; 'Prêmio Museu da Casa Brasileira', Brazilian House Museum, São Paulo; Plasticidades', Casa Shopping, Rio de Janeiro; 2004 'Brazil Faz Design', IBRIT, São Paulo and Milan; 2005 'Design Possível', Brasilartes, Milan; 2006 'Brasil Certificado', Shopping Frei Caneca, São Paulo

Awards
2001 Samsung Prize for Brazilian Design, São Paulo; 2004 Prêmio Atualiadade Cosmética, São Paulo; 2006 VII House & Gifts Design Award, São Paulo

Johannes Norlander
Johannes Norlander Arkitektur AB
www.norlander.se
Born in Sweden, 1974
Lives and works in Stockholm

Education
1995 Architecture, Chalmers University of Technology, Gothenburg; 1996 Crafts & Design, Konstfack University of Arts, Stockholm; 1999 Architecture, Stockholm Royal Institute of Technology

Professional Background
2001 Established own studio

Selected Projects
2000 Waves, pendant lamp, Box Design; 2001 L-Serie, storage units, Asplund; 2003 Kyparn, stacking chair, Nola; 2004 Cano, shelving unit, E&Y; 2005 Post, ashtray, Nola; Ask, chair & table, Collex; 2006 ORI Chair, Johannes Norlander Arkitektur AB

Selected Bibliography
2006 Magnus Larsson, 'Den Ofrivillige Mobel Designern', *Forum*, February, Stockholm

Selected Exhibitions
2006 'Constructs', Forum, Stockholm

Awards
2005 Bruno Mathsson Prize, Värnamo, Sweden

Ken Okuyama
www.kenokuyama.jp
Born in Japan, 1959
Lives and works in Italy and Japan

Education
1982 Advertising Design, Musashino Art University, Tokyo; 1986 Industrial Design, Art Center College of Design, Pasadena, California

Professional Background
1986–1994 Chief Designer, General Motors, Warren, USA; 1990–1991 Senior designer, Porsche AG, Stuttgart; 1996–2006 Design director, Pininfarina S.P.A, Turin; 2001–2003 Department chairman, Art Center College of Design, Pasadena

Selected Projects
2000 Pininfarina Ferrari Rossa, concept car, Pininfarina; 2001 Maserati Quattroporte, production car, Maserati; 2002 Ferrari Enzo, production car, Ferrari; 2003 Nuvo, humanoid robot, ZMP; 2005 Albero, coat hanger, Tendo Mokko; Aliquation Chair, Tendo Mokko; Spiral, light stand/coat hanger, Tada Mokko; Folding Chair/Wall, Tada Mokko; Mayu, teapot, Wazuqu; 2006 Maserati Birdcage 75, concept car, Pininfarina

Selected Bibliography
2002 Wallace Wyss, 'Is Enzo's father Asian?', *Car & Driver*, vol. 48, no.5, November, New York; 2005 Silvia Baruffaldi, 'Sensual Fluidity (Birdcage)', *Auto & Design*, no.152, May/June, Turin; Toshiki Kiriyama, '100 designers in Japan', *Aera Design*, November; Yuji Kotari, 'Interview with Ken Okuyama', *Mac Power*, December, Tokyo; 2006 G. Takeshita, 'Pininfarina Renaissance', *Stile*, vol. 2, February, Tokyo; Flavio Galvano, 'the appeal of power', *Auto & Design*, no.158, June, Turin; Masahiro Kamijo, 'The tractive power of Maison & Objet', AXIS, no.6, June, Tokyo; Yukari Kato, 'Brio Interview', *Brio*, July, Tokyo; Bill Thomas, 'Special Forces', *Top Gear*, October, London; Motoko Saito, 'Story Interview', *Story*, May, Tokyo

OVO
www2.uol.com.br/ovo
Luciana Martins: born Brazil, 1967; Gerson de Oliveira: born Brazil, 1967
Both live and work in São Paulo

Professional Background
1992 Established OVO, São Paulo

Selected Projects
1994 Pigmento, coffee table, OVO; 1995 Cadé, armchair, OVO; 1997 Mientras Tanto, dining table, OVO; 2000 Huevos Revueltos, coat hanger, OVO; 2001 Clip, dining table, OVO; 2002 RGB, light fix-ture, OVO; 2003 Circular, hanging book-case, OVO

Selected Bibliography
2000 Emma O'Kelly, 'São Seers', *Wallpaper**, vol. 27, April, England; Martin Grossman, *Playground*, 4 November, São Paulo; Juliana Mariz, 'A Dupla Vida Dos Objtos Que São O Que Não Parecem', *Jornal Valor*, 28 November, São Paulo; 2001 Malin Zimm, 'En Yta Upprätthallen Av En Serve Händelser', Rum. Tidskriften Orn Arkitektur, *Inredning Och Design*, vol.11, February, Stockholm; 2002 Camila Molina, 'Dupla Subverte Design Em Novo Espaço Expositivo', *O Estado de São Paulo*, 8 November, São Paulo; 2006 Carissa Kowalski Dougherty (ed.), *Young Designers Americas*, May, Cologne, Germany; Tatewaki Nio, 'Brazil and India – A New Current in Design', *AXIS*, Vol.121, May–June, Tokyo

Selected Exhibitions
2000 'Playground', Galeria Brito Cimino, São Paulo; 2001 'Bienal 50 Anos', Fundação Bienal, São Paulo; 2004 'Hora Aberta', Galeria Vermelho, São Paulo; 2004 Biennale Internationale du Design de Saint-Etienne; 2005 'P.A.', Museu de Arte Moderna, São Paulo

Awards
2000 Brazil Faz Design, São Paulo; 2005 Museu da Casa Brasileira, São Paulo

Flávia Pagotti Silva
www.flaviapagottisilva.com
Born in Brazil, 1971
Lives and works in São Paulo

Education
1990 Faculdade de Arquitetura e Urbanismo da Universidade de São Paulo; 1999 MA Design Products, Royal College of Art, London

Professional Background
1994 Zanettini Arquitetura, São Paulo; 1995 Eduardo de Almeida Arquitetos Associados, São Paulo; 2000 Foster and Partners, London; 2001 Self-employed designer, São Paulo

Selected Projects
2001 Curinga Chair, Flávia Pagotti Silva; 2002 Bate-Papo Stool, Flávia Pagotti Silva; 2003 Rondeira Chair, Decameron; Paulistana Chaise, Decameron; Leve Table, Flávia Pagotti Silva; Filé Table, Flávia Pagotti Silva; Pororoca Lamps, Flávia Pagotti Silva; 2004 Excêntrica Table, Flávia Pagotti Silva; Renda Lamp, Flávia Pagotti Silva; Laminado Stool, Monte Azul Wood Workshop; 2005 Laminado Fruit Bowl, Monte Azul Wood Workshop; Laminado Magazine Rack, Monte Azul Wood Workshop; Laminado Saddle Stool, Monte Azul Wood Workshop; Laminado U Stool, Monte Azul Wood Workshop; Laminado Wine Rack, Monte Azul Wood Workshop; 2006 Impressões Table, Flávia Pagotti Silva; Impressões Trays, Flávia Pagotti Silva; Por-Enquanto é Só, shelving and storage pieces, Flávia Pagotti Silva

Selected Bibliography
2001 Carinne Julius, 'Putting talent to the test', *Evening Standard*, 30 May, London; Michael Johnson, 'Reality bites', *Design Week*, 28 June, London; Caroline Roux, 'To die for – Student shows', *Guardian*, 30 June, London; 2002 Roberto Abolafio Jr., 'Invações no front', *Viver Bem*, no.100, January, São Paulo; 2003 Mayra Navarro, 'Parceria afinada entre design e artesana-to', *Casa Claudia*, Year 27, no.5, May, São Paulo; Maria Helena Estrada, 'Design Brasil 2000+3', *Arc Design*, no.33, November/December, São Paulo; 2004 Maria Helena Estrada, 'Sete anos de transformações Design, artesanato, indús-tria, mercado', *Arc Design*, no.38, September/October, São Paulo; Fabiana Caso, 'Desenho a toda prova', *O Estado de São Paulo*, 15 February, São Paulo; Simone Quintas, 'Invenções Animadas', *Casa e Jardim*, no.20, April, São Paulo; Francesca Angiolillo, 'De lo banal a lo industrial: Nuevo diseño em Brasil', *Arquine Revista International de Arquitectura y Diseño*, no.28, Summer, Mexico City; Andrea Codrington, 'Emerging Designers Make Their Mark', *I.D.*, vol. 51, no. 6, New York; 2006 Editorial, 'Um Novo Repertório?', *Arc Design*, no.47, April, São Paulo

Selected Exhibitions
2003 'Uma História do Sentar', Novo Museu, Curitiba, Brazil; 2004 'Oi Novos Urbanos – Chair Exhibition', Jockei Clube, Rio de Janeiro; 'Brasil Faz Design', IBRIT, Milan; '18th Prêmio Design do Museu da Casa Brasileira', Museu da Casa Brasileira, São Paulo; 'Cientoporciento Diseño', Palais de Glace, Buenos Aires; 2006 'I Bienal Brasileira de Design', Oca Ibirapuera, São Paulo

Awards
2002 Brasil Faz Design competition, São Paulo; 2003 IAB Competition for the Santana de Parnaíba Historical Center Renewal (urban furniture), São Paulo; 2004 Brasil Faz Design competition, São Paulo; 18th Prêmio Design do Museu da Casa Brasileira, São Paulo

Zinoo Park
Park Plus
www.zinoopark.com
Born in South Korea, 1973
Lives and works in Seoul

Education
1997 BA Art & Craft, Seoul National University; 2003 MA Product Design, Royal College of Art, London

Professional Background
2003 Internship, Tord Boontje Studio, London; 2003–5 System Appliances Division, Samsung Electronics Co., Seoul; 2006 Director-in-chief, Park Plus, South Korea

Selected Projects
2003 Sunglasses, Alexander McQueen; 2004 Washing Machine, Samsung Electronics; 5 Minute Candles, Park Plus; 2005 Spaghetti Chandelier, Park Plus; Robert Vacuum Cleaner, Samsung Electronics; Clean and white, bathroom accessories, Bumhan; Cutting Edge, designer's bicycle frame, Huffy

Selected Bibliography
2002 British Design & Art Direction, *D&AD 2002 Student Awards Annual*, London; 2003 Maxime Jalou & Marie-Jose Susskind Jalou, *L'officiel 1000 Modèles*, no.3, Paris; Will Dallimore, *The Show 2003 Catalogue*, London; 2004 Brigitte Fitoussi, Reperes; 2005 'Frankfurt als sprongbrett', *Arcade Magazine*, September, Frankfurt; 2006 'Design talk', *Casa Living Magazine*, no.78, September, Seoul

Selected Exhibitions
2003 'Salone Satellite', Salone del Mobile, Milan; 'Biennale de Valencia'; 2004 'Park plus by Zinoo Park', Gallery 'Factory', Seoul; 2005 'Zinoo Park', Space 'Kobo & Tomo', Tokyo; Designers Block, The Nicholls & Clarke building/Atelier Frankfurt, London/Frankfurt; 2006 'Maison & Object – Ugly home edition', Parc des Expositions, Paris; ICFF, New York

Awards
2002 2nd prize, PERGO design competi-tion, London; 2005 Winner, Design leader for next generation, Seoul

Russell Pinch
PINCH
www.pinchdesign.com
Born in England, 1973
Lives and works in London

Education
1993 Ravensbourne College of Art, Kent

Professional Background
1994–9 The Conran Shop/Conran Studio, London; 1999–2004 The Nest, co-founder/creative director, London; 2004 Established own studio

Selected Projects
2004 WAVE, sideboard, Content by Conran; 2005 TWIG, bench and cube, PINCH; Benchmark Range: Maiden, table and stool, Benchmark Furniture; Mrs B, table, Benchmark Furniture; Singer, shelves, Benchmark Furniture; 2006 Avery, chair, SCP; Blackwell, table, SCP; Elsie, sofa, SCP; Leamus, media side-board, SCP

Selected Bibliography
2005 'True Brit', *Elle Decoration*, October, London; 2006 Chris Rochester, 'Furniture Today Interview', *Furniture Today*, January, London; Danusia Osiowy, 'Natural Touch', *Cabinet Maker*, April, London; Oliver Bennett, 'People with passion', *BBC Homes & Antiques*, July, London

Awards

2004 Best newcomer, *Blueprint*/100% Design, London; 2005 Best furniture for the Wave sideboard, Design & Decoration awards, London; Shortlisted for the Wave sideboard, ELLE DECO Furniture Awards, London

POLKA

www.polkaproducts.com
Marie Rahm: born Germany, 1975; Monica Singer: born Austria, 1975
Both live and work in Vienna

Education

Marie Rahm: 2000 Product Design, University for Applied Arts, Vienna; 2000 Product Design, Royal College of Art, London; Monica Singer: 2003 Product Design, University for Applied Arts, Vienna

Professional Background

2004 Established POLKA

Selected Projects

2002 MHT02, men's handbag, One; CUT, light, Innermost; 2004 Tattoofurniture, POLKA; 2005 Short Set, Herend; MINImeter, measure tape for the MINI car, MINI/BMW; POLKAchair, Disguincio; 2006 Josephine, glass carafe, Lobmeyr; Pane, bendable breadbasket, Authentics; DOOSEY, lightshades, Innermost

Selected Bibliography

2004 'En-lightment', *Viewpoint*, no.6, Fall; Jana Semeradova, 'POLKA', *BLOK12*, no.12, October, Prague; 2005 Juland, *Pure Austrian Design*, Vienna; 2006 'POLKA', *Elle Decoration*, February, no.162, London; 'Zeitgemäss', *H.O.M.E*, February, Vienna; Lilli Hollein 'A New Face for Vienna', *Ottagono*, no.189, April, Bologna; Tulga Beyerle and Karin Hirschberger (eds.), *Designlandschaft Österreich*, Vienna; Eichinger oder Knechtl (ed.), UD.A, *Ultimos Disenos*, Vienna, Exhibition Katalogue; Designforum (ed.), *360° Design Austria*, Vienna

Selected Exhibitions

2004 'Salone Satellite', Salone del Mobile, Milan; 'Milky Way', Designblok, Prague; 2005 'Talents', Ambiente, Frankfurt; 'Pure Austrian Design', Kunsthalle project space, Vienna; 'By Invitation', Promosedia, Udine; 'Pure Austrian Design', Prague and Barcelona; 2006 'Ultimos Disenos', Madrid; 'By Invitation', Spazio Orlandi Gallery, Milan; Wallpaper 'Global edit', Salone del Mobile, Milan; '360° Österreichisches Design', MQ Vienna; 'MISS/FIT', MQ, Vienna; 'Pure Austrian Design' 100%east, London; 'Eden ADN', Saint Etienne Design Biennial

Awards

2004 1st prize, Creative Competition by MINI, Vienna; 2005 Best New Product Award, Maison et Objet, Paris; 1st prize, City Council of Vienna: Innovative Christmas Lights,Vienna

ransmeier & floyd

www.ransmeier-floyd.com
Gwendolyn Floyd: born USA, 1979; Leon Ransmeier: born USA, 1979
Both live and work in The Netherlands

Education

Gwendolyn Floyd: 2001 Brown University, Providence, USA; 2005 Design Academy, Eindhoven; Leon Ransmeier: 2001 Rhode Island School of Design, Providence

Professional Background

Gwendolyn Floyd: 2004 Internship, Industrial Facility, London; 2006–present Phillips Design, Eindhoven; Leon Ransmeier: 2002 Industrial design guest lecturer, Parsons School of Design, New York; 2006–present Product design and development, Royal VKB, Zoetermeer, The Netherlands; Both: 2003 Established ransmeier & floyd

Selected Projects

Gwendolyn Floyd: 2005 Printer, prototype; DIY M., lampshade, Droog; Leon Ransmeier: 2005 Worn In, doormat, prototype; 2006 Snow, side tables, limited edition of 20; Both: 2005 Crop, bookshelf, prototype; Gradient, dishrack, prototype; Hood, Range Top Air Purifier, prototype

Selected Bibliography

2005 Joachim Fischer (ed.), *Young European Designers*, Cologne, Daab; Thomas Edelmann, 'Designing Talents' *Form*, July, Neu-Isenburg, Germany; Marcus Fairs, 'Grey Goods', *Icon*, Issue 027, September, London; 2006 Patricia Urquiola (ed.), *International Design Yearbook 2006*, London; Ellen Lupton (curator), *National Design Triennial: Design Life Now Catalogue*, New York; Sam Hecht, 'Focus Design', *ArtReview*, April, London

Selected Exhibitions

Gwendolyn Floyd: 2003 WickerGames, Blindenanstalt, Berlin; Both: 2004 'Salone Satellite', Salone del Mobile, Milan; 2005 'Talents', Ambiente Frankfurt, Frankfurt; 2006 'NEON, Interior Lifestyle', Big Sight, Tokyo; 'National Design Triennial: Design Life Now', Cooper-Hewitt, National Design Museum, New York; Leon Ransmeier: 2004 'Salone Satellite', Salone del Mobile, Milan; 'Spin Off', Museum for Applied Arts, Cologne; 'Das Moma in Berlin, 7x7 museum shop', National Gallery, Berlin; 2005 'colette meets comme de garçons', comme de garçons, Tokyo

Awards

Gwendolyn Floyd: 2004 1st prize, Designboom 100% Tiles Competition, London

Adrien Rovero

www.adrienrovero.com
Born in Switzerland, 1981
Lives and works in Lausanne

Education

2000 Apprenticeship as interior designer, EPSIC, Lausanne; 2004 BA Industrial Design, École cantonale d'art de Lausanne; 2006 Master in Industrial Design, École cantonale d'art de Lausanne

Professional Background

2000 Atmosphère Créative/Interior Designer, La Conversion, Switzerland

Selected Projects

2003 Brush Couple, dustpan and brush, Die imaginäre Manufaktur; Candlelight, Suck UK; 2004 Dis-Order carpet, Galerie Kreo; 2005 Basketstool, prototype, Adrien Rovero; 2006 Dip 1/3, prototype, Christofle; Chain, Belt & Sample, carpets, Tisca France; VD 003, bike rack, Rovagro SA

Selected Bibliography

2003 Yann Gerdil-Margueron, 'Portraits', *Rendez-vous*, no.1, winter 2003–4, Switzerland; 2004 Albert Hill, 'Outside the box', *Wallpaper**, no.70, July, London; Michael Erlhoff, Product Design, Berlin; Silja van der does, 'Planned Disorder', *Form*, no.200, January–February, Neu-Isenburg, Germany; 2005 Henrik Hornung, 'Unordnung mit system', *design report*, 1/05, January, Leinfelden, Germany; Chiara Pilati, 'Caos in casa', *Ottagono*, no.177, February, Bologna; Pierre Keller, *Swiss Design Now*, Shanghai; Thierry Hausermann, 'Portraits', *I.D. Pure*, Hors série, Summer, Lausanne; Charlotte Vaudrey, 'A tidy paradox', *Frame*, no.43, March/April, Amsterdam; 2006 Maroun Zahar, 'Espace Actuel', *Espace Contemporains*, 1/06, January, Morges; Francisco Torres, *INOUT*, Lausanne; Emma Firmin, 'View on Product', *Damn*, no.6, Brussels; Alessandra Paracchi, 'I'd like to park my bike', *Label*, no.22, Summer, Torino; Catherine Geel, *Design Parade*, Summer, Hyères

Selected Exhibitions

2002 'L'Ecal au CCS', Centre Culturel Suisse, Paris; 'Botte-Cul', Centre Culturel Suisse, Milan; 2003 'The Milano Workshops' 2003/ECAL, Galeria La Posteria, Milan; 'Somewhere Totally Else', Design Museum, London; 2004 'Cucina Festiva' Galeria Vinçon, Barcelona; 'DIM', Flux design biennale, Lucerne; 'Dining Design', Salone Satellite, Milan; 'Sélection de diplôme de l'Ecal', ECAL, Lausanne; 2005 'Ex-Ecal', Galeria Care Of, Milan; 'Cristal Clear/Ecal', Galeria Mauro Brucoli, Milan; 'Big-game & Adrien Rovero', Own, Brussels; 'Swiss Design Now', Moca, Shanghai; 2006 'INOUT config. 01', MUDAC, Lausanne; 'Up & Down', Galerie Kreo, Paris;'Christofle/Ecal', Galeria Mauro Brucoli, Milan; 'Sélection de travaux d'étudiants', Lelac, Lausanne; 'YDN showroom', Dutch design week, Eindhoven; 'INOUT config. 01', Galeria Vincon, Barcelona.

Awards

2001 shortlisted, *Intramuros*, Louis Vuitton, Chanel, Concours Le Sac, Paris; 2002 2nd prize, D&AD, Student awards, Nesta product innovation, London; 2004 shortlisted, Les lunetiers du Jura, Paris; 2005 shortlisted, Ferté Vidame, Sur les traces du passé, Paris; 2006 Prix du jury, Design Parade, Hyères, France

Alejandro Sarmiento

www.alejandrosarmiento.com.ar
Born in Argentina, 1959
Lives and works in Buenos Aires

Education

1986 Universidad Nacional de la Plata – Facultad de Bellas Artes, La Plata, Provincia de Buenos Aires

Professional Background

2004 Inow Technologies, Buenos Aires; Museo de Arte Latino Americano, Buenos Aires; 2005 Essen Aluminios S.A., Venado Tuerto, Pcia. de Santa Fe, Argentina; I.D., London

Selected Projects

2003 Ruberta Iron, home furniture, IndustriAL Standard; Cilindros Puf, inflatable seat, IndustriAL Standard; 2004 Nacional Puf, inflatable seat, IndustriAL Standard; Tupac Puf, inflatable seat, IndustriAL Standard; 2005 Querencia Puf, inflatable seat, IndustriAL Standard

Selected Bibliography

2003 Lujan Cambariere, 'Valor Agregado', *Diario Pagina 12 Suplemento M2*, February, Buenos Aires; 2004 Carolina Muzi, 'Diseñador de las Pelotas', *Diario Clarin Suplemento Arquitectura*, 14 September, Buenos Aires; Lujan Cambariere, 'Introducción al Disenismo', *Diario Pagina 12 Suplemento M2*, February, Buenos Aires; 2005 Lujan Cambariere, 'Alejandro Sarmiento', *Elle Déco*, no.138, October, Buenos Aires; Ricardo Blanco, *Cronicas del Diseño Industrial en la Argentina*, December, Buenos Aires; 2006 Bénédicte Duhalde, 'Profils', *Intramuros*, no.123, March, Paris; Francesca Picchi, 'Geodesign', *Domus*, no.891, April, Milan; Winnie Bastian, 'Reciclo Bem-Sucedido', *Arc Design*, no.48, June, São Paulo

Selected Exhibitions

2001 'Circus Stool', Tokyo Designers Block, Tokyo; 2002 'Contenidoneto', Centro Metropolitano de Diseño, Buenos Aires; 2003 'Contenidoneto', Designers Block, London; 'Transformation', Felissimo Gallery/Design House, New York; 2004 'Proyecto Pet/Contenidoneto' British Art Centre, Buenos Aires; 2005 'Panorama Internacional de Design Safety Nest/O Ninho Seguro', Centro Cultural Correios, Rio de Janeiro; 2006 'Mostra Internacional de Design Safety Nest/O Ninho Seguro', Sesc Pinheiros, São Paulo

Awards

2001 Idee Design Award, Marc Newson

Award, Michael Young Award, Tokyo; 2004 Ecologic Design Award, Movelsul, Bento Goncalvez, Brazil; 2005 Zogo Giochi D'Arte Award, Asolo, Italy

Inga Sempé

www.ingasempe.fr
Born in France, 1968
Lives and works in Paris

Education

1993 ENSCI les Ateliers, Paris

Professional Background

1994 Marc Newson, Paris; 1997–9 Andrée Putman, Paris

Selected Projects

2000 Clock, prototype, VIA; 2001 Lamp Plate, Cappellini; 2003 Suspension Plissé, lamp, Cappellini; Brosse, containers, EDRA; Long Pot, Ghaadee France (then Ligne Roset from 2006); 2006 Tables LaChapelle, David design; Table Lunatique, Ligne Roset

Selected Bibliography

2000 Marco Romanelli, 'Horloge VIA', *Abitare*, no.393, March, Milan; 2001 'The Best of Paris', *Frame*, no.20, Amsterdam; 2002 Christoph Radl, 'Inga Sempé', *Interni International*, April, Milan; 2003 'Inga Sempé', *AXIS*, no.105, September/October, Tokyo, Japan; 'Inga Sempé', *Hanatsubaki*, no.640, October, Tokyo; 'Inga Sempé', *Le Point*, 29 August, France; Inga Sempé, 'La passion du bric à brac', *Paris Capitale*, 15 May, Paris; Sophie Djerlal, 'L'autre Sempé', *Vogue*, August, Paris; Bénédicte Duhalde, 'La place d'Inga Sempé', *Intramuros*, no.107, June–July; Anne-Marie Fèvre, 'Sempé toujours simple', *Libération*, 30 June, Paris; Marion Vignal, 'Inga Sempé, rêveuse d'objets', *L'Express, Expressmag*, 16 October, Paris; 2004 Cristina Morozzi, 'Inga Sempé', *The Plan*, June–July, Bologna, Italy; Ian Philips, 'Wizard of Odd', *The New York Times*, New York; Klaus Meyer, '*design report award 2004*', *design report*, August, Heuiedweg, Germany; Tiphaine Samoyault, 'La belleza nace de la comodidad', *Arquine*, no.34, winter, December, Mexico; Natasha Edwards, 'Profile Inga Sempé' *design week*, 13 April, London

Selected Exhibitions

2002 'Tutto Normale', Villa Medici, Rome; 2003 'designer's days', edra, Paris; 'Inga Sempé personal exhibition', Musée des Arts décoratifs, Paris; 2004 'Design d'Elles', VIA, Paris; 2006 'Big Bang', Centre Pompidou, Paris; 'Collection permanente', Musée des Arts décoratifs, Paris

Awards

2003 Grand Prix de la Création de la Ville de Paris, Paris

Wieki Somers

www.wiekisomers.com
Born in The Netherlands, 1976
Lives and works in Rotterdam

Education
1999 Internship, Radi Designers and BLESS, Paris; 2000 The Design Academy, Eindhoven

Professional Background
2001 Established Wieki Somers studio, Schiedam, The Netherlands

Selected Projects
2002 Mattress Stone Bottle, E.J. Kwakkel; 2003 High Tea Pot, porcelain teapot, E.J. Kwakkel; 2004 Trophees, glassware, Arques International; 2005 Blossoms, vase, Cor Unum; Bathboat, t'Vliehout boatbuilders

Selected Bibliography
2002 Joseph Kelly, 'Human intervention', *I.D.*, June, New York; 2003 Ao Ank Trumpie, *Deliciously decadent!*, Rotterdam; 2004 Renny Raymakers, *Simply Droog*, Amsterdam; Francesca Picchi, 'Droog and around', *Domus*, no.872, August, Milan; 2005 Marcel Wanders (ed.), *The International Yearbook*, London; Eva Schaap, 'New Dutch designers', *Frame*, no.46, September, Amsterdam; Liliana Albertazzi, 'The Determination of The Youngest', *Intramuros*, no.121, November/December, Paris; Arjan Ribbens, 'Mijn produkten mag je koesteren', *NRC*, December, Amsterdam; 2006 Cristina Forentini, 'Succeeded in Olanda', *DDN*, no.113, April, Milan; Laila Quintano, 'Dutch Design, Wieki Somers', *Experimenta*, no.54, January, Barcelona; Kristina Raderschad, 'Wleki's Welt', *Decoration*, January, Hamburg; Marcus Fairs (ed.), *Twenty-First Century Design*, London; *Into the Nature*, Die Gestalten Verlag, Germany

Selected Exhibitions
2005 'Wieki Somers s'Winters', Vivid, Rotterdam; 'project Rotterdam', Museum Boijmans van Beuningen, Rotterdam; 'Being in your own bubble', National Glass Museum, Leerdam; 'Grazie dei Fior', Salone del Mobile, Milan; 'New Dutch Designers', Gallery Droog, Amsterdam; 2006 'NLA', Acme, Los Angeles; 'European Ceramic Context 2006', Bornholm

Alexander Taylor
www.alexandertaylor.com
Born in UK, 1975
Lives and works in London

Education
1999 Nottingham Trent University

Professional Background
1999–2001 Proctor: Rihl Architects, London; 2001–2 McDonnell Associates Architects, London

Selected Projects
2003 Antlers, coat hook, Thorsten van Elten; 2005 Fold, light, Established & Sons; HM44 Lounge Chairs, Hitch Mylius; Huskey, bench, E&Y; 2006 Butterfly Table, Zanotta/E&Y coedition

Selected Bibliography
2003 Max Fraser, *Design UK 2*, London; Antonia Ward, 'Unfinished Symphony', *FX*, November, London; 2004 Tom Templeton, 'Movers and Shapers', *Observer Magazine*, 26 September, London; Jenny Dalton, 'Who's Next', *Elle Decoration*, September, London; 2005 Jill MacNair, 'Trophy Hunter', *Observer Magazine*, 6 November, London; 2006 Marcus Fairs (ed.), *Twenty-First Century Design*, London

Awards
2004 *Elle Decoration* 'Best Home Accessory' for Antlers, London; 2005 *Elle Decoration* 'Best in Lighting' for Fold lights, London; *Elle Decoration* 'Young Designer of Year Award', London

Carla Tennenbaum
Born in Brazil, 1979
Lives and works in São Paulo

Education
1998 Faculdade de Arquitetura et Urbanismo – Universidade de São Paulo; 2004 Faculdade de Filosofia, Letras e Ciêcias Humanas, Universidade de São Paulo

Professional Background
2000–2006 Surto Dana Group, São Paulo; 2001–2005 Notechdesign, São Paulo

Selected Projects
2000 Festa Couch/Stools, self-production; 2003 Kraft, chair, self-production; Transoptic Screen, self-production; Undustrial Sculpture, Carla Tennenbaum and Rai Mendes; 2006 Kinetic Tapestry, Carla Tennenbaum and Rai Mendes; Tubete Pencil-Holder, Carla Tennenbaum and Rai Mendes

Selected Bibliography
2000 Vanni Pasca, 'Abitare il tempo – Mostre di Sperimentazione e Ricerca', *Catalog*, Verona; 2002 Vanni Pasca and Maria Helena Estrada, 'Young Design', *Interni*, no.523, July/August, Milan; 2003 Humberto & Fernando Campana, *Campanas*, São Paulo; 2004 Simone Quintas, 'A ordem é reinventas', *Casa e Jardim*, no.589, February, São Paulo; Andrew Wagner, 'What we saw', *DWELL*, vol.4/7, July/August, Harland, USA; 2005 Guilherme Aquino, 'Brasil Reciclado', *Casa e Jardim*, no.604, May, São Paulo; 'Revaloração Artistica', *KAZA*, no.28, September, São Paulo; Bernardo de Aguiar, 'Carla Tennenbaum ganha prêmio no Japão', *Erikapalomino*, September, São Paulo; Marcia Holland, 'Carla Tennenbaum', *Primavera do Design–catalog*, September, São Paulo; Maria Helena Estrade, 'Design Brasil 2000+5', *Arc Design*, no.44, September/October, São Paulo

Selected Exhibitions
2000 'Beyond European Design: Projects from around the World', Abitare il Tempo, Verona; 2001 'Notechdesign', MUBE, São

Paulo; 2003 'LOLAnDA 2003', Latin American Design Foundation, Amsterdam; 2004 'Luz de Nós', Brandia Spot, Lisbon; 2005 'Design Possible – le implicazioni etico-sociali nel tuturo del Design', Spazio Brasilantes, Milan; '"Love/Why?" – 5th International Design Award – Design 21', Hyogo Prefectural Museum, Kobe; Japan/Felissimo Design House, New York/UNESCO Headquarters, Paris; 2006 'Swell – Future Friendly Design', UBC Robson Square, Vancouver; 'Design for Future', Galeria da Restauracão – Olhão City Museum, Olhão, Portugal

Awards
2003 Latin American Foundation hOLAnDA 2003, Amsterdam; 2005 UNESCO/ Felissimo Group Design 21, Kobe, Japan and Paris

TONERICO:INC.
www.tonerico-inc.com
Ken Kimizuka: born Japan, 1973
Yumi Masuko: born Japan, 1967;
Hiroshi Yoneya: born Japan, 1968
All live and work in Tokyo

Education
Ken Kimizuka: 1998 Musashino Art University, Department of Architecture, Tokyo; Yumi Masuko: 1988 Joshibi University of Art and Design, Tokyo; Hiroshi Yoneya: 1988 Musashino Art University, Department of Industrial Design, Tokyo

Professional Background
Ken Kimizuka: 1998 Joined Studio 80, Tokyo; 1999 Contract designer with Hoenkan, Tokyo; 2002 Joined TONERICO:INC., Tokyo; Yumi Masuko: 1987 Book cover designer for Banana Yoshimoto, Tokyo; 1988–97 Worked as coordinator for Ms Setsuko Yamada, Tokyo; 1997 Freelance coordinator, Tokyo; 2002 Joined TONERICO:INC., Tokyo; Hiroshi Yoneya: 1992–2002 Designer at Studio 80, Tokyo; 2002 Established TONERICO:INC., Tokyo; 2005–present Part-time lecturer at Tama Art Univeristy, Tokyo

Selected Projects
2003 TONERICO:FOR TWO, side table and cups, Maki Nakahara; 2004 GUSHA, stool and side chair, Minerva Co.Ltd; EXTO, booth design, CBK Co. Ltd; 2005 MEMENTO, light piece, Yamakin Co.Ltd; MEMENTO-LINK, hanging light, Yamakin Co. Ltd. and Interlux Co. Ltd; 2006 Pon Poco Poon, lighting series, Matsushita Electric Works Ltd

Selected Bibliography
2005 Kazuya Shimokawa, 'Winning the competition', *Nikkei design*, no.223, January, Japan; Virginio Briatore, 'Salone Satellite '05', *Interni*, no.552, June, Italy; 'La Matematica e decorazione', *Case da Abitare*, no.88, June, Milan, *Case da Abitare*; Joachim Fischer, *Young Asian Designers*, September, Germany; Einsame

Hohepunkte, 'Licht-Kunst', *design report*, July/August, Germany; Toshiki Kiriyama, '100 designers of Nippon', *AERA design*, November, Japan; Yudai Tachikawa, '11 designers Japan', *Casa Brutus*, November, Japan; 2006 'Create a Typographic Mood Board for Inspiration' *Elle Decoration*, UK edition, February, UK; Masaaki Takahashi, '50947', *Frame*, no.49, March/April, Amsterdam; Elke Taxis, 'Numeric Shadows', 04.2006, April, Germany

Selected Exhibitions
2002–5 Tokyo Designer's Week, CONTAINER GROUND at Aomi, Shop exhibition at Exto showroom, Tokyo; 2003–5 'Salone Satellite', Salone del Mobile, Milan; 2005 'MEMENTO @ lebain/solo visiting exhibition for MEMENTO', Gallery le bain, Tokyo; 100% Design TOKYO, Meiji Jingu Gaien, Tokyo

Awards
2004 JCD design prize, Tokyo; 2005 1st prize, Salone Satellite 2005 design report award, Milan; JCD design prize, Tokyo; 2006 The Best Store of the Year 2005, Tokyo; JID AWARD 2006 BIENNIAL interior space design award, Tokyo

Peter Traag
www.petertraag.com
Born in The Netherlands, 1979
Lives and works in Antwerp

Education
2001 3-D Design, Hogeschool voor de Kunsten Arnhem, Arnhem; 2003 Product Design, Royal College of Art, London

Professional Background
2004 Established Peter Traag Studio, London; 2006 Established Studio Peter Traag, Antwerp

Selected Projects
2004 Sponge Chair, Edra; 2005 Mummy, beach chair, Edra; 2006 Dip, table, Pallucco

Selected Bibliography
2004 Francesca Picchi, 'An Armchair Generated by Chance', *Domus*, no.871, June, Milan; 2005 Marcel Wanders (ed.), *The International Design Yearbook*, London; Tanya Harrod, 'The Future is Handmade: The Crafts in the New Millennium', *Craft culture*, November, Melbourne; Hannah Booth, 'Well Seated', *Design Week*, vol. 20, no.10, March, London; Nicole Swengley, 'A Graduate Goes, a Master Moves on', *Financial Times*, 9 April, London; Tom Dyckhoff, 'Best of British', *GQ*, no.190, April; 2006 Itsuro Shibata, 'Up and coming creators in the eyes of prominent designers', *Axis*, vol.122, August, Tokyo; Katrin Cosseta, '6 x 6', *Interni*, no.9, September, Milan

Selected Exhibitions
2004 'Design Mart', Design Museum, London; 2005 'Design Tide', Escorter AOYAMA Event Zone, Tokyo; 'My World – The New Subjectivity in Design', Experimenta, Lisbon; 'Great Brits – The

New Alchemists', Paul Smith Headquarters, Milan

Awards
2000 Honourable Mention, HEMA Design Contest, The Hague; 2005 Grant from Esmée Fairbairn Foundation, awarded by the Design Museum, London; 2006 Young Designer of the Year, Classic Design Awards, London

Maxim Velcovsky
Qubus Design
www.qubus.cz
Born in Czechoslavakia, 1976
Lives and works in Prague

Education
1999 Glasgow School of Arts; 2002 Architecture and Design, Academy of Arts, Prague

Professional Background
2002 Founder of Qubus Design Studio, Prague

Selected Projects
1999 Designed for the Messiah, The Cross, One off by Loci for Ambroz Church in Hradec Kralove; 2001 Pure White Cup, D-Sign Inc.; 2002 White Collection, porcelain products, Qubus; 2005 Koi Carp, water heater, Ekotez; Fast Collection Glass Products, Qubus; Maxim Tableware, G. Benedikt

Selected Bibliography
2001 Jiri Macek, 'Revolution for a Can', *Bluk*, 00/01, May, Prague; 2002 Peter Volf, 'Fighting for Better Consumerism', *Reflex*, no.42, October, Prague; 2003 Liz Hoggard, 'Children of the Revolution', *Independent*, 22 March, London; Michelle Hespe, 'Czech Mates', *Object*, no.42, Sydney; 2004 Sophie Davies, 'Crafts Talent', *Elle Deco*, May, London; Zdenek Freisleben, 'White Confrontations', *Glass and Ceramics*, July, Prague; 2005 Lesley Jackson, 'Porcelain Revolution', *Icon*, no.22, April, London; 2006 Alex Wiltshire, 'Poster Boy of Czech Design', *icon*, no.33, March, London

Selected Exhibitions
2001 'My Country', Gallery AAAD, Prague; 'Glued Intimacy', Gallery Jeleni Center for Temporary Art, Prague; 2002 'Prague Session', The Lighthouse, Glasgow; 2003 'Table Tennis at Designblok', Ball-Game Hall, Prague Castle, Prague; 2004 'For all of you and each of you' (with Anna Neborova), Critic's Gallery, Prague; 2005 'Maxim Velcovsky – The Designer', Konsepti Showroom, Bratislava; 2006 'Maxim Velcovsky with Qubus', Gallery for Contemporary Art, Belgrade

voonwong&bensonsaw
www.voon-benson.com
Benson Saw: born Malaysia, 1974;
Voon Wong: born Singapore, 1963
Benson Saw: lives and works in Malaysia;
Voon Wong: lives and works in London

Education
Benson Saw: 1996 BSc Mechanical Engineering, Boston; 2001 MA Product Design, Royal College of Art, London; Voon Wong: 1984 National University of Singapore; 1990 The Architectural Association, London

Professional Background
Benson Saw: 1996 Mechanical Engineer for Massachusetts General Hospital, Boston, USA; 1996–9 Design Engineer for Motorola, Fort Lauderdale, Florida; 1999–2001 Matthew Hilton Studio, London; Voon Wong: 1991–2 Office of Zaha Hadid, London; 1992–3 Rick Mather Architects, London; 1993–4 SAA Architects, Singapore; 1994–6 CF Mollier, London; Both: 1996 Established own practice

Selected Projects
2001 Loop Lamp, Fontana Arte; 2002 Tube Silver Vases, VWBS; Tube Clear Vases, VWBS; 2003 Twig, VWBS; Origami Lamp, Prototype; Slicebox Table, VWBS; 2004 Ribbon Coatrack, VWBS; 2005 Cubist Mirror, VWBS; Landscape Vase, VWBS; 2006 Bone China Lighting, VWBS

Selected Bibliography
2004 Tom Dixon (ed.), *International Design Yearbook*, London; Toby Wiseman (ed.), 'Worldly Possessions, Details Section', *GQ*, July, London; Giovani Progettisti, 'Function vs poetry', *Abitare*, July–August, Italy; Talib Choudhry, 'Boudoir Punk', *The Sunday Times*, 5 December, London; 2005 Kyoko Nakajima, 'Establishing Brands Without Going to Corporations', *AXIS*, no.113, February, Japan; 2006 Michael Mazzeo, 'Guiding Lights, image', *Surface*, no.57, February, USA; 'Big in Milan', *Design Week*, no.21, 30 March, London; Alex Bagner, 'Wong to Watch', *Wallpaper**, no.86, March, London; Devin Jeyathurai, 'Va Va Voon!', *Home Concepts*, April, Singapore; Laura Maggi and Piera Belloni, 'Design Week 2006, DestinazioneMi, Brera-Garibaldi', *Elle Decoration*, April, Italy; 'Design bordering on art', *Monitor*, no.36, April–May, Russia

Selected Exhibitions
2001 'Spin-Off', Cologne; 2001–3 Designers Block, London and Milan; 2003 100% Design, London; 2004 Salone Satellite, Milan; 2005 'Confetti', Space Furniture, Singapore; 2006 'Light Last', Galleria Ciovasso, Milan

Awards
2002 OXO Peugeot Award Lighting, for Elma Vases, UK; 2004 Shortlisted for the Loop Lamp, Compasso d'Oro

Dominic Wilcox
www.dominicwilcox.com
Born in England, 1974
Lives and works in London

Education
1997 BA (Hons) Visual Communication, Edinburgh College of Art; 1998 Dip. Visual Communication, Edinburgh College of Art; 2002 MA Product Design, Royal College of Art, London

Professional Background
2002–5 Mosley meets Wilcox, London; 2006 established Dominic Wilcox, London

Selected Projects
2002 Nimbus, self-production; Ivy Shelf, self-production; The Glove, glove, self-production; White Elephant, giant lamp-shade/space, self-production; Foot Print, shoe, self-production; 2002–6 War Bowl, Thorsten van Elten; Honesty Stamps, ink stamps, Thorsten van Elten; 2003 Ghost Table, Mosley meets Wilcox; 2004 Nose Light, self-production; 2004–5 Rock Table Series, Mosley meets Wilcox; Rock Plates Series, Mosley meets Wilcox/Figgjo; Plectrum Light, Mosley meets Wilcox; 2005 Bowie Vase Set, Mosley meets Wilcox

Selected Bibliography
2002 Katy Greaves, 'Your Flexible Friends', *Blueprint*, no.202, December, London; 2004 Luis Royal, 'Design 100%', *Oindependente*, 16 April, Lisbon;' Weijie Yang, 'When Mosley meets Wilcox', *Design*, no.13, June, Beijing; Sarah Brownlee, 'People Profile', *FX*, October, London; 2005 Jerome Rivolta, 'Mosley meets Wilcox meets Artravel', *Artravel*, no.3, March, Paris; Aric Chen, *Spoon*, October/November, New York; 2006 Brendan L. Koener, 'Caring enough to stamp I Love You', *New York Times*, 19 March, New York

Selected Exhibitions
2002 'Small Step Exhibition', Aram Gallery, London; 2004 'Mosley meets Wilcox meets Rock', Hanbury Gallery, London; 'Brilliant', Victoria & Albert Museum, London; 2005 'Touch Me', Victoria & Albert Museum, London; 2005-6 'Rock n'Roll Icons: The Photography of Mick Rock', Urbis, Manchester

Awards
2001–2 RCA, Pergo design, London; 2003 Peugeot Design Awards, London

WOKmedia
www.wokmedia.com
Michael Cross: born England, 1979; Julie Mathias: born France, 1978
Both live and work in London

Education
Michael Cross: 2002 BA Hallam University, Sheffield, UK; 2004 MA Product Design, Royal College of Arts, London; Julie Mathias: 2002 École des Beaux-Arts de Saint-Etienne; 2004 MA Product Design, Royal College of Arts, London; 2005 Royal College of Arts Research, Hele Hamlyn Research Centre, London

Professional Background
Michael Cross: 2004–present Freelance designer, London; Julie Mathias: 2004–present Established WOKmedia with Wolfgang Kaeppner; creative director of WOKmedia, London

Selected Projects
2004 Flood, light installation; Sprinkle, rug; Blow, fan; 2005 Lunuganga, shelf; Knot, toilet; Hidden, table

Selected Bibliography
2005 *Well Done*, Tapei; 'Great Brits', *Wallpaper**, July, London; 'Dangerous Idea', *Casa International*, no.34, October, Japan; Ham Capon, 'When Michael met Julie', *Trend*, no.49, November, Israel; Grant Gibson, 'Commissions', *Crafts*, no.197, November/December, London; 2006 'Antennae', *The World of Interiors*, January, London; 'Mission Mash', *Blueprint*, no.238, January, London; 'Seven Up', *Time Out*, January, London; 'Great Brits', *Elle Décor*, no.82, February, Japan; Michael Cross and Julie Mathias, *Muse*, London; 'Experimentalism', *Viewpoint 18 & 19*, London

Selected Exhibitions
2004 'Design Mart', Design Museum, London; 2005 'Great Brits', Paul Smith Headquarters, Milan; 'My World'/Experimenta, Botanical Garden, Lisbon; 'Well Done', Taiwan Museum of Contemporary Art Tapei, Tapei, Taiwan; 'Great Brits', Travelling, Tokyo/Singapore/Australia; 'Best of', Sprada Bank, Münste; 'British Design', Droog Gallery, Amsterdam

Shunji Yamanaka
Leading Edge Design
www.LLEEDD.com
Born in Japan, 1957
Lives and works in Japan

Education
1982 School of Engineering, University of Tokyo

Professional Background
1982–7 Exterior designer, Design Center of Nissan Motor Company, Tokyo; 1991–4 Associate professor, University of Tokyo

Selected Projects
1989 O-product, compact camera, Olympus; 2001 INSETTO, wristwatch, Issey Miyake; 2002 Cyclops, humanoid installation, Leading Edge Design Corp./Nichinan Corp.; morph3, humanoid robot, Leading Edge Design Corp./Kiano Symbiotic Project, The Japan Science and Technology Corporation; Hallucigenia01, robotic vehicle, Leading Edge Design Corp./Future Robotics Technology Center of the Chiba Institute of Technology; 2005, tagtypeGK, Japanese character keyboard, takram engineering; 2006 OXO Kitchen Tools, OXO International

Selected Bibliography
2004 Tom Dixon (ed.), *The International Design Yearbook*, London; 2005 Katsutoshi Ishibashi, 'Cover Interview', *AXIS*, vol.116, 1 July, Tokyo; Kaz Yuzawa, 'Design Drives Technology', *Casa Brutus*, vol. 68, 1 November, Tokyo; 2006 Shunji Yamanaka, *Projections of Function, Shunji Yamanaka works, 1988–2006*, 13 May, Tokyo; Kenji Hall, 'Yamanaka: A Free Design Spirit', *Business Week Online*, 9 June, New York

Selected Exhibitions
2001 'Workspheres', MoMA, New York; 2002 permanent collection, Ars Electronica Center, Linz; 2005 'Chubu Community for Millennial Symbiosis', EXPO 2005 Aichi Japan, Nagoya; 'Leading Edge Design exhibition "MOVE"', Spiral Gallery, Tokyo; 2006 'Projection of Function', AXIS Gallery, Tokyo

Awards
2001 iF design award, Hanover; 2005 '2004 Designer of the Year' Mainichi Newspaper Co., Ltd, Tokyo; 2006 Good Design Award 'Gold Prize 2006', Tokyo.

Noriko Yasuda
Born in Japan, 1975
Lives and works in Tokyo

Education
1996 Nihon University, Tokyo; 1998 Kanazawa International Design Institute, Ishikawa, Japan; 2001 Parsons School of Design, New York

Professional Background
Writer, designer, translator working in Tokyo

Selected Projects
2000 Teacuptea/Ring/Wan, porcelain objects, self-production; 2001 Bulbshell, light, self-production; 2003 Plastibear, teddybear made from reused plastic bags, self-production; Anouk Lepere Earing Display Tile, self-production; 2004 Cremated Teddy Bear, self-production

Selected Bibliography
2002 Emma O'Kelly, Laura Houseley and Albert Hill, 'Intelligence Report', *Wallpaper**, June, London; *Inview*, no.20, September, Paris; 2003 Ken Mori, *Giant Robot*, no.28, Spring; 2004 Aaron Betsky, *Simply Droog*, Amsterdam

Selected Exhibitions
2001–4 'Droog Design Presentation', Milan; 2002 'Milan in a Van', Victoria & Albert Museum, London; 'Fragile', Namori Gallery, Osaka; 2003 'Tokyo Designers Week (Droog Presentation)', UN, Tokyo; 2005 'Touch Me', Victoria & Albert Museum, London

zuii
www.zuii.com
Alana Di Giacomo: born Australia, 1979; Marcel Sigel: born Australia, 1976
Both live and work in Melbourne

Education
Alana Di Giacomo: 2000 BA Visual Communications, Curtin University, Perth; Marcel Sigel: 2001 BA Product Design, Curtin University, Perth

Professional Background
Alana Di Giacomo: 2000 Graphic Designer, Sydney; Marcel Sigel: 2001 Product Designer, Perth; Both: 2003 Established zuii, Melbourne

Selected Projects
2004 Woodland, table lamp, self-production; Swoon, candleholder, self-production; Trace, table light, self-production; 2005 Carbon Sapphire, chair, handmade by designer; Wall Brooch, coat rack, self-production; Henry's Collar, fruit bowl, self production; 2006 Pewter Range, Royal Selangor

Selected Bibliography
2004 Gunda Siebke, 'Talentshow in Mailand: zuii', *Schoner Wohnen*, no.8, August, Germany; Fleur Watson, 'Design Atlas: Zuii – Musical Chairs', *Monument*, no.62, August/September, Sydney; 2005 Anne-Celine Jaeger, 'Best Young Designers', *Wallpaper**, January/February, London; Jacqueline Khiu, 'Fresh Picked: Antipodean Antics', *Surface*, no.53, April, San Francisco; Luis Royal, 'Objectto de Desejo; Woodland', *O Independente*, June, Lisbon; Maggie Ma (ed.), 'zuii – ology', *Casa International*, no.31, July, Beijing; Yenny Kim, 'Living Idea of Next Generation 2535 – zuii', *Design*, no.325, July, Seoul; D.P., 'La Sinergia Creativa Di Zuii', *Ottagono*, no.183, September, Bologna; 2006 Daniela Patane, *Young Asian Designers*, Cologne; Joachim Fisher (ed.), *Pure Design Now*, Barcelona; Brian Parkes, *Freestyle*, Sydney

Selected Exhibitions
2004 'Fringe Furniture', Melbourne Museum, Melbourne; AIFF, Sydney Exhibition Center, Sydney; 'Young Designer of the Year Exhibition', Powerhouse Museum, Sydney; 2004 & 2005 'Salone Satellite', Salone del Mobile, Milan; 2005 'Bombay Sapphire: Sapphire Inspired', Bathhouse, New York; 'zuii: Somewhere Between', Object Gallery, Sydney; 2006–8 'Freestyle: New Australian Design for Living', Melbourne Museum, Object Gallery, QUT Art Museum, Art Gallery of South Australia, Triennale Milano, Victoria & Albert Museum, Melbourne, Sydney, Brisbane, Adelaide, Milan, London

Awards
2004 Young Designer of the Year: *Sydney Morning Herald*, Sydney; Design Excellence Award: Fringe Furniture, Melbourne Museum, Melbourne; 2005 Best Young Designers: *Wallpaper**, London

Index

Page numbers in *italic* refer to the illustrations

5.5 designers 10–13
Aalto, Alvar 412, *412*
Aboriginal people 419, *419*
Acriresinas *298–300*
Adal 237
Aebi, Aurel 22
AIDEC 201
Akahoshi *196–7*
Albini 416
Alessi 58, *58–60*, 98
ANTEEKSI 14–17
Antenna Design 18–21
Antonelli, Paola 266, 290, 354
Antwerp
   Peter Traag *378–81*
Apple Computers 286
Aquamass 278
Arad, Ron 42, 258, 302, *305*
Arco 180
Arnhem
   Ineke Hans *178–81*
Arp, Jean 416
Arques International 362, *365*
Artecnica 346
Artek 412, *412*
Arteluce 416, *416*
Artemide 58, *258*
Artquitect 182, *182–3*, 274
Asplund, Gunnar 318
Atelier Oï 22–5
Attitudes Art Gallery 162
Authentics 346

B&B Italia 22, 162
Baas, Maarten 26–9
Baccarat 358
Baci Perugina 34
Bae, Suyel 30–3
bags
   4 Zipper Purse *64*
   Black Hole 316
   Dilly Bag 419, *419*
   Pady Bag 202, *204*
   SoftBox 282
Bandeira, Manuel 34–7
Baranger, Vincent 10
Barbosa, Tatiana *50*
Barford, Barnaby 234
Barry Friedman Gallery 150
Bastos, André 298
bathrooms
   AQHayon Bathtub 182, *183*
   AQHayon Sink and Mirror 182, *182*
   Bathboat *354–5*, 362
Bauhaus 412, 414
BD Ediciones de Diseño *183, 185*
beds
   Flip Down Crib *95*
   Rest 257
   Voyage 110
   Wogg 24, 66
Béhar, Yves 38–41
Behnisch, Günter 242
Belgioioso 415
benches *see* chairs
Benchmark Furniture *338*
Benetton Group 182, 226
Bengtsson, Mathias 42–5
BENTEK 260
Berlin
   BLESS 62–5

FUCHS + FUNKE *154–7*
Bernabei, Rina 46
bernabeifreeman 46–9
Bernades, Mana 50–3
Bernades, Sergio *50*
Bernardaud 10, *13*
BIC 54
bicycles
   Abri-vélo bike stand *280*
   Banc-vélo bike stand/bench *280*
   Bicycle Brake sketches *270*
   Schwinn Sting-Ray 417, *417*
Big-game 54–7
Bill, Max 414, *414*, 420
Birkenstock 38, *39*
Blaess, Steven 58–61
Blanc, Jean-Sebastién 10
BLESS 62–5, *138*
Blomqvist, Malin 16
BMW Group 39
Boner, Jörg 66–9
bookmarks
   Hotel Butterfly 118, *120*
Borka, Max 278
Bosa 185
Bouroullec, Roman 162
bowls and dishes
   Black Honey fruit bowl 266, *266–7*
   Bowls & Spoons 181
   Clamshell Bowls 97
   Hands bowl 376
   Henry's Collar fruit bowl 408
   Newton Fruit Bowl 162, *163*
   Rust 240
   Serve-Bowl 176
   Solitaire Olive Dish 234, *234–5*
   War Bowl 390, *391*
Braun 420, *420*
bridges
   Air Bridge 218, *218–19*
Brisbane
   Alexander Lotersztain 274–7
British Council 381
Brízio, Fernando 70–3
Brussels
   Big-game 54–7
   Xavier Lust 278–81
Buenos Aires
   Alejandro Sarmiento 354–7
Burks, Stephen 74–7
Buxton, Sam 78–81

cable tidy *33*
Caccavale, Elio 82–5
cafetières
   Aroma 200
Calder, Alexander 416
calendars
   Everyday Calendar 30
   One-a-Day Calendar 393
Campana brothers 50, 90, 162, 326
Campbell, Louise 86–9
candles
   5 Minute Candles 334, *337*
   Black Gold candleholder 178
   Burn the Candle at Both Ends 230, *233*
   Candlelight 352
   Candlelight Dinner for One 282
   Digit table piece 226, *226–7*
   Flames 214, *217*
   Mistic 269

Norwegian Wood Candleholder 385
   Turner candlestick 281
Capote, Leo 90–3
Cappellini 74, *308*, 358, *358*
Carpenter, Jennifer 94–7
carpets and rugs
   Belt/Chain carpet 350, *350–1*
   Kinetic Spiral rug 372
   Sprinkle rug 396
cars
   Maserati Birdcage 75th concept car 322
Cassina 29, 210
Castiglioni, Achille 34, 415, *415*
Castiglioni, Pier Giacomo 415, *415*
cat accommodation shelter
   House 213.6 *262–3*
Cebu City
   Kenneth Cobonpue 110–13
Ceramic Japan 376
ceramics
   Broken White 186, *189*
   Cremated Teddy Bear 405
   Journey 70, *71*
   Mattress Stone Bottle 362, *364*
   Moneybox 385
   Painting a fresco with Giotto 70, *70*
   Short Set 334
   Stakhanov 248
   see also bowls and dishes; cups and
     saucers; plates
chairs
   'A 100 Chairs in 100 days' 158,
     158, 160
   Aeron chair 417
   Armchair 41 'Paimio' 412, *412*
   Avery chair 341
   Balou seating 113
   Banc-vélo bike stand/bench *280*
   Beer Set 179
   Billy goes Zen 88
   BoCu 66–7
   Cadé armchair 327
   Carbon Chair 406, *407*
   Carioca Chair 297
   Chainsaw Chair 152
   Cilindro Puf 356
   Clay Chair 26, *26–7*
   Clone Chaise 78, *80*
   Collage 260
   Cow Bench 'Antonia' 270, *272*
   Cup Chairs 74, *76*
   D/side Series 162
   Disk Chair 90, *90*
   Dragnet lounge chair 110, *113*
   Envelope Chair 294, *294–5*
   Festa Pill 370, *370*
   FlightCaseChair 16
   Folding chair 323
   FormForm Chair 15
   Grand Pouf de Paille 238, *238*
   GUSHA side chair 376
   Gwanju chair 174
   Hidden bench 395
   Husky bench 366, *368*
   Iron Chair X 190
   Jeesus Furniture Chair 16
   Jolly Jubilee easy chair 178, *180*
   Kraft Chair 370, *370–1*
   Kyparn chair 321
   Levita Chair 37

Lidl by Lidl Chair 17
Lissajous 169
Low Chair *194–5*
LTD Collection 381
Mandi Chair 237
Meccano Chair 236
Mono 154, *157*
Mummy beach chair 378
ORI Chair 318, *319*
Papton 156
Patapouf 132
PicNik 279
POLKAchair 342, *345*
Poltrona Mole (Sheriff) armchair 413, *413*
Poltrona with cover 183
Polytope Seating Landscape 105
Prince Chair 86–7
Quattropolka 335
Relax Set 179
Rodrock 106
Rubber Chair (ANTEEKSI) *16*
Rubber Chairs (Peter Traag) 378, *380*
Ruberta Iron Home furniture 357
Scheluw 249
Smoke ZigZag chair 28–9
Solid C1 210, *213*
Spade Chair 91
Sponge chair 378, *379*
Spoon Chair 90, *93*
Spun Carbon Fibre Bench 42–3
Spun Carbon Fibre Chair 44
Stair Chair 36
Sturdy Straws Chair 174, *176*
Tailor Chair 214, *217*
Takata 04-neo child car seat 198
Tangram Chair 317
Terceira 326
Tobogã 328
Treasure Dining Chair 29
Tree Log 158
Tuoli Mummilleni Chair 17
TWIG bench and cube 338, *338*
Twig Concrete Bench 274
Twin chairs 146, *148*
Vague à l'Âme carpet seat *130–1*
Veryround 86, *89*
Waiting for You 124
Water Bench 138, *141*
Who's gonna drive you home tonight? 68
Yin & Yang armchairs 110, *112*
YODA 111
CHARPIN, PIERRE 98–101
Chermayeff, Ivan 94
Chirnside, Lucas 102–5
Christofle 162, 250
Cigar Case 250
cigarette butt case
   EXT Square 60
Cinna 146
ClassiCon 66
cleaning machine
   Lovely Polisher X 191
clocks and watches
   Cuff 256
   Magnet Clock 221
   MINI_motion watch 38
   PI World Time Clock 102, *102–3*
   Public Clock for the City of Geneva 250, *253*

Seconds Tick Away *121*
smlwrld Desk Clock 102, *104*
Timepiece *80*
Word Clock 78, *81*
Close, Chuck 94
clothing
Bodywork 254, *254*
Mangastrap 202, *202*
coat hangers
Albero coat hanger *325*
Antlers coat hook *366*
Forest for the Trees coat stand *179*
Point de Suspension peg *133*
Ribbon Coatrack 386, *389*
Sasae coat stand 306, *307*
Spiral coat hanger *325*
UPON wall coat hanger *127*
Cobonpue, Betty 110
Cobonpue, Kenneth 110–13
Coca-Cola 162, 334
Cocksedge, Paul 114–17
Collex 318
computers *see* electronic goods
Conran 338, 340
Conran, Terence 338
Contrast 201
Contrasts Gallery, Shanghai 394
Copenhagen
Cecilie Manz 282–5
Louise Campbell 86–9
Cor Unum 365
Costa, Lucio 413
Covo 274
Cross, Michael 394
cups and saucers
Biscuit en Couvercle *133*
Rivulet of the Heart 118, *120*
Teacuptea 402, *402*
TONERICO: FOR TWO *377*
Waltz 118, *120*
cutlery
Bowls & Spoons *181*
Cameo *200*
Sugarcubes *226*
SUNAO series *173*

D-BROS 118–21
Danese 38, 346
David design 358
de Graauw, Judith 122
De Padova 308
DEMAKERSVAN 122–5
Derlot.com 277
Desalto 22
Design By project 150
Design Trust for Public Space 96
desks
Desk – Crisis collection *145*
Solitaire workstation *275*
d'Esposito, Martino 146–9
Di Giacomo, Alana 406
Diez, Stefan 126–9
diwill 258, *260*
Dixon, Tom
Alexander Taylor 366–9
ANTEEKSI 14–17
Armchair 41 'Paimio' by Alvar Aalto *412*
Doshi Levien 134–7
FUCHS + FUNKE 154–7
Johannes Norlander 318–21
Kenneth Cobonpue 110–13

Louise Campbell 86–9
Russell Pinch 338–41
Stephen Burks 74–7
voonwong&bensonsaw 386–9
DMA *139*, 264
Doléac, Florence 130–3
Doshi, Nipa 134
Doshi Levien 134–7
Draft 118
Driade 278
Driade Kosmo 278, 280
Droog B.V. 249
Droog Design 186, 214, 216, 246, 270, 346, 362, 402, *403*
Dubai
Steven Blaess 58–61
dubois, david 138–41
Ducasse, Alain 210
Duchamp, Marcel 34, 258
Dupont Corian 394

E&Y 74, 318, *319*, 366, *367*, 368
Eames, Charles 26
East Japan Railway Company 398
Easy Track 314
Eau de Paris 98, *100*
ECAL 350, *353*
Edag Future *105*
Edra 58, 378, *378*
Eek, Piet Hein 142–5
Eindhoven
Maarten Baas 26–9
ransmeier & floyd 346–9
electronic goods
Blob mobile phone 314–15
Bluetooth headset 40
Bordeaux Project – LCD TV and Home
Theater System 222, 222–3
Carregador Solar de Pilha e Celula
battery charger *296*
dB/Elements Collection 262, *263*
Fujitsu 23/6 Mobile PC 20
Hub Commissioning Centre 78–9
L7 Project (50" DLP TV and Home
Theater System) 222, *224*
Luon Modual PC 230, *233*
Minimizing Einstein radio *122*
O/Elements Collection – oxygen
generator 262, 264–5
One Laptop Per Child 40
Pocket Imager 222, 224–5
Telephone 510, *198*
Toshiba Laptop Transformer 40
TP1 (T4 & P1) by Dieter Rams 420
Elixir 278, *280*
ElmarFlötotto 237
Epstein, Mitch 94
ESA of Bergamo 413
Established & Sons 369
Estée Lauder 74
Estrada, Maria Helena
Alejandro Sarmiento 354–7
Carla Tennenbaum 370–3
Flavia Pagotti Silva 330–3
Leo Capote 90–3
Mana Bernardes 50–3
Manuel Bandeira 34–7
Marcelo Moletta 294–7
Nada Se Leva 298–301
Nido Campolongo 310–13
Nódesign 314–17

Poltrona Mole (Sheriff) armchair by
Sergio Rodrigues 413
Extremis 278

Fabrica 182, 226
fencing
How To Plant a Fence *125*
STRETCH Fence *96*
Fermob 132
Ferrari 322
Figueira, Suelen 50
Fiorini 300
Fitzgerald, Chris 304
Flaubert, Stefanie 242
Flavin, Dan 416
FLOOD© 265
Flos 415
Flower Robotics 286, 286–9
Floyd, Gwendolyn 346
FMG.SA 149
Fontana Arte 386
Footrest Pebbles/Grass 402, *403*
Foscarini 22
Foster and Partners 330
Fredericia Furniture 285
Frederik, Crown Prince of Denmark 86
Freeman, Kelly 46
Fritz, Al 417, *417*
FRONT 150–3
FRP 172
Fuchs, Wilm 154
FUCHS + FUNKE 154–7
Fujitsu 20
Fukazawa, Naoto 421
Funke, Kai 154
furniture
3/6 Series 170
Entre Console 329
garden furniture 10–11
Graft Kit *12–13*
Multileg Cabinet *185*
Scrap Wood Cabinet *144*
Sketch Furniture *151*
Tattoofurniture 342, *342*
Treteau 56
UPON Series *126*
Wave Sideboard 338, *340*
see also chairs; sofas; stools; tables

Gaia & Gino 268
Gaillard, Alexandre 146
Galeria do Papel Gráfica Editoria 311–13
Gamper, Martino 158–61
garden accessories
Blockcrack 228
Furniture to Garden 10–11
Grow Bag 146, *149*
Gaudard, Sébastien 132
Georgacopoulos, Alexis 162–5
Gian 210
GK design 421
glassware
Josephine water jug *343*
Minima glassware 282, *284*
Mistic *269*
Trophees *365*
Goodrum, Adam 166–9
Google Cab *20–1*
Goretti, Nicola 354
graf:decorative mode no.3 170–3
Grcic, Konstantin 318

Groningen
Lambert Kamps 218–21
Gugelot, Hans 414, *414*
Gur, Tal 174–7
Guzzini, Fratelli 61

Habitat 154, *154–5*, 178
Hans, Ineke 178–81
Harvey Nichols 226
Hattori, Shigeki 170
Hay 86, *86*
Hayon, Jaime 182–5
Heal's 338
Heijdens, Simon 186–9
Heiss, Desiree 62
Helsinki
ANTEEKSI 14–17
Herend 335
Hildinger, Paul 414, *414*
Hofit 174–5
Holmegaard 285
Hong, Jackson 190–3
household goods
Brush Couple 350, *353*
Carafe Eau de Paris 98, *100*
Drink'Tray 146, *146*
Fan 137
Fur Hammock 2 62, *64*
GENIO cookware 128–9
Gio clothes dryer 77
HERITAGE IN PROGRESS 54, 56
Inflatable Bottle Cooler 162, *162*
Marli Apribottiglie bottle opener
58, *58*
Marli Ice Tongs 58–9
Moose 56–7
PACK, SWEET PACK 54, *54–5*
Paso Doble umbrella stand 278
Shaker 61
Spider Picture Holder 90, *92*
Spoon Simple Picture Holder *92*
Tutoring Flowerpot holder 138, *140*
Worn-In doormat 346, *347*
see also electronic goods; kitchen
equipment; tableware
Hutten, Richard 162
Hyrkäs, Johanna 14

Idee 74, 274, *277*
Ikea 22, 26
Die Imaginaire Manufactur (DIM)
164, *353*
In Africa Community Foundation 274
IndustriAL Standard 354, *356*
installations
Querencia Dodecahedron
modular system 354, *355*
Interior Crafts of the Island, Inc.
110, *110–13*
interior design
Casa do Leitor (Reader's House)
310, *312*
Model for a Paper House 310, *313*
Restaurant Flood© 265
Ishigami, Junya 194–7
ISM Objects 206, *209*
Istanbul, Meriç Kara 226–9
Istanbul_cosmetics 226
Ivry sur Seine
PIERRE CHARPIN 98–101
Iwasaki, Ichiro 198–201

ixi 202–5

Jansen, Trent 206–9
Jeanmonod, Grégoire 54
jewellery
Brazil Necklace *52*
Fan Necklace *52*
Glass Ball Necklace 50, *51*
Magnetic Jewellery 45
Messenger Band 33
No. 26 Cable Jewellery collection
62, *62–3*
Nokia Sphere mobile phone pendant
50, *50*
Serpent Necklace *53*
Shaman necklace *266*
Stirrer Necklace *53*
Stylefree Signature Pin *64*
Urban Indian Bracelet *51*
You're Something On Me Necklace
50, *53*
JL Laboratory 249
Johnson & Johnson 58
Jouin, Patrick 210–13
Judd, Donald 190
jugs
Ceramic Jugs *142–3*
Josephine water jug *343*

Kaag, Ines 62
Kabel, Chris 214–17
Kagaya, Mr *421*
Kalliopuska, Jussi 17
Kamps, Lambert 218–21
Kang, Yun-je 222–5
Kara, Meriç 226–9
Kartell 210, *213*
Keller, Pierre
Alexis Georgacopoulos 162–5
Atelier Oï 22–5
Big-game 54–7
Florence Doléac 130–3
Jörg Boner 66–9
Martino d'Esposito 146–9
Nicolas Le Moigne 250–3
Paul Cocksedge 114–17
Peter Traag 378–81
Ulmer Hocker by Max Bill *414*
Yves Béhar 38–41
Kenison Furniture 36
Kenji Nakachi *205*
keyrings
Baby Holder Keyring 30–1
Kim, Bosung 230–3
Kim, Sang-kyu
Arik Levy 266–9
Bosung Kim 230–3
ixi 202–5
Jackson Hong 190–3
Khashayar Naimanan 302–5
Sang-jin Lee 258–61
Stella stool by Achille and Pier
Giacomo Castiglioni *415*
Suyel Bae 30–3
Tal Gur 174–7
Yun-je Kang 222–5
Zinoo Park 334–7
Kimizuka, Ken 374
Kitano, Hiroaki 286
kitchen equipment
Cookuk2 cooking studio *69*

Egouttir (colander) 360
Fruit-Tree 309
Gradient dish rack 346, *348*
Little Kettle 256
Matlo drinking vessel 136
Mosaic range for Tefal 134, *134–5*
OXO Kitchen tools 398, *398–9*
PET-Pick Bottle Carrier 270, *271*
Rape à Fromage (cheese grater) 360
Sink 138
Stainless Steel Kettle 198, *199, 201*
see also household goods
Klauser, André 234–7
K.M. Hardwork 242
Knecht, Tetê 238–41
KODAS 230
Kohama, Izumi 202
Kohonen, Otto 412
Korban, Janos 242
korban/flaubert 242–5
Kovanen, Erika 14, *15*
Kram, Reed 106–9
Kreo 98
Krzentowski, Didier
   Adrien Rovero 350–3
   Chris Kabel 214–17
   david dubois 138–41
   Gino Sarfatti 416
   Ineke Hans 178–81
   Inga Sempé 358–61
   Julia Lohmann 270–3
   Lampadaire '1063' by Gino Sarfatti 416
   Mathieu Lehanneur 262–5
   Nori Yasuda 402–5
   PIERRE CHARPIN 98–101
   Wieki Somers 362–5
Kurosaki, Teruo 274
Kwakkel, Erik Jan 178, *365*

la Fontaine, Christophe de 126
La Neuveville
   Atelier Oï 22–5
Laarman, Joris 246–9
Lacroix, Christian 138
Lagerkvist, Sofia 150
Lasky, Julie
   5.5 designers 10–13
   Antenna Design 18–21
   Cecilie Manz 282–5
   Dominic Wilcox 390–3
   Jennifer Carpenter 94–7
   Lambert Kamps 218–21
   Maarten Baas 26–9
   Mathias Bengtsson 42–5
   Patrick Jouin 210–13
   Schwinn Sting-Ray by Al Fritz 417
   WOKmedia 394–7
Lausanne
   Adrien Rovero 350–3
   Alexis Georgacopoulos 162–5
   Big-game 54–7
   Martino d'Esposito 146–9
   Nicolas Le Moigne 250–3
   Tetê Knecht 238–41
Ldesign 266
Le Corbusier 110
Le Moigne, Nicolas 250–3
Leading Edge Design Corp. *400*
LeAmon, Simone 254–7
Lebossé, Anthony 10
Lee, Sang-jin 258–61

Lehanneur, Mathieu 262–5
Lensvelt BV 179
Levien, Jonathan 134
Levy, Arik 266–9
lighting
   Adventure installation 245
   Animal Things 153
   Avestruz Lamp 310
   Balladeuse 203
   Boab Light 166–7
   Bookmark 258
   Bulbshell 402, *404*
   Burstlight 242, *242–3*
   Candlelight 352
   Caravaggio lamp 283
   Collage 89
   A Composition for Cords 22, *22–3*
   Deco Lamp 172
   Eash light 175
   EL-butterfly 258
   Ema Pendant Light 46, *49*
   Ether 210–11
   Flood 394, *394*
   Fold lamp 366, *369*
   Football Lamp 145
   Funghi lamp 184
   Grande Lampe Plissée 358, *359*
   Hanabi lamp 306, *306*
   Head Light 256
   Hotel Butterfly Light 118, *120*
   Iltama Chandelier 15
   In Vitro 210–11
   Iron Lamp 92
   K/Elements Collection daylight
receiver/transmitter 262, *264*
   Knock Down Lamp 154, *157*
   Lace Pendant Light 46–7
   Lady Light 259
   Lamp Tome 24–5
   Lampadaire '1063' 416, *416*
   Lead LED table light 38, *41*
   Ligero Ceiling Lamp 298–9
   Light As Air 114, *114*
   Light Frame 74
   LightButton Lamp 14
   Lightweeds 187
   MEMENTO-LINK 374, *374–5*
   Mesofär lamp 22
   Metamorphosis Chandelier 230, *230–1*
   Milky Pendant Lamp 16
   Mummy Vessel 408–9
   NeON lights 114, *117*
   Nomadic Eye Chandelier 230, *232*
   Optic 212
   Origami Light 388
   Pinlight 216
   Pixelated Peacock Light 178
   Rising Slowly Chandelier 186
   Ruminant Bloom 270, *273*
   Shaali Light Scarf 46, *46*
   Skinny table/floor light 406
   Sound System 72–3
   Spaghetti Chandelier 334, *334–5*
   Straw Light 259
   Styrene Lampshades 116
   Supernova lamp 354
   Suspension plissée lamp 358, *358*
   Tellut Pendant Lamp 15
   Topple Lamp 206, *208*
   Tree 188
   Waist lamp 154, *154–5*

Watt? 114, *115*
Weblight 244
White October Light 17
Woodland table lamp 406, *406*
Yam Yabasha light 174, *174*
Zipper Light 258, *261*
Ligne Roset 56, 146, *146*, 149, *361*
Lima de Silva, Juliana 52
Limoges 10
LinBrasil *413*
Lindgren, Anna 150
Linley 338
Lionni, Pippo 266
Lisbon
   Fernando Brízio 70–3
Lobmeyr *343*
Lohmann, Julia 270–3
London
   Alexander Taylor 366–9
   Dominic Wilcox 390–3
   Doshi Levien 134–7
   Elio Caccavale 82–5
   graf : decorative mode no.3 170–3
   Julia Lohmann 270–3
   Khashayar Naimanan 302–5
   Martino Gamper 158–61
   Mathias Bengtsson 42–5
   Matthias Aron Megyeri 290–3
   Paul Cocksedge 114–17
   Russell Pinch 338–41
   voonwong&bensonsaw 386–9
   WOKmedia 394–7
Lotersztain, Alexander 274–7
Louis, Armand 22
Lust, Xavier 278–81

Mackintosh, Charles Rennie 26
Magnussen, Erik 406
Manz, Bodil 282
Manz, Cecilie 282–5
Margiela, Martin 62
Martins, Luciana 326
Maserati 322, *322*
Masuko, Yumi 374
Materialise.MGX 210, *213*, 266
Mathias, Julie 394
Matsui, Tatsuya 286–9
Maxray Inc. 172
MDF Italia 278
medicine
   One-a-Day Calendar 393
   Therapeutic Objects 262, *262*, 265
Megyeri, Matthias Aron 290–3
Meia Pataca *413*
Melbourne
   Lucas Chirnside 102–5
   Simone LeAmon 254–7
   zuii 406–9
Memphis Group 98
Mendes, Rai *373*
Metalarte 185
Miller, Herman 38, *41*
Minerva Co. 376
Miró, Joán 416
mirrors
   Fountain Mirror 250, *250–1*
   Platform Medium White wall mirror 98
Missoni 74
Mituisi 194–5
Miyata, Satoru 118
mmmg (MILLIMETER MILLIGRAM.INC)

30–3
Modus 74
Moeslinger, Sigi 18–21
Moeve 234
Moholy-Nagy, László 415
Moletta, Marcelo 294–7
Mollino 26
Montina 98, *101*
Moooi 26, *152*, 214, 216
Moormann, Nils Holger 66
Morris, William 46
Morrison, Jasper 234, 318
Mosley, Steve 390, *393*
Mosley meets Wilcox/Figijo *393*
Moulin, Xavier 202
Moulinex 34
Moura Guedes, Guta
   André Klauser 234–7
   Fernando Brízio 70–3
   Jaime Hayon 182–5
   Maxim Velcovsky 382–5
   Meriç Kara 226–9
   OVO 326–9
   Paperclip by Johan Vaaler 418
   POLKA 342–5
   ransmeier & floyd 346–9
   Tetê Knecht 238–41
   Xavier Lust 278–81
Mouvements Modernes Paris 25, 268
Movement 8 110
MUDAM *139*, 140
MUJI 421, *421*
Munich
   Clemens Weisshaar & Reed Kram
106–9
   Khashayar Naimanan 302–5
   Stefan Diez 126–9
Murano Due 210–11
Murray Moss 26
MUTECH 198, *198*

Nada Se Leva 298–301
nails
   Ça Ne Vaut Pas un Clou – Hidden
Wealth Project 305
Naimanan, Khashayar 302–5
Nakagawa, Masahiro 206
Nakahara, Maki 376
Neal's Yard 338
Nelson, Nirith 174
nendo 306–9
New York
   Antenna Design 18–21
   Jennifer Carpenter 94–7
Newson, Marc 358
Nichinan Corp. *400*
Nido Campolongo 310–13
Niemeyer, Oscar *413*
Nissan Motor Company 398
Nódesign 314–17
Nokia 50, *50*
Nola *321*
Norlander, Johannes 318–21
Notech Design 238
Novus 34
Nurseryworks 94
Nymphenburg *302–3*

Oca *413*
Oiva, Vesa 14

Ojala, Pauli 17
Okinawa
   ixi 202–5
Okuyama, Ken 322–5
Olin, Johan 16
Oliveira, Gerson 326
Oluce 254
Omega Electronics 250, *253*
Osaka
   graf:decorative mode no.3 170–3
Oshima, Nagisa 334
OVO 326–9
OXO International 398, *398–9*
Pagotti Silva, Flávia 330–3
Paimio Sanatorium 412
Panton, Verner 42
Paola C. 226
paperclip 418, *418*
parasols
   Shadylace 214, *216*
Paris
   5.5 designers 10–13
   Arik Levy 266–9
   Big-game 54–7
   BLESS 62–5
   david dubois 138–41
   Florence Doléac 130–3
   Inga Sempé 358–61
   Mathieu Lehanneur 262–5
   Patrick Jouin 210–13
Park, Zinoo 334–7
Park Plus 334, *337*
Parkes, Brian
   Adam Goodrum 166–9
   Alexander Lotersztain 274–7
   bernabeifreeman 46–9
   Dilly Bag 419
   korban/flaubert 242–5
   Lucas Chirnside 102–5
   Sam Buxton 78–81
   Simone LeAmon 254–7
   Steven Blaess 58–61
   Trent Jansen 206–9
   zuii 406–9
pen clips 32
Pesce, Gaetano 378
Petit, Elric 54
Picchi, Francesca
   BLESS 62–5
   Clemens Weisshaar & Reed Kram
106–9
   DEMAKERSVAN 122–5
   Dieter Rams 420
   Elio Caccavale 82–5
   FRONT 150–3
   Joris Laarman 246–9
   Junya Ishigami 194–7
   Martino Gamper 158–61
   TP1 (T4 & P1) by Dieter Rams
420
   Simon Heijdens 186–9
   Stefan Diez 126–9
Pinch, Russell 338–41
Pininfarina, Carrozzeria 322
Pinkus, Kendra 58
PINKUSBLAESS 58
Planex 274
plates
   Drop 200
   Mon Cirque Collection 184
   Rock Plates 392

POLKA 342–5
Poltrona 413
postcard stand 14
Poulsen, Louis 86, 89
Prague
    Maxim Velcovsky 382–5
Prosomedia 106, 234
Putman, André 358
PYPEX 198, 199, 201

Qubus 382, 382, 385
Radi Designers 130
radiators
    Gary Cooper 264
    Heatwave 246, 246–7
radios see electronic goods
Rahm, Marie 342
Raleigh 417
Rams, Dieter 420, 420
Ransmeier, Leon 346
ransmeier & floyd 346–9
Readymade Projects 74
Réanim project 10, 12
Reebok 78
Renard, Claire 10
Reymond, Patrick 22
Ribeiro, Guilherme 298
Rietveld, Gerrit 28–9
Rio de Janeiro
    Mana Bernades 50–3
    Marcelo Moletta 294–7
robots
    Cyclops 398, 400–1
    Masked Citizen X 190, 193
    Nubo 322, 322
    P-noir 288
    Palette 286, 287
    Platina 289
    Posy 286, 289
    Upper Torso Palette 286
Rodrigues, Erivelto 52
Rodrigues, Sergio 413, 413
Roethlisberger 22
Rogers Marvel 94
Rosenthal 126, 128–9
Rotterdam
    Chris Kabel 214–17
    DEMAKERSVAN 122–5
    Joris Laarman 246–9
    Simon Heijdens 186–9
Rovero, Adrien 350–3
Royal Selangor 406
Royal VKB 178, 180, 214
rugs see carpets and rugs

Sadler, Marc 34
Samsung Electronics 222, 222–5
San Francisco
    Yves Béhar 38–41
Santi 415
São Paulo
    Carla Tennenbaum 370–3
    Flávia Pagotti Silva 330–3
    Nada Se Leva 298–301
    Nido Campolongo 310–13
    Nódesign 314–17
    OVO 326–9
Sarfatti, Gino 416, 416
Sarmiento, Alejandro 354–7
Sato, Oki 306
Sävström, Katja 150

Saw, Benson 386
Schiedam
    Wieki Somers 362–5
Schwinn 417
Scott de Martinville, Augustin 54
SCP 341
screens
    Armour Screen 169
    Cellscreen 242, 245
    Kinetic Tapestry 372
    Transoptic Screen 372
    Yuki 308
sculpture
    MIKRO products 78, 78
security
    Billy B. old English padlock 290
    CityCatTV 293
    Heart to Heart chain 290
    Mr. Smish & Madame Buttly razor wire
        290, 293
    Mrs Welcome, Nottingham lace curtain
        292
    R. Bunnit, Peter Pin and Didoo,
        Victorian railings 291
    Reactive Hair X 192
Sejima, Kazuyo & Associates 194
Sempé, Inga 358–61
Seoul
    mmmg (MILLIMETER MILLIGRAM.INC)
        30–3
    Sang-jin Lee 258–61
    Zinoo Park 334–7
Serralunga 162, 250
shelving
    Cano 318, 318
    Cardboard Rings Shelf 312
    Crop bookshelf 346
    Feriado bookcase 327
    Ivy Shelf 390
    Lunuganga 394, 397
    Parallel Shelving System 74–5
    Porta-Tudo – Por Enquanto é Só 333
    Steel Shelving Unit 421, 421
shoes
    Birkis Slip-on clogs 38, 39
    Sabot 238, 240–1
Sigel, Marcel 406
signs
    Light-Emitting Rooftiles 220
Siitonen, Tuomas 17
Singer, Monica 342
Singh, Dayanita 94
SIOS 80–1
Sisley 226
Smith, Paul 290, 298
sofas
    Crusoe Sofa 276
    Soft Sofa 274, 276
SOGO furniture Co. 170
Somers, Wieki 362–5
Song, Aamu 15
Sony 198
Sottsass, Ettore 26
Sowden, George 98
Spade, Kate 94
speakers
    Loudspeakers (glass) 150
sport
    Robin Board 304
    Winter Wonderland® Walkerhill
        ice rink 336

Sputnik 274
stamps
    Honesty Stamps 390, 393
Starck, Philippe 210, 420
Stelton 86, 88
Stockholm
    Clemens Weisshaar & Reed Kram
        106–9
    FRONT 150–3
    Johannes Norlander 318–21
stools
    Arnold Circus Stool 158–9
    Basketstool 353
    Bate-Papo Stool 330, 332
    Bau Stool 332
    BOX stool 54
    Disk Stool 90
    Glass Stool 140
    GUSHA stool 376
    KADA 38, 38
    Laminado Stool 333
    Lost&Found 122
    Maiden stool 339
    Monza Bar Stool 107
    Mushroom Stools 172
    Nauplius 171
    Oneshot 210, 213
    Poche-Cul 25
    Popo Stool 17
    Ruberta Iron Home furniture 357
    Sign Stool 206, 206–7
    Spade Stool 91
    Sella stool 415, 415
    Ulmer Hocker 414, 414, 415
    Wax Stool 300
    X Stool 311
storage
    Ambrogio magazine rack 77
    Box in the Box 326, 328
    Chéne-a-velo bicycle rack 138, 139
    Euripedes Hanger 36
    Home Work 88
    Laminado Magazine Rack 333
    Red x storage system 209
    Rigicordes 24–5
    Up storage unit 202, 205
    see also shelving
street furniture
    Antenna Design 18–19, 21
    Atelier Oï 22
    DEK range 277
    Vortex drain cover 146, 149
Street and Garden Furniture Co. 277
Studio 80 374
Studio Nova 97
Suhonen, Janne 16
Swarovski 178, 186
Swatch 22, 54
Swedese 306
Sweet Dreams Security™ 290, 290–3
Sydney
    bernabeifreeman 46–9
    korban/flaubert 242–5
    Steven Blaess 58–61
    Trent Jansen 206–9
Szekely, Martin 138

Beam table 320
Beer Set 179
Breeding Tables 106, 108–9
Bull's Eye Table 301
Butterfly Table 367
Chab-table 308
Cinderella Table 122, 123
Cloth Table 46, 48
Cone Table (Nido Campolongo) 313
Cone Tables (André Klauser) 237
Disco coffee table 329
Earthquake Table 146, 146–7
Hidden table 395
Impressoes Table 331
KADA 38, 38
Ligero Small Side Table 300
Maiden table 339
Marble Table 137
A Matter of Taste 177
Meteo side table 201
Micado side table 285
MutsisoliFaijas Table 14
Out Door Table 168
PicNik 279
Platform Coffee Table 98–9
Polar nesting table 307
Relax Set 179
Rock coffee table 268
Ruberta Iron Home furniture 357
SIOS Table 81
Snow side table 346, 349
Table 194, 196–7
Table Lunatique 358, 361
Tables LaChapelle 358
Tambour Table 94
TONERICO: FOR TWO 377
tableware
    5.5 designers 13
    Breadbasket 342, 343
    Incognito - Rococo & Empire Hidden
        Wealth Project 302–3
Tada Mokko 322, 325
Talka, Maria 15
Tallon, Roger 378
Tange, Kenzo 286
Tatil Design 50
Taylor, Alexander 366–9
teaching aids
    Comforting Device 84–5
    Low-resolution Pig TV 85
    MyBio dolls 82–3
    Toy Communicator 84–5
Team by Wellis 66, 67
teapots
    Harp 200
    Mayu 324
Tefal 134, 134–5
Tennenbaum, Carla 370–3
Tergoprint 300
TG 230, 233
Thomas 128–9
Thonet, Gebrüder 306
Tisca France 350
Toivonen, Tuomas 16
Tokyo
    graf : decorative mode no. 3 170–3
    Junya Ishigami 194–7
    nendo 306–9
    Noriko Yasuda 402–5
    Shunji Yamanaka 398–401
    Tatsuya Matsui 286–9

TONERICO:INC. 374–7
TONERICO:INC. 374–7
Toolsgalerie 132, 215, 216
Toscani, Oliviero 182
Toshiba 40
toys
    Cremated Teddy Bear 405
    Design Products: Ron Arad with Little
        Albert 302, 305
    Plastibear 402, 405
Traag, Peter 378–81
trays
    Animal Things 153
    CMYK trays 162, 164–5
    Impressöes Trays 330
    O-Bon 200
TRUCK Product Architecture 94–7
Tsuboi, Nene 17
Tsubame Shinko Co. 173
Turin
    Ken Okuyama 322–5

Uchida, Shigeru 374
Udagawa, Masamichi 18–21
Uehara, Ryosuke 118
Ultradesign 34, 36
underwear dryer
    Sequinha® 34–5

Vaaler, Johann 418, 418
van Elten, Thorsten 234, 366, 390, 393
vases 185
    Air Vase 384
    Black Money Vase (BMV) 214, 214
    Blossoms 362, 364–5
    Flower frame 387
    Hope Forever Blossoming 118, 118–19
    Landscape Vase 386
    Mistic 269
    Oil 238–9
    Platinum Money Vase (PMV) 215
    Shirtvase 229
    Stando 164
    Triplo Vase 98, 99
    Waterproof 382–3
Vauxhall Motors UK 78
Velcovsky, Maxim 382–5
Venini 98, 99
Verhoeven, Jeroen and Joep 122
VIA (Valorisation de l'innovation dans
    l'ameublement) 264–5, 358
Viceversa 250
Vienna
    POLKA 342–5
Viganò 415
Vitória, Fernando 52
Vitra 74, 414
Vlaemsch 56
't Vliehout boatbuilders 354
von der Lancken, Charlotte 150
voonwong&bensonsaw 386–9
Vuitton, Louis 286

wallpaper
    Moving Wallpaper 187
    Wallpaper by Rats 153
Wanders, Marcel 26, 206
Watanabe, Yoshie 118
watches see clocks and watches
watering spout
    Verso Diverso 250, 252

Wazuqu *324*
Weisshaar, Clemens 106–9
Wellcome Trust 134, *137*
Wilcox, Dominic 390–3
Wittgenstein, Ludwig 186
Wogg 24, 66
Wohnberdarf (Zurich) 414
WOKmedia 394–7
Wong, Voon 386
Worldwide Co. *78*

Yamakin Co. *375*
Yamanaka, Shunji 398–401
Yasuda, Noriko 402–5
Yoneya, Hiroshi 374
Yoshiie, Chieko
    D-BROS 118–21
    graf:decorative mode no.3 170–3
    Ichiro Iwasaki 198–201
    Ken Okuyama 322–5
    Matthias Aron Megyeri 290–3
    nendo 306–9
    Shunji Yamanaka 398–401
    Steel Shelving Unit 421
    Tatsuya Matsui 286–9
    TONERICO:INC. 374–7

Zanotta 86, 89, 366, 367, 414, 416
Zanuso *415*
Zeltec 270
ZMP/Seiko Instruments *322*
zuii 406–9
Zurich
    Jörg Boner 66–9

Photo Credits

Antenna Design
Ryuzo Masunaga: p20; Bruce Pringle: p21 – renderings of Civic Exchange

Atelier Oï
André Bucher: p22 – Bank Plus

Maarten Baas
M. van Houten: all photography

Suyel Bae, MILLIMETER MILLIGRAM.INC
Hyunsuk O: all images; except Myounghee Kí: p30, Everyday

bernabeifreeman
Dieu Tan: p46 – Lace Pendant, p48–9

Mana Bernardes
Mauro Kury: all photography

Big-game
Milo Keller: all photography

Steven Blaess
Marcus Ehrenblad: p58 – Marli Apribottiglie, p61 – Shaker

Jörg Boner
David Willen: p67; Niklaus Spoerri, p68; Milo Keller: p69

Fernando Brízio
António Nascimento: p70–1

Stephen Burks
Marcos Bevilacqua: p74–5 – Parallel Shelving System, p76 – Cup Chairs

Sam Buxton
Eugene Franchi: p78 – MIKRO-House; Dave Willis: p78 – MIKRO-Man Jungle; Peter Mallet: p80 – Clone Chaise, p81

Elio Caccavale
Ingrid Hora: all photography

Louise Campbell
Erik Brahl: p88 – Billie goes Zen

Leo Capote
Damiano Leite: all photography

Jennifer Carpenter,
    TRUCK Product Architecture
Tom Francisco: p97 – Clamshell Bowls (7)

PIERRE CHARPIN
Morgan Le Gall: p98 – Platform Medium White, p98 – Platform Coffee Table; Simone Maestra: p100 – Carafe Eau de Paris; Pierre Antoine: p101 – Basket-Up, p101 – Ceram X Vases

Lucas Chirnside
Robert Colvin: p102–3 – PI World Time Clock, p104 – smlwrld Desk Clock

Clemens Weisshaar & Reed Kram
Mathias Ziegler: p107 – Monza (3);

Frank Stolle
p108 – Breeding Tables (5), p109 – Breeding Tables (6); Florence Böhm: p109 – Breeding Tables (9)

Kenneth Cobonpue
Conrado Velasco: all photography; except Katya Zialcita: p111

Paul Cocksedge
Richard Brine: all photography; except Jean François-Carly: p114 – Watt? (detail)

DEMAKERSVAN
Raoul Kramer: p122 – Lost&Found, p123 – Cinderella Table, p124 – Waiting For You; Hans van de Mars: p122 – Minimizing Einstein; Raoul Kramer and Bas Helbers: p125 – How To Plant a Fence

Doshi Levien
John Ross: all photography

david dubois
Morgane Le Gall: p134

Martino d'Esposito
Actimage: p146 – Drink'Tray, p149 – Grow Bag; Pierre Fantys/ECAL: p146 – Earthquake Table; Emilie Muller and Yan Gross: p149 – Vortex

FRONT
Katja K: p150 – Loudspeakers, p152 – Chainsaw Chair; Anna Lönnerstam: p153 – Wallpaper by Rats; Tommy Bäcklin: p153 – Animal Thing

FUCHS + FUNKE
Samuel Pietsch: p157 – Knock Down Lamp (5)

Martino Gamper
Anders Kjaergaard and Sue Parkhill: p158 – Tree Log; Edward Horsford: p158, p160 – all photographs from '100 Chairs in 100 Days', p158–9 – Arnold Circus Stool; åbäke: p161 – Martino Gamper's studio

Alexis Georgacopoulos
Milo Keller: p162 – Chair – D/side Series, p162 – Inflatable Bottle Cooler (2), p165 – CMYK; Anoush Abrar: p163 – Newton Fruit Bowl/Green Apples; D.I.M.: p164 – Rainmat

Adam Goodrum
Blue Murder Studios: p169 – Lissajous (7); Paul Pavlou: p169 – Armour Screen (8)

graf:decorative mode no.3
Yasunori Shimomoura: p162 – 3/6 Series, p164 – Deco Lamp, p163 – SUNAO series

Tal Gur
Naomi Yogev: p174 – Yam Yabasha (3), p175 – Eash (5); Yoram Reshef: p174 – Yam Yabasha (4); Carlo Draisci: p175 – Eash (6); Boaz Drori – p175 – Eash (7); Venetzian Verhaptig: p177

Ineke Hans
Corne Bastiaansen: p178 – Black Gold; Swarovski: p178 – Pixelated Peacock; Alphons ter Avest: p179 – Relax Set, p179 – Beer Set; Ron Steemers: p180, p181 – Garlic Crusher; Royal VKB: p181 – Bowls & Spoons

Jaime Hayon
Nienke Klunder: p184 – Mon Cirque Ceramic Plates

Simon Heijdens
Daniel Nicholas: p187 – Moving Wallpaper

Jackson Hong
Sang-tae Kim: all photography

Ichiro Iwasaki
Hiromasa Gamo: p198 – Takata 04-neo, p198 – Telephone 510, p201 – Meteo; Mutsumi Kaneko: p200 – Drop, p200 – O-Bon, p200 – Aroma; Mie Morimoto: p200 – Cameo, p200 – Harp

Trent Jansen
Alex Kershaw: all photography

Patrick Jouin
Varianti: p210–1 – Ether; Thomas Duval: p213 – Solid C1; Richard Perron: p213 – Oneshot

Chris Kabel
Toolsgalerie: p215–6 – Pinlight (7); Droog Design: p216 – Shadylace (8); Daniel Klapsing: p216 – Shadylace (10); Maarten van Houten: p217 – Flames (13)

Lambert Kamps
Bjorn Eerks: p220 – Light-Emitting Roof Tiles (7); Christophe Brocoru, p220 – Light-Emitting Roof Tiles (5)

Meriç Kara
Zak Swenson/Fabrica: p226 – Sugarcubes, p228–9; Fabrica: p226 – Digit

Bosung Kim
Chungwa Park: all photography; except KODAS: p233 – Lluon Modular PC

André Klauser: Ed Carpenter
p237 – Cone Tables (8)

Tetê Knecht
Andrés Otero: p238–9 – Oil, p240–1

korban/flaubert
Sharrin Rees: p242–3 – Burstlight (2), p244 – Weblight, p245 – Cellscreen (7)

Joris Laarman
Anita Star: all photography

Nicolas Le Moigne
ECAL/Milo Keller: p250 – Cigar Case; INOUT/Yan Gross & Emille Muller: p250–1 – Fountain Mirror; Anoush Abrar: p252

Simone LeAmon
Jeremy Dillon: p257 – Supersystem

Mathieu Lehanneur
Véronique Huyghe: p262 – Medicine
by the Centimetre, p263 – dB/Elements
Collection, p264 – K/Elements Collection,
p265 – O/Elements Collection, p265 –
Third Lung–Therapeutic Objects Collection

Julia Lohmann
Studio Bec: all photography

Alexander Lotersztain
Florian Groehn: p275

Xavier Lust
Driade/Adelaïde Astori: p278, p281;
EXTREMIS: p279; ELIXIR sprl: p280

Cecilie Manz
Jeppe Gudmundsen-Holmgreen: p282 –
SoftBox (4–6); Lightyears A/S: p283;
Fredericia Furniture A/S: p285

Tatsuya Matsui
Flower Robotics Inc.: p286; Masao
Okamato: p287–9

Matthias Aron Megyeri
Felix Amsel, Matthias Aron Megyeri:
all photography

Marcelo Moletta
Rudy Hühold: all photography

Nada Se Leva
Daniel Pinheiro: p299

Nendo
Masayuki Hayashi: p306, p307 – Polar;
Gebrüder Thonet: p307 – Sasae; De
Padova: p308 – Chab-table; Daici Ano:
p308 – Yuki

Nido Campolongo
Eduardo Barcelos: p310 – Avestruz
Lamp, p312 – Casa do Leitor, p313 –
OCA; Mária Daloia: p311; Alex Salim:
p313 – Cone Table

Nódesign
Ioram Finguerman: p316

Johannes Norlander
Rasmus Norlander: p318 – Cano (2),
p321; Johan Ödmann: p318–9 – ORI
Chair (5–7), p320

Ken Okuyama
ZMP: p322 – Nuvo (3); Noboru Murata:
p323–5

OVO
Rômulo Fialdini: p326 – Terceira (2),
p326 – Box in the Box, p328 – Box
in the Box; Cacá Bratke: p327 – Cadé &
Feriado in situ (5 & 9); Fernando Laszlo:
p327 – Cadé (6–8); Amilcar Packer:
p328 – Disco; p329 – 2.88, p329 – Entre;
Marcelo Zocchio: p329 – In Vitro

Flávia Pagotti Silva
Andres Otero: p332 –
Bate-Papo Stool (6–7)

Zinoo Park
Yun Sukmu: p335

Russell Pinch
David Clerihew: p338, p340; Beth Evans:
p339; John Ross: p341

POLKA
Lisi Gradnitzer: p342 – Tattoofurniture;
Maurizio Maier: p342 – Josephine, p343,
p344; Marie Jacel: p345 (17–8)

Adrien Rovero
Philippe Jarrigeon: p350–1

Alejandro Sarmiento
Violeta Villar: p354 – Supernova (1–2);
Valeria Pedelhez, Fabio Scrugli: p355;
Valeria Pedelhez, Marcos Lopez: p356–7

Wieki Somers
Elian Somers: p362–3; Frank Kouws:
p364 – Mattress Stone Bottle; Ramak
Fazel: p365 – Trophees (11–2)

Alexander Taylor
Marino Ramazzotti: p367; E&Y: p368;
Leon Cohen: p369 – Fold (6); David
Sykes: p369 – (7 & 9)

Carla Tennenbaum
Chico Rivero: p370, p372 – Kinetic
Tapestry (6–7); Dante: p371, p372 –
Kinetic Tapestry (4–5), p373

TONERICO:INC.
Nacása & Partners Inc: p374–5; Kentaro
Kamata: p376–7

Peter Traag
Leon Chew: p381 – LTD Collection (21)

Maxim Velcovsky
Marek Novotny: p382–4, p385 –
Moneybox; Gabriel Urbanek: p385 –
Norwegian Wood

Dominic Wilcox
Mick Rock/Dominic Wilcox: p392

WOKmedia
Alessandro Evangelista: p394 –
Flood (3)

Shunji Yamanaka
Yukio Shimizu: all photography

Courtesy Credits

5.5 designers
Bernardaud Foundation/Hélène Huret,
p13 – Creamer Casting

Mana Bernardes
Suelen Figueira, all images

Big-game
ECAL, p57, Moose

Stephen Burks
Halvor Thorsen, design assistant for
Gio and Ambrogio

Sam Buxton
Elumin8 Ltd, p80 – Clone Chaise,
p81 – Word Clock and SIOS Table

PIERRE CHARPIN
Galerie Kreo, p98 – Platform Medium
White, p98–9 – Platform Coffee Table;
Venini, p99 – Triplo Vase

david dubois
MUDAM, p135, p137

Martino d'Esposito
Ligne Roset, p146 – Drink'Tray, p149 –
Grow Bag; ECAL, p146 – Earthquake
Table, p148 – Twins; InOut, p149 – Vortex

Martino Gamper
åbäke, for graphic design, p161

Alexis Georgacopoulos
Matteo Gonet, glassblower:
p164 – Stando

Adam Goodrum
Tom Rowney and Alexandra Chambers
(glass blowers), p166–7 – Boab
Light; Gunnersen Inspirations, p169 –
Armour Screen

Jackson Hong
Jung-hyun Cho, Dae-jae Kim, Chang-
kyoung Kim, Yang-ho Kin, Nam-hyeong
Kim, Jae-Yup Lee: p190 – Iron Chair X,
p192 – Reactive Hair X

Junya Ishigami
Gallery Koyanagi: p196–7

Chris Kabel
Toolsgalerie: p215–6 – Pinlight (7),
Droog Design: p216 – Shadylace (8);
Daniel Klapsing: p216 – Shadylace (10);
mooi: p217 – Flames (13)

Yung-je Kang
Samsung Electronics: all imagery

Meriç Kara
Fabrica/shoedust – Markus/Zak & Andy:
p226 – Sugarcubes, p228–9; Fabrica
with Paola C.: p226 – Digit

Bosung Kim
TG: p233 – Lluon Modular PC

Tetê Knecht
Workshop with Paul Cocksedge:
p240 – Rust

Joris Laarman
Droog Design/Jaga: p246–7 – Heatwave
(1&2); Droog Design: p248, Arco: p249

Nicolas Le Moigne
Nelly Moneger-Glayre (donor): p253

Simone LeAmon
Digital team – Daryl Munton & Dave
Morison: p254 – Bodywork (2–4), p256,
p257 (except Supersystem)

Mathieu Lehanneur
FLOOD, Paris©: p265 – Restaurant
FLOOD, Paris©

Arik Levy
MGX.Materialise: p266; Gaia & Gino: p269

Julia Lohmann: Verolution, London:
p270; Zeltec: p271; Alma Home,
London: p272

Xavier Lust
EXTREMIS: p279

Cecilie Manz
Holmegaard: p284

Tatsuya Matsui
Flower Robotics, Inc.: all photography

Marcelo Moletta
Celsos Santos, Ilana Abramoff: p294–5;
Celsos Santos, Mauro Schwanke: p296;
Carlos Eduardo Ribeiro Leitáo, Lula: p297

Adrien Rovero
Galerie Kreo: p350 – Chain (4);
ECAL: p352–3

Alejandro Sarmiento
Milanesa: p355, p357

Wieki Somers
Galerie Kreo: p362–3; EKWC
(European Ceramic Work Centre),
s'Hertongenbosch: p364 – Mattress
Stone Bottle

Dominic Wilcox
Mick Rock: p392

WOKmedia
Dupont Corian: p395; British Council:
p397 – Lunuganga (9)

Noriko Yasuda
Droog Design: p403 – Footrest
Pebbles (4–5)

10 Curators' choices

Armchair 41 'Paimio'
p412 – image courtesy of Artek

Poltrona Mole (Sheriff) armchair
p413 – image courtesy of LinBrasil

Ulmer Hocker
p414 – image courtesy of Zanotta

Lampadaire '1063'
p416 – image courtesy of Galerie Kreo

Sella stool
p415 – image courtesy of Zanotta

Schwinn Sting-Ray
p417 – image courtesy of The Bicycle
Museum of America

Paperclip
p418 – image courtesy of ACCO
Brands Inc.

Dilly Bag
p419 – image courtesy of Maningria
Arts & Culture

TP1 (T4 & P1)
p420 – image courtesy of Dieter Rams

Steel Shelving Unit
p421 – image courtesy of MUJI/Ryohin
Keikaku Co. LTD